International Perspectives on Teaching
and Learning with GIS in Secondary Schools

Andrew J. Milson · Ali Demirci · Joseph J. Kerski
Editors

International Perspectives on Teaching and Learning with GIS in Secondary Schools

Foreword by Roger Tomlinson

 Springer

Editors
Andrew J. Milson
University of Texas at Arlington
601 South Nedderman Drive, Box 19529
Arlington
Texas 76019
USA
milson@uta.edu

Joseph J. Kerski
Environmental Systems Research Institute
1 International Court
Broomfield
Colorado 80021-3200
USA
jkerski@esri.com

Ali Demirci
Fatih University
Department of Geography
Buyukcekmece, Istanbul
Turkey
ademirci@fatih.edu.tr

ISBN 978-94-007-2119-7   e-ISBN 978-94-007-2120-3
DOI 10.1007/978-94-007-2120-3
Springer Dordrecht Heidelberg London New York

Library of Congress Control Number: 2011937059

Printed on acid-free paper

Springer is part of Springer Science+Business Media (www.springer.com)

Foreword

Geographic Information Systems (GIS) are in use worldwide – over 400,000 institutions in over 125 countries use them on a daily basis (based on the actual licenses sold by major GIS software manufacturers). Millions of people work in the geospatial industry, estimated at $64 billion per year by the US Department of Labor, and growing at 20% per year. However, in my estimate, the current production of GIS trained students from all educational institutions, secondary and tertiary, does not keep up with the growing demand. In fact, the global uptake of this productive and exciting technology is determined by the supply of trained people able to use it effectively.

Geographic Information Systems are contributing to every imaginable activity across the administrative, military, and scientific spectrum. The technology has demonstrated significant benefit to those organizations, countries, and cultures that use it. Clearly education is a critical component in the future of the nations involved. GIS education in secondary schools is the foundation of this progress.

This book tells stories from many countries describing their experience with bringing GIS education into the high school. It is a description of real-world experience from which lessons can be extracted. It brings the realization that GIS education is far more than developing skills needed in the workforce. It shines a light on the importance of inquiry-based teaching, where GIS is the enabling tool that allows students to engage in meaningful issues about their environment, time, and place. GIS education leads to critical thinking in a wide range of disciplines, and is fundamental to the creation of decision-makers. This emerges as a persistent theme in the experiences that are related from widely different institutional settings and cultural backgrounds.

This is an important book, of value not only to students, parents, and teachers, but also to the leaders in pedagogy, curriculum, and organizational policy in the world of education.

Tomlinson Associates Ltd, Ottawa, Canada   Roger Tomlinson

Acknowledgments

The editors are thankful to Springer for publishing this long overdue global perspective on GIS in education. We wish to express our sincere gratitude to Bernadette Ohmer from Springer for her support of and assistance with this project. We also appreciate the external reviewers selected by Springer who provided extremely helpful feedback. The authors have gained tremendous respect for one another in the process of producing this volume. Each has been grateful to work with such a diligent team of co-editors. Finally, we salute the visionary educators who have introduced GIS to young people around the world and who have contributed their valuable time to share these stories. We believe that the students they have worked with around the world will be well-equipped through their work with GIS to enable a brighter future for the planet and its people.

– Andrew J. Milson, Ali Demirci, and Joseph J. Kerski

Contents

Contributors

Svein Andersland Akvator, Stord, Norway, sa@akvator.no

Maria Attard University of Malta, Msida, Malta, maria.attard@um.edu.mt

Jim Ayorekire Makerere University, Kampala, Uganda, jayorekire@arts.mak.ac.ug

György Borián Danube-Drava National Park, Hungary, gy.borian@freemail.hu

Gregory Breetzke University of Canterbury, Christchurch, New Zealand, gregory.breetzke@canterbury.ac.nz

Mohamed R. Bualhamam United Arab Emirates University, Al Ain, United Arab Emirates, mbualhamam@uaeu.ac.ae

Theodore Burikoko Kigali, Rwanda, buritheo@yahoo.fr

Alfredo del Campo Instituto Geográfico Nacional (IGN), Madrid, Spain, adelcampo@fomento.es

Joan Capdevila Instituto Geográfico Nacional (IGN), Catalonia, Spain, joan.capdevila@mpr.es

Victoria Castro De la Rosa Education Ministry, Santo Domingo, Dominican Republic, victoriacastro@gmail.com

Chris Charman Huron Heights S.S., Kitchener, ON, Canada, chris_charman@wrdsb.on.ca

Che-Ming Chen National Taiwan Normal University, Taipei, Taiwan, jeremy@ntnu.edu.tw

Ali Demirci Fatih University, Istanbul, Turkey, ademirci@fatih.edu.tr

Natalia Andrea Diaz Vega Universidad Distrital F.J.C., Bogotá, D.C. Colombia, nataliadiazv@gmail.com

Pinliang Dong University of North Texas, Denton, TX, USA, Pinliang.Dong@unt.edu

Stephanie A. Eddy GISMAPED, Auckland, New Zealand,
stephanie@gismaped.co.nz

Sanet Eksteen University of Pretoria, Pretoria, South Africa,
sanet.eksteen@up.ac.za

Mary Fargher Institute of Education, University of London, London, UK,
m.fargher@ioe.ac.uk

Tim Favier VU University, Amsterdam, The Netherlands, t.favier@ond.vu.nl

Martina Forster Esri Deutschland, Kranzberg, Germany, m.forster@esri.de

Sylvain Genevois Institut National de Recherche Pédagogique, Lyon, France,
sylvain.genevois@inrp.fr

Iain Greensmith Esri Canada, Toronto, ON, Canada,
igreensmith@esricanada.com

Niem Tu Huynh Association of American Geographers, Washington, DC, USA,
nhuynh@aag.org

Yoshiyasu Ida University of Tsukuba, Tsukuba, Japan, ida@human.tsukuba.ac.jp

Thomas Jekel Institute of GIScience, Austrian Academy of Sciences, Salzburg,
Austria, thomas.jekel@oeaw.ac.at

Torben P. Jensen Langkær Gymnasium and HF, Tilst, Denmark, tpj@tpj.dk

Tino P. Johansson University of Helsinki, Helsinki, Finland,
tino.johansson@helsinki.fi

Thierry Joliveau Jean Monnet University of Saint Etienne, Saint Etienne, France,
thierry.joliveau@univ-st-etienne.fr

Joseph J. Kerski Esri, Broomfield, Colorado, USA, jkerski@esri.com

Minsung Kim Texas A&M University, College Station, TX, USA,
minsungkim@neo.tamu.edu

John C. Kinniburgh The King's School, Sydney, NSW, Australia,
jck@kings.edu.au

Arne Frank Knudsen Laksevåg Upper Secondary School, Bergen, Norway,
arnknu@hfk.no

Alfons Koller University of Education, Linz, Austria, kol@ph-linz.at

María Luisa de Lázaro Universidad Complutense de Madrid (UCM), Madrid,
Spain, mllazaro@ghis.ucm.es

Sang-Il Lee Seoul National University, Seoul, South Korea, si_lee@snu.ac.kr

Peiying Lin Capital Normal University, Beijing, China, pylin2000@263.net

Yan Liu University of Queensland, St. Lucia, Queensland, Australia, yan.liu@uq.edu.au

Andrew J. Milson University of Texas at Arlington, Arlington, TX, USA, milson@uta.edu

Gustavo Moreira-Riveros Ministry of Education, Santiago, Chile, gmoreirar@gmail.com

Madalena Mota Pinhal Novo Secondary School, Pinhal Novo, Portugal, madalenamota@gmail.com

Osvaldo Muñiz-Solari Texas State University, San Marcos, TX, USA, o.muniz@txstate.edu

José Antonio Nieto Instituto de Cartografía de Andalucía, Seville, Spain, jantonio.nieto@juntadeandalucia.es

Albert Nsengiyumva Workforce Development Authority, Kigali, Rwanda, ansengiyumva@wda.gov.rw

Benjamin Ofori-Amoah Western Michigan University, Kalamazoo, MI, USA, ben.ofori@wmich.edu

Anne F. Olsen Chilton Saint James School, Lower Hutt, New Zealand, anne@gismaped.co.nz

Joseph R. Oppong University of North Texas, Denton, TX, USA, Joseph.Oppong@unt.edu

Quinta Ana Pérez Sierra Instituto Tecnólogico de las Américas, Santo Domingo, Dominican Republic, qperez@gmail.com

Erika Pretorius University of Pretoria, Pretoria, South Africa, erika.pretorius@up.ac.za

David Rayner Institute of Education, University of London, London, UK, D.Rayner@ioe.ac.uk

Luz Angela Rocha Salamanca Universidad Distrital F.J.C., Bogotá, D.C. Colombia, lrocha@udistrital.edu.co

Jan Ketil Rød Norwegian University of Science and Technology, Trondheim, Norway, jan.rod@svt.ntnu.no

Concepción Romera Instituto Geográfico Nacional (IGN), Madrid, Spain, cromera@fomento.es

Eric Sanchez Université de Sherbrooke, Sherbrooke, Québec, Canada, eric.sanchez@usherbrooke.ca

John A. Schembri University of Malta, Msida, Malta,
john.a.schembri@um.edu.mt

Henk J. Scholten Geodan, Amsterdam, The Netherlands, henk@geodan.nl

Bob Sharpe Wilfrid Laurier University, Waterloo, ON, Canada, bsharpe@wlu.ca

Alexander Siegmund University of Education Heidelberg, Heidelberg, Germany,
siegmund@ph-heidelberg.de

Hans-Jörg Stark University of Applied Sciences Northwestern Switzerland,
Muttenz, Switzerland, hansjoerg.stark@fhnw.ch

Josef Strobl Center for Geoinformatics University of Salzburg, Salzburg, Austria,
josef.strobl@sbg.ac.at

Geok Chin Ivy Tan National Institute of Education, Nanyang Technological
University, Singapore, Singapore, ivy.tan@nie.edu.sg

Vinod Tewari TERI University, New Delhi, India, vinodt@teri.res.in

Chetan Tiwari University of North Texas, Denton, TX, USA,
Chetan.Tiwari@unt.edu

Jean Tong Esri Canada, Toronto, ON, Canada, jtong@esricanada.com

Carmen Treuthardt Association of Swiss Geography Teachers, Switzerland,
Carmen.Treuthardt@edulu.ch

Revocatus Twinomuhangi Makerere University, Kampala, Uganda,
levtwin@arts.mak.ac.ug

Joop van der Schee VU University, Amsterdam, The Netherlands,
j.vanderschee@ond.vu.nl

Kathrin Viehrig University of Education Heidelberg, Heidelberg, Germany,
viehrig@ph-heidelberg.de

Xi Xiang National Institute of Education, Nanyang Technological University,
Singapore, Singapore, shellyxiang1203@gmail.com

Rawan Yaghi Nabil Adeeb Sleiman Secondary Public School, Bednayel-North
Bekaa, Lebanon, rawan.yaghi@gmail.com

Minori Yuda University of Tokyo, Tokyo, Japan, minori@csis.u-tokyo.ac.jp

List of Figures

List of Tables

About the Editors

Ali Demirci is an associate professor at the Department of Geography, Fatih University in Istanbul, Turkey. He received his Ph.D. degree in 2004 from Marmara University in Turkey and studied secondary school geography education in the USA as his dissertation subject. A part of his dissertation was published in a Turkish book entitled *Education System in the USA with a special focus on Secondary School Geography Education*. His major research interests are geography education in general and GIS and its applications in geography education in particular. He has published many articles in national and international journals in a wide range of topics including curricular developments, the use of GIS and other spatial technologies in geography education, and the application of GIS and remote sensing in different areas. He is the author of the book *GIS for Teachers* published in Turkish with the support of Esri-Turkey in 2008. This book is the only source in Turkey that provides secondary school teachers with sufficient theory and practice on GIS along with digital data, GIS software licensing, and many GIS-based applications to be implemented in geography lessons. He is also co-editor of two books: *Geography,* which is a textbook written for Grade 9 geography lessons and *Methods and Approaches in Geography Education*, which is used as a textbook for a graduate course offered at universities in Turkey. Over the past ten years, he has been involved in a number of activities such as organizing conferences, workshops, and courses to introduce GIS to secondary and higher education institutions in Turkey. He is currently working in a committee in the Ministry of Education to revise the secondary school geography curriculum in Turkey. One of his recent projects is entitled "Using GIS to develop social sensitivity among students: Implementation of GIS-based projects in secondary school geography lessons" supported by The Scientific and Technological Research Council of Turkey (TÜBİTAK).

Joseph J. Kerski serves as education manager on the education team for Environmental Systems Research Institute (Esri) in Colorado, USA. Prior to joining Esri, he served for seventeen years as geographer in the Education Program at the US Geological Survey, and for four years as geographer at the US Census Bureau. He has taught as adjunct instructor of GIS at Sinte Gleska University on the Rosebud Sioux Reservation, at the University of Denver, in primary and secondary schools, and in online courses. Joseph holds three degrees in geography and thus is rather enthusiastic about studying biomes, population, landforms, hazards, and other topics using technology, maps, and imagery. Passionate about all aspects of spatial learning, Joseph seeks and fosters educational partnerships, promotes GIS in education and society through service and scholarship, and conducts training in geotechnologies for instructors, students, and administrators in a wide variety of educational levels and disciplines, internationally. He creates curricula focused on spatial thinking that uses geotechnologies as its chief inquiry-based tool, and conducts research in the effectiveness and implementation of these technologies in formal and informal educational settings.

Andrew J. Milson is a Professor of social science education and geography with faculty appointments in the College of Education and Health Professions and the College of Liberal Arts at the University of Texas at Arlington, USA. He teaches courses in human geography, GIS, and social studies education. He earned his Ph.D. in 1999 from the University of Georgia (USA), where he studied social science education and geography. Prior to pursuing his doctorate, he taught secondary school history and geography near Dallas, Texas. Andrew conducts research on geographic education and the use of geospatial technologies in educational environments and has published numerous journal articles and book chapters on this work. His co-edited books include *Digital Geography: Geospatial Technologies in the Social Studies Classroom* (Information Age, 2008) with Marsha Alibrandi. He is an elected member of the Executive Board of the National Council for Geographic Education and serves as an associate editor of the *Journal of Geography*.

About the Authors

Svein Andersland is Associate Professor at Stord/Haugesund University College, Department of Teacher Education and Cultural Studies in Norway. He teaches geography, geography didactics, in-service GIS courses, development studies, and other geography-related subjects within the broader social science study in teacher education. His research interests are related to the usage of ICT and GIS in lower education. Currently, he is participating in a national survey within the research program Education, Curriculum and Technology, focusing on teachers' usages and attitudes toward the use of ICT in various school subjects. He has presented and published GIS-related papers in IGU-CGE-held conferences, national and international ESRI user conferences, national geography conferences, and conferences and workshops held by the HERODOT network. He has also written GIS-related chapters in two education-related books

Maria Attard holds the position of lecturer in Geography as well as Director of the Institute for Sustainable Development at the University of Malta. Her specializations are transport geography and GIS. She has coordinated the Geographic Information Systems Laboratory and the teaching of GIS at the university since 1997, mostly at undergraduate but also as part of various postgraduate courses. She has been involved in a number of training programs and the institute will see the development of further GIS study programs at the University of Malta, thus increasing the skills in the available and future workforce. She has published in the areas of transport and GIS education.

Jim Ayorekire is a Lecturer at in the Department of Forestry, Biodiversity and Tourism, College of Agricultural and Environmental Sciences, Makerere University. He attained his doctorate in environmental and geographical sciences from University of Cape Town and holds a master of arts degree in land use and regional development, Makerere University. His teaching and research interests are in regional geography and the application of GIS in land use planning with a bias to tourism planning and management.

György Borián is senior manager of environmental education and ecotourism for the Danube-Drava National Park in Hungary. He has authored or co-authored numerous materials and teachers' manuals for environmental education including five secondary school textbooks.

Gregory Breetzke is a Lecturer of geographical information systems (GIS) at the University of Canterbury, New Zealand. He has been involved in teaching post-graduate GIS programmes in his native South Africa and has now transferred that knowledge to assist in the development of a new master's in GIS programme at the University of Canterbury. His research focuses broadly on geospatial analysis and on different pedagogical approaches to GIS education.

Mohamed R. Bualhamam is a professor in the geography department at the UAE University and head of the Geographical Information Systems committee in the Emirate of Ras al Khaimah. He received his Ph.D. in environmental dynamics and GIS from the University of Arkansas. He is also a chair of authorship and review of the social studies curriculum for Grades 5 and 6 level committees at the Ministry of Education. His research interests include GIS in education, GIS for project management, and environmental applications of GIS.

Theodore Burikoko assists the "GIS in Secondary Schools" project in various activities and serves as instructor for teacher trainings. He has taken part in the project since its early days when he taught GIS to his classmates of the technical school ETO Gitarama. He currently continues his studies of civil engineering at Kicukiro College of Technology in Kigali.

Alfredo del Campo is a geographer engineer at the *Instituto Geográfico Nacional* (National Geography Institute), Madrid, Spain. He is in charge of the *Área de Cartografía Temática y Atlas Nacional* (thematic cartography and national atlas) of the Cartography Section in the General Directorate of the IGN. He studied forest engineering at the Polytechnic University of Madrid (UPM) and completed doctoral studies in the Department of Cartographic Engineering, Geodesy and Photogrammetry: Graphic Expression. He has also obtained the titles of Master of Geographic Information Systems (UPM), Manager of Information Systems and Communications and Information Technologies (UPM) and Senior Public Management at the *Universidad Complutense de Madrid* (UCM). He has taught and worked on speeches and presentations on thematic cartography and the National Atlas of Spain in specialized forums and outreach activities in teaching.

Joan Capdevila directs the regional service of the *Instituto Geográfico Nacional* in Catalonia. He studied physics and geography at the University of Barcelona. He is a member of *Consejo Superior Geográfico* (Geographic High Council) and works on the Committee on spatial data infrastructures and the Committee on Geographical Names. He publishes the blog IDEE, the *Boletín SobreIDEs*, and directs the working groups SDI Observatory and Cartographic Heritage on SDI. In all these fields, he performs tasks related to the diffusion and dissemination of geographic information in education.

Victoria Castro De la Rosa holds a degree in computer science from the *Universidad Autónoma de Santo Domingo* (1998) and a master's in educational technologies from *Pontificia Universidad Católica Madre y Maestra* (PUCMM), Dominican Republic. Currently, she is developing her PhD thesis in Information and Knowledge Society from the *Universidad Pontificia de Salamanca, Madrid* (UPSAM), Madrid, Spain. She works for the Education Ministry in the technology integration department, Dominican Republic. Her research has focused on information technology for education in middle schools in the Dominican Republic.

Chris Charman began his teaching career in 1997 in Waterloo, Ontario. In 2001 he completed his MA in urban/economic geography at Wilfrid Laurier University and became involved in GIS and GPS in the classroom. He found it curious that there was more support for geo-technology from the information technology world than there was from the geographic lobby. This led to questions and research into the difficulties of GIS implementation and the role of geographic education in general. Over the years these questions lead Chris to begin working on a doctorate, focusing on geographic literacy and how students perceive the space around them. In his current role as Geography Department Head in a new secondary school he has sought to build the Geography department with likeminded individuals, who embrace technology while still balancing the need for out of classroom field based experiences for our students.

Che-Ming Chen is Associate Professor of Geography at the National Taiwan Normal University. He has 18 years of experience in geography teacher training at the secondary level. His research and teaching interests include geography education, mobile learning, and spatial information technologies. Che-Ming has approximately 40 scientific publications in peer-reviewed journals, books, proceedings, and other outlets. His current research focuses on mobile learning for high school fieldwork. In the recent 5 years, he held over 60 workshops helping secondary geography teachers to use GPS, GIS, and Google Earth in their classroom.

Natalia Andrea Diaz Vega graduated from the Department of Geodesy and Cadastre Engineering at *Universidad Distrital Francisco José de Caldas* in Bogotá. In 2010, she joined as a professional on the geographic information systems group at the Mobility Secretary Bogotá. Prior to this she worked in a private enterprise company "Ingetec S.A.," managing the geographic information system and developing engineering projects. Also she worked at the Institute for Urban Development of Bogotá on feasibility projects for valorization. She is a member of the students research group SDI (Spatial Data Infrastructure) and has won an award for the project regarding a GIS application for teaching geography in basic secondary education in Colombia. Currently, she is doing her postgraduate studies in local development and environment at the *Universidad Distrital Francisco José de Caldas.*

Pinliang Dong is an Associate Professor at the Department of Geography, University of North Texas, Denton, Texas, USA. He has been doing research and teaching in areas of geographic information systems and remote sensing. His research interests include spatial analysis and modeling, GIS application development for emergency response and animal tracking, GIS and remote sensing for land cover mapping, biomass estimation, and population estimation. He has taught

Introduction to GIS, Intermediate GIS, Advanced GIS, GIS Programming, Remote Sensing, and Special Topics. He has published over 20 journal papers and book chapters.

Stephanie A. Eddy is the Director GISMAPED. She served as Head of Social Sciences at Botany Downs Secondary College in New Zealand from 2003 to 2006, Deputy Principal of Botany Downs Secondary College in 2005, and Head of Geography Macleans College, New Zealand from 2000 to 2003. In 2000, she introduced GIS into the Social Sciences Department at Macleans College, and in 2001 was the recipient of the Macleans College staff scholarship that allowed her to visit schools in the USA. She has attended numerous international workshops and training sessions on GIS in education and was recently awarded the Royal Society of New Zealand Teacher Fellowship to study GIS in New Zealand secondary schools.

Sanet Eksteen is GIS lecturer in the Department of Geography, Geoinformatics and Meteorology, at the University of Pretoria, South Africa. She teaches courses in introductory and advanced GIS, and her research focuses on the introduction of GIS in secondary school as well as the use of artificial intelligence in GIS. She was involved in the development of a Paper GIS to enable schools to teach GIS without the use of computers.

Mary Fargher is a Lecturer in Geography Education at the Institute of Education (IoE), University of London. After over twenty years working in schools as a geography teacher and subject leader, she was awarded an ESRC studentship to study for her PhD at IoE in 2005. Since then, she has taken part in a number of research projects using GIS with teachers and students in schools (including the Geographical Association's 'Spatially Speaking' project) and is regularly involved in geography education and GIS conferences (including ESRI). Mary is currently writing up her PhD thesis on place and GIS.

Tim Favier is a PhD student at the Free University Amsterdam. His research focuses on how geography teachers can use GIS to stimulate progression in students' geographic literacy. In his research, he has developed several successful GIS-supported inquiry-based geography modules with teachers from different schools. Tim Favier also works in the EduGIS project, which aims to stimulate the diffusion of GIS in secondary education in The Netherlands.

Martina Forster is the project coordinator for "GIS in Secondary Schools" at the GIS and RS Centre of the National University of Rwanda (CGIS-NUR). The program is designed to promote GIS integration through all levels of secondary education in Rwanda and inter-institutional collaboration in the field of GIS for Development.

Sylvain Genevois works at the National Institute for Educational Research (INRP), Lyon, France. His PhD work in geography and science education examines the questions posed by the introduction of geomatic tools in French secondary schools. His research interests also concern localized games and augmented reality.

Iain Greensmith has been with ESRI Canada since September of 2007 in the capacity of Technical Solutions Specialist, Education. Iain is able to draw from his technical experience in previous roles as well as his 12 years of coaching experience to ensure that educators benefit from the training and resource development, and direction he provides. Iain is a graduate of McMaster University with a degree in geography that specialized in GIS and spatial analysis, and is a current M.Sc. GISc candidate through Birkbeck College, University of London.

Niem Tu Huynh joined the Association of American Geographers (AAG) staff as Senior Researcher in June 2011. Prior to joining AAG, Niem taught secondary level science in Canada and was an Assistant Professor in the Department of Geography at Texas State University – San Marcos. Her research interest is how spatial thinking and geographic skills influence problem solving with geospatial technologies, particularly with GIS. This interest stems from her dissertation, which was completed in Fall 2009 at Wilfrid Laurier University.

Yoshiyasu Ida is a professor at Graduate School of Comprehensive Human Sciences (Institute of Education), University of Tsukuba. He is the Chair of Education Sciences and School Education in Graduate School, University of Tsukuba. He obtained his Ph.D. from the University of Tsukuba. He is interested in curricula in every county and GIS education in secondary school. He has published articles on GIS, geography, and environmental education in journals such as *Tsukuba Journal of Educational Studies* and *Geographiedidaktische Forschungen*. He is also the author of *Social Studies Education and Region*.

Thomas Jekel is senior scientist at the Institute of GIScience, Austrian Academy of Sciences, Salzburg, Austria. He is currently responsible for research and education projects for GI use in secondary schools. He studied geography and communication science at the University of Salzburg for MSc and PhD.

Torben P. Jensen is a Teacher of Geography and History at Langkaer Gymnasium and HF, Aarhus, Denmark. He earned his M.Sc. in Geography and History from the University of Aarhus, Denmark. He is the former Chairman of the Danish Association of Upper-Secondary Teachers of Geography and former President of the European Association of Geography Teachers' Associations. He is co-author and co-editor of *Geografi: Natur - Kultur - Menneske*, Geografforlaget 1992. Geography: Nature–Culture–Man and *GO, Naturgeografi, Jorden og mennesket*, Geografforlaget 2006 (GO, Geoscience, Earth, and Man); both widely used upper-secondary Geography course books. He is also co-author and co-editor of *Bæredygtig energi, Natur og Viden*, Geografforlaget 2010 (Sustainable Energy), an interdisciplinary basic introductory upper-secondary science course on sustainable energy including biology, chemistry, geography, and physics.

Tino P. Johansson received his Ph.D. from the University of Helsinki in Finland in 2009. He works as a research coordinator at the Department of Geosciences and Geography and has specialized in educational use of Geographical Information Systems. He has organized several pre- and in-service teacher courses on the educational use of GIS in Finland. He has been actively involved in national and European projects that have introduced GIS into schools and developed ways to integrate these interdisciplinary teaching and learning tools into the daily activities of teachers and pupils. He is currently coordinating the Climate Change Impacts on Ecosystem Services and Food Security in Eastern Africa (CHIESA) Project in Nairobi, Kenya.

Thierry Joliveau is professor of geography and geomatics at Jean Monnet University of Saint-Etienne (France) where he is in charge of the Master "GIS and Space Management" program. He is member of the CNRS research laboratory EVS (*Environnement, Ville, Société*) and director of the CNRS collaborative research network MAGIS (Models and Applications of Geographical Information Science). His research interests include several topics relating to the use of geographical information in society, education, environmental management, and landscape planning.

Minsung Kim is a Ph.D. candidate in the Department of Geography at Texas A&M University in College Station, Texas, USA. His primary research interests include spatial literacy, geo-spatial technologies, and spatial concept development. He received his master's degree in geography education from Seoul National University in South Korea and was involved in projects developing geography education materials.

John C. Kinniburgh is the Head of Geography at The King's School, a comprehensive independent boarding school in Sydney, Australia. Having taught at the school for the past thirteen years, John first introduced students to Geographic Information Systems (GIS) in 1998 and since then has integrated GIS into the Geography curriculum from years 7 to 12. He is a leader in the use and application of GIS in geographical education and a key advocate for its role in fostering authentic learning environments for boys. He has conducted numerous workshops on the use and application of GIS in the Geography classroom and has spoken at a number of conferences including the ESRI International Education User Conference in San Diego in 2003. He is currently completing a PhD at Macquarie University by investigating how, and to what extent, GIS enhances the conceptual understanding of geography students in New South Wales. His current research investigates the way in which GIS supports constructivist learning environments, particularly those that adopt instructional frameworks that incorporate problem-based learning (PBL).

Arne Frank Knudsen is a senior high school teacher in geography, history, and Norwegian language. He teaches geography, geoscience, and Norwegian language in Laksevåg Upper Secondary School. From 1984 to 2006 he taught geography didactics at the Department of Teaching Education, University of Bergen. He has been a practice teacher at the same place since 1984. For several years, he has had a special interest in teaching geography and has participated in a pilot project "GIS in Schools," initiated by and under the leadership of Svein Andersland.

Alfons Koller is a high school teacher at the Petrinum High School, Linz, Austria. He coordinates geography and economics within the Austrian Innovations in Mathematics and Science Teaching (IMST) program. He is also involved with several EU projects aiming at the inclusion of GI in teacher training and comanages the annual Learning with GI conference at Salzburg. He studied geography and mathematics at the University of Salzburg and has been active in vocational training of geography teachers since the early 1990s.

María Luisa de Lázaro is an Associated Professor of Human Geography at the *Universidad Complutense de Madrid* (UCM), Spain. She has been member of the board of the *Real Sociedad Geográfica* of Spain since 1992 and President of the Spanish work group in geography didactics (*Didáctica de la Geografía, Asociación de Geógrafos Españoles*) since November 2008. She has published GIS-related papers in the journal "*Didáctica Geográfica*" (Focus on Teaching Geography - for which she has been editorial secretary since 2010), for the meetings of HERODOT, EUROGEO, and several congresses. Her research interests are GIS and ICT for learning, researching and teaching geography in higher education and sustainability development in cities and in the Spanish countryside.

Sang-Il Lee is an associate professor in the Department of Geography Education at Seoul National University. He received his Ph.D. degree from Ohio State University. His research interests include GIS, spatial data analysis, and cartography. He has been responsible for running a branch of the National GIS Education Center in South Korea. He has also been involved in various teacher retraining programs with special focus on GIS for K-12 geography education.

Peiying Lin is a Professor at the Department of Geography, College of Resources, Environment and Tourism, Capital Normal University, Beijing, China. She has been teaching undergraduate and graduate courses in Geography Pedagogy, Geography Education, Computer Aided Geography Teaching, Modern Teaching Technologies in Geography, Environment Education, and English for Geography Majors. She has worked as Principal Investigator for six research projects on information technology for geography education in Chinese secondary schools. She has also edited and co-edited 10 books, and published 27 papers.

Yan Liu is Senior Lecturer in Geographical Information Sciences at the School of Geography, Planning and Environmental Management of the University of Queensland, Australia. Her research interests include GIS applications, spatial analysis and modeling, as well as learning with GIS in schools. She was an Assistant Professor at the National Institute of Education, Nanyang Technological University in Singapore when she conducted this research project. She is a member of the Surveying and Spatial Sciences Institute in Australia.

Gustavo Moreira-Riveros earned the master's in education from Universidad de Chile and public administration from the same higher education institution in Chile. He is a full-time professional working for the Ministry of Education, Santiago, Chile. His responsibilities are focused on information and communication technologies (ICT). As a specialist in public administration he coordinates activities related to the electronic government and the design and development of GIS for regional development in Chile. GIS development in school environments is his main focus of research. He has attended and participated in four ESRI International Conferences.

GIS applied to territorial information system for education and teaching with GIS in K–12 have been his main topics of applied research.

Madalena Mota is a geography teacher, teaching middle school and high school in Portugal since 1994. She has a M.Sc. degree in "Science and GIS" (2005) from the *Instituto Superior de Estatística e Gestão de Informação*, New University of Lisbon (ISEGI-UNL). Since 2004, she has used GIS in her classrooms and is involved in organizing events like GIS-Days with students. She also worked as the pedagogical coordinator of the ConTIG project at ISEGI-UNL, during a sabbatical year from teaching, in 2008–2009.

Osvaldo Muñiz-Solari, earned his Ph.D. in Geography from the University of Tennessee, Knoxville, and MA in Geography from Michigan State University, East Lansing. He is an associate professor in the Department of Geography and Associate Director of the Gilbert M. Grosvenor Center for Geographic Education at Texas State University. His initial academic preparation was in Pedagogy in Geography, History, and Civic Education at the University of the North, Antofagasta, Chile. As a geographer, he held academic positions in several universities and developed private consultant work in Argentina, Brazil, Colombia, Ecuador, Peru, and Chile. His specialties in geography education are focused on new technologies for international collaboration, one of which is GIS applications. Online learning methods and scientific diffusion through electronic networking are also important specialties that concentrate his publications. He is a member of the Steering Committee for the International Geographical Union's Commission on Geographical Education representing countries in Latin America.

José Antonio Nieto is a technician at the Institute of Cartography of Andalusia in Seville, Spain. He coordinates the department working on the elaboration of didactic and educative material related to cartography and geographic knowledge. He is also responsible for Didact-ICA, which is the section of the Web page of the Institute of Cartography of Andalusia that incorporates didactic material and links. He is member of the Group of Didactics of the Association of Spanish Geographers and is doing research on the geography of the population.

Albert Nsengiyumva serves as the Coordinator ICT in Education and also coordinates the establishment of the Rwanda Education and Research Network (RwEdNet) that aims at providing affordable Internet connectivity to tertiary education in Rwanda. He is also a network partner of the Research ICT Africa Network (RIA!) and the Deputy Chair of Ubuntunet Alliance for Research and Networking in the Eastern and Southern African region. He is a multidisciplinary professional with more than 12 years of experience. He is a former director of the National University of Rwanda Computing Center and has worked in computer networking projects as well as in the area of ICT research particularly in policy and regulation.

Benjamin Ofori-Amoah is a Professor of Geography and Chair of the Department of Geography at Western Michigan University, Kalamazoo, Michigan, USA. He holds a PhD degree from Simon Fraser University, Canada (1990), an MA (higher education administration) from the University of Exeter, UK (1984), an MSc (planning) degree from the University of Science and Technology, Kumasi, Ghana (1980), and a BA (geography with statistics) degree from the University of Ghana (1977). Prior to entering the university Ben taught as an elementary school teacher, from 1972 to 1977 in his native Ghana, after completing a four-year teacher training program. After obtaining his baccalaureate degree, he went back to teach geography at high school from 1977 to 1979. From 1980 to 1983, he worked as assistant registrar at the Kwame Nkrumah University of Science and Technology, in Kumasi, Ghana. In 1983, he left Ghana for further studies in the UK and Canada and eventually arrived in the USA in 1991, when he received appointment as Assistant Professor of Geography at the University of Wisconsin-Stevens Point. In 1999, Ben served as the Administrative Associate to Dr. David J. Ward, the then Senior Vice President for Academic Affairs of the University of Wisconsin System. He was the Chair of the Department of Geography and Geology at Stevens Point from 2001 to 2006, before he moved to Western Michigan University in 2006. Ben is an economic geographer and a regional planner with expertise in economic development, location analysis, urban and regional planning, and geographic information systems (GIS), and research interests in development theory, development planning, technological change, and human factor development. He has given scholarly presentations at many professional conferences and published in these areas of expertise. His most recent publication is an edited volume titled *Beyond the Metropolis: Urban Geography as if Small Cities Mattered.* University Press of America Lanham, MD: University Press of America, 2006.

Anne F. Olsen is the head of social sciences at Chilton Saint James School in New Zealand and director of GISMAPED. She has attended numerous international workshops and training sessions on GIS in education. In 1999, she introduced GIS into the Social Sciences Department at Chilton Saint James School. In 2001, she received a Royal Society of New Zealand Teacher Fellowship to study "The integration of GIS into Secondary School Geography" and visited schools in the USA, Canada, and Australia to observe the use of GIS. In 2002, she established GISMAPED with Stephanie Eddy to promote GIS in schools by providing training and resources. Recently, she served as the lead author on New Zealand GIS lessons.

Joseph R. Oppong is a Professor of Geography and Associate Dean of Toulouse Graduate School at the University of North Texas, Denton, Texas, where he has been teaching since 1992. He has a BA in geography from the University of Ghana and MA and PhD degrees from the University of Alberta, Edmonton, Canada. Prior to coming to UNT, Dr. Oppong taught at the University of Iowa. He currently serves as the US representative to the International Geographers Union Commission on Health and Environment and has served as chair to the Medical Geography Specialty Group and also the Africa Specialty Group of the Association of American Geographers. Dr. Oppong's research centers on medical geography, particularly the application of spatial analysis and GIS techniques to health care issues in Africa and Texas. Dr. Oppong has published articles on HIV/AIDS in Africa, teen HIV/AIDS in Dallas County, and Tuberculosis Genotypes in Tarrant County, Texas.

He coedited the volume *HIV-AIDS in Africa: Beyond Epidemiology* (Blackwell Publishers) and a special issue in Social Science and Medicine titled *HIV/AIDS, gender, agency and empowerment issues in Africa*. His latest research focuses on computational epidemiology, mapping tuberculosis genotypes, and racial/ethnic disparities in tuberculosis and HIV/AIDS in Texas and Africa. As associate dean of Toulouse Graduate School, Dr. Oppong is responsible for UNT's Responsible Conduct of Research policy formulation and implementation.

Quinta Ana Pérez Sierra earned a degree in computer science (2000) and a master's degree in telecommunication (2003) from the *Universidad Autónoma de Santo Domingo* (UASD), Dominican Republic. She also holds a Master's degree in Geographical Information Systems (GIS) and is currently developing her PhD thesis in GIS from the *Universidad Pontificia de Salamanca, Madrid* (UPSAM), Madrid, Spain. Her research has focused on (1) GIS technology for disasters management focused on crisis mapping, (2) GIS for the prevention of flood disasters with particular emphasis on simulation of hydrological processes, and (3) GIS technology for education. Currently, she is research professor for *Instituto Tecnológico de Las Américas* (ITLA), TI Professor for the *Universidad Iberoamericana* (UNIBE), and founder and director of the Geographic Information Systems School in the Dominican Republic.

Erika Pretorius is Technical Assistant at the Department of Geography, Geoinformatics and Meteorology, at the University of Pretoria, South Africa. She was previously an educator and a part-time contributor to the online Earthwire News Portal. Her research interests include an integrated approach to teaching geography with GIS as well as the use of GIS in developing suitability indices for the positioning of Fog Water Collection Systems.

David Rayner is a Lecturer in Geography Education at the Institute of Education (IoE), University of London. After thirty years working in schools as a geography teacher and subject leader, he took on the role of National Subject Lead (NSL) for KS3 Geography, leading a team of 25 teachers and advisors in supporting the government roll-out of the New Secondary Curriculum in England. David has a long-standing interest in GIS and technology in education. He was involved in one of the first UK local authority GIS initiatives in 2002 and wrote the online GIS teacher guidance for the Royal Geographical Society website. Alongside his work at the IoE, David continues his role as NSL running GIS workshops and conferences to support teachers wishing to embed GIS into the secondary curriculum.

Luz Angela Rocha Salamanca graduated from the Department of Geodesy and Cadastre Engineering at *Universidad Distrital Francisco José de Caldas* in Bogotá and received her Msc degree in geoinformation systems from ITC, The Netherlands, in 1997. She worked for almost nine years at the National Geographic Institute of Colombia "Agustin Codazzi" at the Cartography Department, where she gained a lot of experience in map production. In 2001, she joined the Universidad Distrital at the Faculty of Engineering, Cadastral and Geodesy program, as an assistant professor teaching cartography and geographic information systems. Currently she is a researcher of the research group NIDE (Research Nucleus in Spatial Data). She is also a team director of the students research group SDI (Spatial Data Infrastructure) and belongs to the editorial committee of the scientific journal "*UD y la Geomatica.*" Presently, she is continuing her PhD studies at the Universidad Nacional of Colombia in the Department of Geography.

Jan Ketil Rød is Associated Professor in Geographical Information System and Science at Department of Geography, Norwegian University of Science and Technology. He teaches statistics, cartography, GIS, and remote sensing. His research is on applications of GIS in urban planning, studies of civil armed conflicts, vulnerability to environmental hazards, and the implementation of GIS in upper secondary schools. He has published in *Environmental Planning B: Planning and Design, Political Geography, Journal of Conflict Resolution, International Studies Quarterly, International Organization, Conflict Management and Peace Science, Norwegian Journal of Geography,* and *Cartographica and Cartographic Perspectives.*

Concepción Romera is a geographer engineer at the *Instituto Geográfico Nacional*, Madrid, Spain. She is in charge of the CARTOSOPHIA Project and leads the Web page "*Cartografía para la enseñanza*" (Cartography for teaching). She deals with the management of knowledge and with training in thematic cartography. She is the vice president of cartography in the *Sociedad Española de Cartografía, Fotogrametría y Teledetección* (Spanish Society of Cartography, Photogrammetry and Remote Sensing).

Eric Sanchez is professor at the faculty of education of the University of Sherbrooke, QC (Canada), and assistant professor at *École Normale Supérieur de Lyon* (France). His research work concerns the uses of information and communication technology for educational purposes (elearning, simulation, serious games).

John A. Schembri is a senior lecturer in Geography at the Mediterranean Institute of the University of Malta. He has a B.A. from the University of Malta in contemporary Mediterranean studies and history and later obtained an M.A. in the geography of the Middle East and the Mediterranean and a Ph.D., both from Durham University, UK. He is a fellow of the Royal Geographical Society. John is coordinator of the geography division at the Institute and lectures mainly in human geography. His main research interests and publications are in populations of walled towns, development in ports and harbors, and historical heritage along urban coastal areas. John is a regular contributor to courses organized by the International Ocean Institute in Malta and is involved in the local national geography examinations, the development of syllabi in geography for ordinary, intermediate, and advanced levels and has been involved in the development of geography at the university.

Henk J. Scholten studied mathematics and geography at the Vrije Universiteit Amsterdam and obtained his PhD in 1988. Since 1990 he is professor in spatial informatics at the Faculty of Economics of the Vrije Universiteit Amsterdam and Director of the Spinlab (www.spinlab.vu.nl).

Prof. Scholten is founder and CEO of Geodan, one of the largest European companies specialized in Geospatial Information Technology (www.geodan.nl). Prof. Scholten is advisor for the European Union and several ministries in the field of disaster management and GIS. He has written more than 100 articles and 9 books on GIS. In July 2009, the 'Lifetime Achievement Award'was given to Prof. Scholten by Jack Dangermond, founder of ESRI. This award is given to a person who has contributed significantly to advancing the science and technology of GIS throughout his career.

Bob Sharpe is an Associate Professor and Associate Dean of Arts (Academic Development) at Wilfrid Laurier University in Waterloo, Ontario. His primary teaching and research interests are at the intersection of human geography and geomatics. He has taught and supervised students for twenty years at both the undergraduate and graduate levels in the fields of urban and economic geography, geographic information systems, cartography, and geographic education. Whenever possible, he incorporates fieldwork and geotechnologies into student learning experiences. Among Bob's research interests are issues of change and conflict in the city, and the application of geotechnologies to geographic education.

Alexander Siegmund currently is a full professor and head of the Department of Geography at the University of Education Heidelberg, as well as an honorary professor at the Institute for Geography at the University of Heidelberg. He has set up the "Klaus-Tschira-Competency Center for Digital Geomedia in Schools" and leads the Research Group for Earth Observation (rgeo), which focuses on research in geography and geography didactics, the development of learning environments using geospatial technology and in- and pre-service teacher training. His PhD, completed in 1997, focused on the regional climate of the Baar region in Germany. He studied geography, economics, and education at the University of Mannheim and was a teacher at a trading school for several years. He also held a professor position at the University of Education in Karlsruhe (Germany), a researcher position at the University of Mannheim as well as lecturer positions at the University of Karlsruhe, the University of Applied Science Karlsruhe, and the University of Heidelberg.

Hans-Jörg Stark is a professor at University of Applied Sciences Northwestern Switzerland, School of Architecture, Civil Engineering and Geomatics, Institute of Geomatics Engineering. He has been teaching geographic information systems and science since 2004. Before that he had worked for more than twelve years in the GIS industry, mainly as GIS project manager and in the development of GIS applications. His research interests are in the field of Volunteered Geographic Information (VGI), collaborative mapping, Open Source GIS, and Geo-Marketing. Besides the initiation of the "Map your World" project for secondary level students, he is also the founder of the VGI Project OpenAddresses.org. Currently he is involved in many national research projects in the aforementioned fields.

Josef Strobl is head of the center for geoinformatics at the University of Salzburg, Austria. He studied geography, meteorology, and geology with an MSc and PhD in Geography from the University of Vienna, Austria. Since 1985, he has taught and conducted research in computer-assisted cartography, remote sensing, statistical methods, and geographical information systems at the University of Salzburg.

Geok Chin Ivy Tan is an Associate Professor of Humanities and Social Studies Education Academic Group and Sub-Dean, Diploma and Practicum, Office of Teacher Education at the National Institute of Education, Nanyang Technological University, Singapore. She has taught as a geography teacher and has been Head of the Humanities Department in secondary schools. She has also been a Gifted Educational Specialist (Geography) in the Ministry of Education, Singapore. Presently she is a Commission Member of the International Geographical Union – Commission on Geographical Education and also serves as a regional representative of the International Association for the Study of Cooperation in Education. She is an executive member of the South East Asian Geography Association and the Geography Teachers' Association, Singapore.

Vinod Tewari is chair professor for National Capital Region Studies at TERI University, New Delhi. He has more than four decades of national and international experience working on various spheres of urban development planning, management, and governance. His specific research, consultancy, and teaching interests are in the areas of urban policy, urban institutional reforms, urban infrastructure management and financing, and urban poverty alleviation. He has taught courses in postgraduate- and PhD-level teaching programs and short-term training programs related to urban policy, urban management, quantitative urban modelling, public policy, geographic information systems, and research methodology. He has published a number of books/reports (with publishers such as Johns Hopkins University Press, Concept publishers, Allied publishers, and World Bank) and a number of research papers on topics related to his areas of interest in refereed journals including *Regional Development Dialogue, International Regional Science Review, Environment & Planning B: Planning & Design, IIMB Management Review, Geographical Review of India, Problems de Croissance Urbaine dans le Monde Tropical, CEGET, Sankhya*, and *Analytical Geography*, and in the reports of the National Commission on Urbanisation, Government of India, World Bank, Washington, and UNCRD, Nagoya. For his pioneering work in the area of GIS and service planning in India during the 1980s, he received Geodyssey Award of Geodyssey Environmental GIS Research Initiative instituted by Autodesk Inc., California, USA, and Environmental Systems Research Institute (ESRI), Inc., Redlands, California, USA in March 1993.

Chetan Tiwari received his PhD in geography from The University of Iowa in 2008. He is currently an assistant professor of geography at the University of North Texas in Denton, Texas. He is also affiliated with the Center for Computational Epidemiology and Response Analysis (CeCERA) at UNT. His research interests are in the applications of Geographic Information Science and spatial analysis methods for problems in public health. Specifically, he is interested in using GIS for mapping disease burdens and for environmental health surveillance. He has previously collaborated with several health departments across the USA and the National Cancer Institute. He is also the primary developer for an open-source disease mapping framework called WebDMAP. He currently teaches GIS, medical geography, location modelling, and human geography courses to undergraduate and graduate students at the University of North Texas and the University of North Texas Health Science Center.

Jean Tong has been with ESRI Canada since May 2007 in the Education Industry Manager's role. Prior to this, she taught Grades 7–10 Social Sciences in Toronto. Her primary role at ESRI Canada is to introduce educators and administrators to Geographic Information Systems (GIS) by demonstrating how GIS can be used both in the classroom and for campus management in areas such as bus routing and planning. She is actively involved in presenting to educators and conducting training about the use of GIS in education. In addition, she is responsible for many other aspects of the education program such as creating resources and working with museums. Jean is a graduate of Canterbury University and McMaster University with degrees in Education and Geography.

Carmen Treuthardt studied geography at the University of Zurich. Since 1998 she has taught geography and political education at the Lucerne High school. She is co-author of the GIS book *Geografische Informations Systeme*, which will be used often in upper secondary education and is the first Swiss GIS training book for Grades 7–12 in German. This book was honored with "Special Achievement in GIS Award" at the Esri International User Conference 2007. Carmen Treuthardt is also the president of the Association of Swiss Geography Teachers (VSGG).

Revocatus Twinomuhangi is a Lecturer in the Department of Geography, Informatics and Climatic Sciences, College of Agricultural and Environmental Sciences, Makerere University. He received his doctorate from Makerere University. His current research interests are in GIS Applications in waste management, natural resources management, and climate change adaptation and mitigation.

Joop van der Schee is professor for geography education at the Center for Educational Training, Assessment and Research at VU University Amsterdam, The Netherlands. He teaches human geography, geography didactics, and in-service courses, and is coach for research projects of master students and a group of geography education PhD students. His research interests are related to the use of maps and GIS and thinking skills in geography education. He has written scientific articles as well as books for teachers about geography in education, thinking skills, and the use of digital maps. He is co-chair of the International Geography Olympiad and member of the IGU Commission on Geographical Education.

Kathrin Viehrig currently is a researcher at the Department of Geography at the University of Education Heidelberg in Germany. Her PhD research focuses on the effects of GIS use in secondary schools on student achievement in geography, particularly with regard to systemic and spatial thinking. She also works within the research project "HEIGIS", which in its first phase aims at developing a geographically, didactically, and psychometrically sound assessment instrument for geographic system competency. Kathrin graduated with a first state exam for teachers, an M.A. in European Bilingual Education and an additional qualification for intercultural education/ foreigner pedagogy from the University of Education in Karlsruhe, Germany. She has studied for a semester at the University of Auckland, New Zealand.

Xi Xiang is a doctoral student at the National Institute of Education, Nanyang Technological University, Singapore. Her research interests include using geospatial technologies to develop spatial thinking of students, and designing teaching and learning resources for geography learning.

Rawan Yaghi has been a teacher for 16 years and is now the coordinator of English language and extracurricular activities in Nabil Adeeb Sleiman Secondary Public School in Bednayel-North Bekaa, Lebanon. She is also the educational director of the nationwide project, Teach Women English (TWE). She is a teacher trainer with the British Council in a project funded by HSBC and in a partnership with the Ministry of Education.

Minori Yuda is an assistant professor at the Center for Spatial Information Science, the University of Tokyo. She obtained her Ph.D. in utilization of GIS in school education from Kanazawa University. Her research interests are utilization of GIS in school education in Japan and other countries, and development of applications and courses using GIS and WebGIS in school education. She has published articles on GIS in education in Japan and Finland.

Chapter 1
The World at Their Fingertips: A New Age for Spatial Thinking

Andrew J. Milson, Joseph J. Kerski, and Ali Demirci

1.1 Introduction

As the first decade of the twenty-first century comes to a close, it is clear that young people around the world have opportunities to learn in ways that are quite different from those of their parents and grandparents. The *New York Times* columnist Thomas Friedman has famously described this technologically interconnected world as flat (Friedman, 2005). Yet, geographer Harm de Blij cautions us to pay close attention to the differences that exist between places despite the forces of globalization (de Blij, 2008). Both authors, along with many others, have captured the numerous ways in which the Internet and personal computers have altered the acquisition and diffusion of information, along with patterns of commerce and culture. New terms have been coined in an attempt to describe this altered landscape. Young people who were born since the rise of personal computers, the Internet, and handheld digital devices are referred to as the "net generation" (Tapscott, 1998) or "digital natives" (e.g., Palfrey & Gasser, 2008), while the gap between the technological haves and the technological have-nots is summed up as the "digital divide" (e.g., Norris, 2001). One important question that has arisen for educators and citizens around the world is, "How do we effectively educate digital natives, while also working to narrow or eliminate the digital divide, in an era of globalization?" Any attempt to respond to that question requires an acknowledgment of the spatial relationships at play. "Where?" and "Why there?" become fundamental questions for understanding the twenty-first century world. These questions are the essence of the study of geography, and the importance of these questions in the twenty-first century has brought renewed attention to the geographer's primary tool: the map.

Maps have always been the most powerful tool in human history for understanding, analyzing, and managing the physical and human characteristics of the Earth's surface. Early maps drawn on paper were primarily used to locate features such as rivers, resources, roads, and settlements across the world. These maps helped

A.J. Milson (✉)
University of Texas at Arlington, Arlington, TX, USA
e-mail: milson@uta.edu

A.J. Milson et al. (eds.), *International Perspectives on Teaching and Learning with GIS in Secondary Schools*, DOI 10.1007/978-94-007-2120-3_1,
© Springer Science+Business Media B.V. 2012

the map reader to answer the most essential question asked in geography: "Where are things located?" However, the developments in science and technology, especially since the 1950s, have changed not only the shape of maps but also our understanding of maps in terms of their meaning, creation, and uses. Although cartographers throughout history produced maps on stone tablets, cloth, and paper, maps have changed from static geographic depictions to dynamic reference materials with the advent of advanced spatial technologies such as Geographic Information Systems (GIS), Remote Sensing (RS), and Global Positioning Systems (GPS). With these technologies, maps today enable users to observe, modify, and analyze spatial data and information. Today's map reader can more effectively investigate questions beyond "where?" such as "why are things located where they are?" "how and why do places differ from one another?" and "how do people interact with the environment across time and space?"

The birth of GIS in the 1960s was due to the convergence of a variety of factors such as improved database and graphics technologies, theoretical developments in spatial statistics, advancements in satellite remote sensing, and the need among government agencies to efficiently manage geographic data. The Canada Geographic Information System, developed by Roger Tomlinson and his colleagues in the early 1960s, is widely cited as a pioneering moment in the development of the technology and the use of the term GIS. During the 1960s, the Harvard Laboratory for Computer Graphics and Spatial Analysis developed the first raster GIS, and a Harvard student, Jack Dangermond, went on to create the Environmental Science Research Institute (Esri) in 1969. These technical and commercial advancements were fueled by academic advancements in the field of quantitative geography that outlined GIS concepts and theoretical frameworks for spatial analysis (e.g., Berry, 1968; McHarg, 1995). By the early 1980s, the lower cost of hardware and the advancements in computing and software development made GIS a cost-effective tool for managing spatial data and decision making for a variety of industries and government agencies. During the following two decades, it also became apparent that GIS was more than a set of computer technologies, but that it involved a distinct set of methods and ways of looking at the world, in essence, a science unto its own, or Geographic Information Science (Goodchild, 1992).

Instruction in GIS around the world was largely confined to universities from the 1960s through the 1980s. During the 1990s, however, a group of enterprising secondary educators began using it as an instructional tool. Since then, as evidenced by this book, GIS has spread around the world as educators have discovered its value in connecting students to issues ranging in scale from their local communities to global concerns, to outdoor education, to careers and skills demanded in the workplace, to rich content, and to inquiry-based teaching and learning.

Working with GIS involves the examination of spatial information to solve problems using computer hardware and software. The "G" part of GIS is a two-dimensional or three-dimensional map, such as a topographic map, a satellite image, or a map of population, biomes, or any other theme. Objects in a GIS represent real-world phenomena and are stored as points, lines, polygons, or images. The "I" part of GIS is the database behind the map, full of information about water quality, mineral resources, ecoregions, or a whole host of other variables. The "G" and "I" are integrated into the "S" part of GIS – the system, so that asking a question on the map simultaneously asks the same question of the database, and vice versa.

The transition from map reader to map analyst and interpreter is made easier with new spatial technologies, but these technologies do not substitute for users' spatial thinking skills. On the contrary, new spatial technologies require skills of spatial thinking that have often been neglected in formal schooling (NRC, 2006). Fortunately, the spatial thinking skills that are vital for understanding the world can be taught and strengthened through the use of GIS and related spatial technologies (e.g., Huynh, 2009; Lee & Bednarz, 2009). The emergence of GIS has revolutionized the way people explore and understand the world around them. The ability to capture, manage, analyze, and display geographic data and information has enabled GIS users to make decisions and solve problems as diverse as designing routes for buses, locating new businesses, responding to emergencies, and researching climate change. Each of these decision-making and problem-solving applications of GIS has the common focus on investigating the "whys-of-where." Students and teachers around the globe are using GIS in the secondary school classroom to investigate the whys-of-where as they study social and scientific concepts and processes, engage in problem solving and decision making about local and global issues, and broaden their skills in using this significant emerging technology. Undoubtedly, GIS has an important role to play in the education of digital natives for global citizenship and in the shrinking of the global digital divide.

The chapters in this book offer an international perspective on the use of GIS for teaching and learning in secondary schools in 33 countries (see Fig. 1.1). The global terrain of GIS in education cannot be characterized as flat. There are important differences between and within countries in terms of the context of schooling, the technological infrastructure, and the recognition of GIS as a vital tool for teaching and learning. One important thread that runs through the chapters, though, is that the pedagogical, curricular, and technological context is important – but not deterministic – of the success of GIS in education. The editors asked each author to guide the reader through an overview of the context of secondary schooling in the country, an explanation of the status of GIS in the country's schools, a case illustrating the use of GIS for teaching and learning in the country, and a discussion of the prospects for GIS in the country's schools in the future. To focus the description of the context of secondary education and GIS in the country, each author was asked to describe factors such as the structure of schooling, the level of centralization and standardization of the curriculum, the process of teacher preparation, and the status of GIS in the curriculum. The second major section of each chapter focuses on a model or case that illustrates how secondary school teachers and students in the country are using GIS. The authors tell the story of the use of GIS in these schools and classrooms and explain whether or not this case represents the norm in the country or is an example of innovations toward which others are striving. Some authors provided multiple cases to represent the diversity of approaches to GIS in education in their country. In a few cases where discernable use of GIS is limited or nonexistent in the secondary school setting, the editors asked the authors to identify and reflect

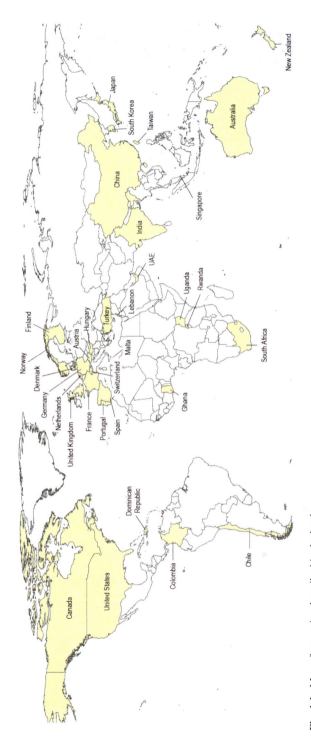

Fig. 1.1 Map of countries described in the book

upon the contextual factors in the country that influence the limited implementation of GIS in schools. Finally, each chapter closes with a discussion of the opportunities and challenges evident in the country in terms of infusing GIS into secondary schools.

The book as a whole represents a partial snapshot of how GIS is used in secondary education around the world. As with any attempt to capture an international perspective on educational practice, we were confronted with numerous questions. First, which countries should be included? At the outset, we conducted a detailed review of the literature and consulted with personal contacts to identify countries where GIS is used in secondary education. We also sought to include countries from every world region if possible. We are pleased with the number and diversity of countries that this process yielded, but we do not claim that the use of GIS in secondary education is limited to those countries covered in the book. We also made the decision to include a few countries where GIS use in secondary education is limited to provide a perspective on the prospects for growth in these countries. Second, how should we handle the variety of terms and phrases that are used in different countries to refer to GIS-related work? The reader will notice a variety of phrases used for GIS in the book. These include Geographic Information Systems, geomatics, geoinformation, geographic information technologies (GIT), geotechnologies, geographic information science and technology (GIS & T), WebGIS, and others. Rather than standardizing them, we respected the traditions that gave birth to the different phrases used around the world by leaving them intact. No matter what term is used, they all include GIS, remote sensing, GPS, and Web mapping with the goal to foster critical thinking through spatial analysis. Third, how should we group the countries in the book? Although various typologies for grouping the chapters of this book were considered, we decided that an alphabetical sequence suited the purpose of surveying the international landscape without imposing an artificial hierarchy. For the reader who wishes to focus on particular topics or structures, Tables 1.1 and 1.2 are offered as a guiding framework for the chapters in the book. A glance at Table 1.1 reveals that only a few countries, such as Taiwan and South Africa, have included GIS as a component of their national curriculum, while educators in many countries have established partnerships between universities or industries and secondary schools. Many of the chapters also describe projects that involve students in investigations of their local environment. Table 1.1 is not intended to display every topic within each chapter and the information displayed is based only on what was provided by the authors of these chapters. Similarly, Table 1.2 provides a summary of the case studies from each chapter to give the reader an overview of the projects described within each chapter. Finally, what does all of this mean for educators around the world? In the final chapter, we offer a synthesis of the emerging international trends that are apparent across the chapters in the book. It is our hope that you will find the work described by these authors as inspiring and intriguing as we do. We welcome your feedback.

Table 1.1 Overview of selected topics within chapters

Country	Formal national curriculum	GIS in national curriculum[a]	Industry/University partnerships	Local investigations	WebGIS
Australia	■			■	
Austria	■		■	■	■
Canada			■	■	
Chile	■		■	■	■
China	■	■	■	■	■
Colombia	■		■		■
Denmark			■	■	■
Dominican Republic	■		■	■	
Finland	■	■	■	■	
France			■	■	■
Germany			■	■	■
Ghana	■				
Hungary	■		■	■	■
India	■	■	■	■	
Japan	■		■	■	■
Lebanon	■		■		
Malta	■				
Netherlands	■		■	■	■
New Zealand	■		■	■	
Norway	■	■	■	■	■
Portugal	■		■	■	■
Rwanda	■		■	■	
Singapore	■		■	■	
South Africa	■	■	■	■	
South Korea	■				
Spain	■		■		■
Switzerland			■	■	■
Taiwan	■	■	■	■	■
Turkey	■	■	■	■	
Uganda	■				
United Arab Emirates	■		■	■	
United Kingdom	■	■	■	■	
United States			■	■	

[a]Includes elective courses

Table 1.2 Overview of case studies

Country	Case study	Description
Australia	The King's School	Multiyear GIS program in geography, Year 7–12, plus other disciplines and in Senior Geography Project studying coastal processes
Austria	Schools on Ice	Local effects of climate change; fieldwork in glaciated mountains
	Location-based games	"Applications on the move" – 6 games
	Geovisualization in participatory decision-making processes	Effects of geovisualization in decision-making processes; spatial planning development and evaluation
Canada	Hants East Rural High School, Nova Scotia	Geomatics course
	Neelin High School, Manitoba	International baccalaureate program
	St Michaels University School, British Columbia	GIS through experiential geography program in multiple grade levels
Chile	ArcReader and open-source platform	Use of learning management system to study gold mining in Andes
China	Santiago study by female students in Providencia	Study and encourage use of bike trails in the community
	Experimental high school projects for arts majors	GIS in geography class: agricultural planning
	Experimental high school projects for science majors	Locating optimal routes to get to school and optimal living location in community
Colombia	Distrital university and GIS professional society project to create WebGIS	Understanding the economic and physical geography of Colombia
Denmark	Silkeborg Gymnasium	Freeway impact study

Table 1.2 (continued)

Country	Case study	Description
Dominican Republic	Instituto Tecnologico de Las Americas partnership with Boca Chica students	Examination of the Haitises section of Hispanola
Finland	Jaakola Lower Secondary School	Geocaching project
	PaikkaOppi project	Web-based GIS curriculum and development
France	Game research and development	Land use plan associated with high-speed train
	Game research and development	Sustainable energy development
Germany	Thadden Gymnasium, Heidelberg	Tree mapping project
Ghana (no case study)		
Hungary	Biotic index at secondary education level	Over 70 schools; courses and fieldwork
India	Almora district, Uttaranchal	Mapping the neighborhood, participatory mapping project
Japan	Tsukuba City, Ibaraki	Junior High School neighborhood safety project
	Takasaki High School	Cell Phone GIS project; mulberry land use change
Lebanon	Nabil Adeeb Sleiman Secondary Public School	GIS for teaching natural hazards and environmental problems
Malta (no case study)		
Netherlands	Merewade High School, Gorinchem	Services and customers survey and mapping project for local businesses
New Zealand	Botany Downs Secondary College	Tsunami and resources studies; creating an island; community atlas
	Chilton Saint James School	Recreation facility planning; vineyard study; cyclone analysis

Table 1.2   (continued)

Country	Case study	Description
Norway	Map in the school	Web Atlas for Norway for multiple disciplines
Portugal	Google Earth in education	Visualization and mapping
	ConTIG project	Four schools; field trips; ArcPad and mobile mapping; bird studies
Rwanda	Ecole Technique Officiel (ETO) Gitarama Technical Secondary School	Campus mapping; land use mapping
Singapore	Raffles Institution	Volcano science and hazards
South Africa	St David's Marist Inanda private school	Variety of projects over multiple years including a research project
	Afrikaans Hoër Seunskool public school	Paper-based GIS exercises; guest lectures
	Hoërskool Sanctor School	Use of transparencies and other noncomputer tools to teach GIS concepts
South Korea	GIS implementation study of 9 schools	Modified from Audet and Paris
Spain	Instituto Geográfico Nacional	Developed e-learning courses and online atlas
	Instituto Cartográfico de Andalucía	Data portal; WebGIS; curriculum development
Switzerland	Matura School, Lucerne	Volcano study; sea level change study; field collection; tourist map

Table 1.2 (continued)

Country	Case study	Description
Taiwan	National Yilan Senior High School	3D campus mapping; urban heat island; GPS survey; viewshed analysis for landscape pavilion; sea level rise simulation
Turkey	Anatolia Teacher High School	City minibus lines study
	The Scientific and Technological Research Council of Turkey	3 high schools; 9 GIS-based projects, including solid waste, air quality in school, noise and marine pollution
Uganda	Rainbow International School	WebGIS for land use and settlement pattern study
United Arab Emirates	Al-Ain Al-Namothajia secondary school	Planning new school locations
United Kingdom	Bishop's Stortford College	Presenting, processing, analyzing, inputting, and editing data
	Spatially speaking GIS support project	Professional development, curriculum development, and research
United States	Piner High School, California	Articulation with junior college; career technology focus; stream studies; surveying

References

Berry, B. (1968). *Spatial analysis: A reader in statistical geography*. New York: Prentice Hall.

De Blij, H. (2008). *The power of place: Geography, destiny, and globalization's rough landscape*. London: Oxford.

Friedman, T. L. (2005). *The world is flat: A brief history of the 21st century*. New York: Farrar, Straus, and Giroux.

Goodchild, M. F. (1992). Geographical information science. *International Journal of Geographical Information Systems, 6*, 3–45.

Huynh, N. (2009). The role of geospatial thinking and geographic skills in effective problem solving with GIS: K-16 education. Ph.D. dissertation, Wilfrid Laurier University, Canada. Retrieved November 9, 2010, from Dissertations & Theses: Full Text. (Publication No. AAT NR54258).

Lee, J. W., & Bednarz, R. (2009). Effect of GIS learning on spatial thinking. *Journal of Geography in Higher Education, 33*(2), 183–198.

McHarg, I. (1995). *Design with nature* (2nd ed.). New York: Wiley.

National Research Council. (2006). *Learning to think spatially: GIS as a support system in the K-12 curriculum*. Washington, DC: National Academies Press.

Norris, P. (2001). *Digital divide: Civic engagement, information poverty, and the Internet worldwide*. Cambridge: Cambridge University Press.

Palfrey, J., & Gasser, U. (2008). *Born digital: Understanding the first generation of digital natives*. New York: Basic Books.

Tapscott, D. (1998). *Growing up digital: The rise of the net generation*. Columbus, OH: McGraw-Hill.

Chapter 2
Australia: Inquiry Learning with GIS to Simulate Coastal Storm Inundation

John C. Kinniburgh

2.1 Introduction

The study of geography is vital to the education of every young Australian as it provides students with a holistic view of the world, combining the natural and social sciences. Students of geography gain the understanding, knowledge, and skills to make sense of complex issues such as climate change, drought, aging populations, urban growth, ethnic conflicts, and globalization (National Committee for Geography, 2007, p. 2). Geography studies have long been part of the Australian school curriculum and this is certain to continue with a national curriculum to be implemented from 2011 onward. After at least two decades of discussion and debate about what Australian children should and should not learn, the Australian Curriculum, Assessment and Reporting Authority (ACARA) was established in May 2009 after an act of parliament in December 2008. The functions of ACARA are to develop and administer a national school curriculum, including content of the curriculum and achievement standards for school subjects. The development of the geography curriculum will take place from July 2010 to June 2011 with implementation soon after.

As the new national geography curriculum takes shape, a unique opportunity exists to examine the role of Geographic Information Systems (GIS), within the teaching and learning of geography in Australia. Since the emergence of GIS in education in the early 1990s (in the United States), proponents have identified the technology as a powerful tool that supports research-based investigations and encourages inquiry learning. Significant interest in GIS has arisen due to its ability to examine real-world issues and the fact that it encourages a diverse range of students to actively engage in their own learning; hence it is supportive of the constructivist paradigm.

This chapter describes the way in which GIS is perceived within secondary schools in Australia. In particular, it will provide a general outline of its use at the

J.C. Kinniburgh (✉)
The King's School, Sydney, NSW, Australia
e-mail: jck@kings.edu.au

A.J. Milson et al. (eds.), *International Perspectives on Teaching and Learning with GIS in Secondary Schools*, DOI 10.1007/978-94-007-2120-3_2,
© Springer Science+Business Media B.V. 2012

secondary level and it will present a case study of a GIS-based learning activity conducted at The King's School, Parramatta, in Sydney. The chapter will conclude with perspectives about the future role of GIS within Australian geography classrooms.

2.2 The Context of Secondary Education in Australia

Australia follows a three-tier model of schooling that includes primary education, secondary education, and tertiary education. Secondary schooling in Australia covers the school Years 7 or 8 through to Year 12. Education is compulsory in Australia between the ages 5 and 15 to 17, depending on the state or territory, and date of birth. The minimum leaving age is generally at the end of Year 10. The key learning areas in Australian secondary schools are English, math, studies of society and the environment, science, foreign languages, technology, health, and physical education. Year 12 students can obtain a Senior Secondary Certificate of Education (referred to differently in each state and territory) that is nationally recognized for the purpose of admission to universities, Technical and Further Education (TAFE) colleges and technical institutes. Other students may choose to attend schooling that is complemented by vocational training in order to better prepare them for the job market.

Education in Australia is primarily the responsibility of states and territories with each respective government providing the funding and regulations for the public and private schools within its governing area. Government schools (often referred to as "public schools") are free for Australian citizens, while Catholic and Independent schools charge fees. Irrespective of whether a school is part of the government, Catholic, or independent systems, they are expected to adhere to the same curriculums of their state or territory.

As each state or territory provides its own secondary education curriculum, there is inconsistency in what is taught. For example, both New South Wales and Victoria have structured curriculum documents that compel educators to teach content-specific detail. Others, including Queensland, have broader curriculum frameworks from which lesson content is administered. As a result of these differences, the quality of teaching varies due to the nature of learning that is conducted in the classroom in each jurisdiction.

2.2.1 The Nature of Geography Education in Australian Schools

In recent years, significant attention has been drawn to issues surrounding the study of geography in Australian secondary schools with a number of concerning trends emerging, including a decline in student numbers. Hutchinson and Pritchard (2006, pp. 16–18) refer to the dilemma in New South Wales and highlight that the number of students taking Higher School Certificate Geography (Year 12) is less than one third of its level of 15 years ago.

The media has also shown some interest in the debate, especially in terms of curriculum content and the way in which geography is taught in schools. "Naïve syllabus neglects basics" (Ferrari, 2006a, p. 1 in *The Australian* newspaper) and "The Geography Wars" (Ferrari, 2006b [online] in *The Australian* newspaper) provide an insight into the media's agenda. Secretary of the International Geographical Union's Commission on Geographical Education, Professor John Lidstone [cited in Ferrari, J. (2006a) *The Australian* (September, 25), p. 1] suggested that the Geography "syllabus lacks coherence and tends to become issues-based" and that "you're asking kids to solve problems that adults and politicians can't solve."

A number of factors are likely to have contributed to the decline in the teaching of geography in Australia up to this point. Some relate to curriculum structures and others more generally to teaching and learning, school organization, and teacher professional development (Erebus International, 2008, p. 36). Obstacles include:

– Problems caused by amalgamation of geography into studies of society and environment (SOSE);
– Loss of status for geography and geography teachers;
– Loss of priority for geography in an overcrowded curriculum;
– The shortage of suitably qualified geography teachers;
– Failure to engage students; and,
– Lack of incentives for the study of geography in senior years (pp. 36–44)

Generally, there exists a strong expectation in Australian state and territory curriculum documents that in the compulsory years of schooling, young people will be exposed to key concepts of place and space, specific skills such as map reading, and the development of higher order skills of analysis and interpretation of data to assist them to understand the world they live in (Erebus International, 2008, p. 58). Whilst this view continues to be important, there has been significant variation in what is taught in geography between each state and territory. This is due to the fact that each region has control over what is taught in primary and secondary schools.

Currently, New South Wales is the only state or territory which requires all students to complete geography as a core subject before the end of their compulsory years of school (end of Year 10). Two hundred hours of geography are mandated over Years 7–10 whilst in other states and territories, students cover a combination of social science areas, together with history, within the compulsory studies of society and the environment (SOSE) learning area. At the senior secondary level, geography is often available to students as an elective subject in most states and territories.

2.2.2 Tertiary Training and Post-university Support of Geography Teachers

In Australia, individuals seeking a teaching qualification are required to complete at least four years at a tertiary institution. They also need to hold a Bachelor's degree and a one or two year professional qualification, for example, a Graduate Diploma or Masters in Education. Once completed and the individual enters the classroom, geography departments or faculties within individual schools play an important role in supporting teachers. Much of this support comes from national geography organizations (for example, the Australian Geography Teachers Association, the Institute of Australian Geographers, and the Royal Geographical Society of Queensland) as they are well placed to provide opportunities for teachers to broaden their knowledge, enthusiasm, and expertise beyond a local context through such avenues as national conferences. Also, state and territory education departments and curriculum and assessment authorities provide appropriate structures to ensure that teachers are well prepared to teach geography within their system (National Geography Curriculum Steering Committee, 2009, p. 84).

2.3 The Use of GIS in Australian Geography Classrooms

Houtsonen (2003, p. 57) states, "GIS presents geography education with one of its greatest opportunities – and its greatest challenges." Indeed there has been continued debate about the effective integration of GIS to education including Kerski (2000) and Bednarz (2004). Others, including West (2006, p. 467), have noted that there is not a single empirical study that categorically answers the question of whether GIS should be used in schools. The challenge for geography teachers is to effectively use resources such as GIS in constructive and innovative ways to promote student learning.

It can be said that generally the adoption of GIS in Australian secondary schools is variable, except within Queensland and South Australia, where it seems that GIS has been more widely accepted. The reasons for this appear consistent with the broader impediments to the adoption of GIS, but one important factor is likely to be the limited exposure that undergraduate education students receive whilst completing their teaching qualification. Generally, very little is taught to undergraduate education students about the merits of using GIS as an educational tool within the geography curriculum area. As noted by Hutchinson (personal communication, September 2010), there is some work being undertaken at the University of New South Wales and the Australian Catholic University and also at Macquarie University where one or two sessions, of approximately 3 hours duration, are conducted each semester. A similar situation exists in Victoria where, according to Kriewaldt (personal communication, September 2010), students learn the basics of GIS applications in the classroom in a two-hour seminar during the latter part of their teaching qualification. Most students are unfamiliar with commercial GIS software unless they have previously used GIS in their undergraduate geography degree. It is highly likely that a similar situation exists at other major universities around the country. Bliss (personal communication, September 2010) suggests that university lecturers are either not trained in GIS, have limited IT skills, or are simply not interested in showcasing its merits.

Following is a brief outline of the extent to which GIS is utilized within selected states.

South Australia – South Australia has been a success story in Australia in terms of the wider adoption of GIS in schools. In 1996, the Spatial Industry introduced an Australia wide "Australasian Urban and Regional Information Systems Organisation (AURISA)" GIS in schools competition. South Australian schools were offered training between 1996 and 2001 and during these years, 113 teachers from 33 schools received free GIS training and access to the GIS software. Also involved in this project was Malcolm McInerney, a geography teacher from Findon High School and now an internationally regarded advocate for, and exponent of, the educational use of GIS in the classroom. In an effort to introduce GIS into classrooms around Australia, Malcolm developed a "GIS Skill Development Course" for secondary students and a range of across the curriculum resource materials. Largely as a result of the efforts of Malcolm McInerney, South Australia is recognized around Australia as being a leader in the educational use of GIS. Between 1999 and 2001, the Senior Secondary Assessment Board of South Australia (SSABSA) initiated the integration of GIS into the curriculum. In 2002, the SSABSA released the new 2003 SACE Stage 1 Geography course that incorporated GIS into the methodology and assessment components. Mandating the use of GIS at a curriculum level has encouraged even wider adoption of the technology throughout the state. The Geography Teachers' Association of South Australia (GTASA) and the Technology School of the Future (TSoF) developed GIS teacher training resources and by the end of 2002, 125 teachers had introductory GIS training. Since 2007, the GTASA, in partnership with the Department of Education and Children's Services (DECS) and Esri Australia, have conducted online training for teachers using the CENTRA program. In 2009 and 2010, GTASA employed external consultancy Contour Education (operated by Mick Law) to conduct staff training. Diffusion of GIS, however, continues to be largely the result of the efforts of motivated and innovative individuals including Malcolm McInerney and Ross Johnson. One of the leading schools in South Australia integrating GIS-based learning is Glenunga International High School where some excellent work is being undertaken by a number of teachers led by Paul Ridge, Roy Croft, and Ross Cameron.

Queensland – Some excellent examples of GIS use have also been recorded in the state of Queensland and a number of the early adopters, including Bryan West and Brett Dascombe, were responsible for promoting its early use. The Queensland Studies Authority (QSA) has also advocated the use of the technology in their junior SOSE and senior geography courses. The appointment of Meegan Maguire as Geography Project Officer for the Queensland government resulted in further promotion of GIS use in schools by encouraging teachers to use different approaches to develop spatial literacy skills. This role also nurtured partnerships and cooperation between schools and industry and also acquired data and software for teachers. Meegan Maguire also developed the Spatial Technology in Schools (STiS) project during this time and coordinated a number of successful conferences promoting spatial technologies in education. A spatial technology education consultant Mick Law (Countour Education), has provided professional development training in the form of workshops and in-school training as well as resource support to teachers throughout the state. Over the past two years, the Geography Teachers' Association of Queensland (GTAQ) has worked actively to connect regional teachers to industry so that they are able to gain training in the use of GIS. A number of key educators have also provided localized and wider support through conference presentations and workshops including Tony Dawson, Neil Gray, Mark Camman, Mike Railton, and Bec Nicholas.

Tasmania – Whilst the adoption of GIS technology in Tasmanian schools has not occurred on a large scale, there are a number of excellent examples of GIS being used in the classroom. In particular, Devonport High School has a well-developed program for students in Years 7–10. Students in the junior years learn the basics of GPS navigation as well as developing their own GPS treasure hunts and GIS virtual tours. Those in Years 9 and 10 can study a Spatial Technology elective in which they learn about all aspects of GPS, GIS, remote sensing, and surveying. Developed by teacher and spatial education consultant Darren Llewellyn (who was awarded the National Technology Teacher of the Year in 2008 for his spatial initiatives), the school is a leader in community-based GIS inquiry activities. One notable project completed by students involved them teaming up with industry experts and volunteers from the Tasmanian Arboretum (a 140-acre living tree museum) to help update and develop the site's GIS. This involved the installation of 37 survey reference points and the mapping of the entire site's infrastructure, including 5,000 trees. There have also been some schools at the primary level which have utilized GPS technology. Year 5 and 6 students at Miandetta Primary School have completed beach profile surveys using survey equipment to investigate the effects of coastal erosion. There are also some schools at the pre-tertiary level (Years 11 and 12), which also use GIS in their geography courses. These include public schools such as Hobart and Elizabeth College, and private schools such as Launceston Church Grammar and Hobart Collegiate.

Victoria – Whilst there appears to be variability in the uptake of GIS within geography classrooms in Victoria, a number of schools are acknowledged as having integrated GIS to some extent into their curriculum. These include Presbyterian Ladies' College Melbourne and Camberwell High School. Also, the Geography Teacher's Association of Victoria (GTAV) conducts GIS professional learning programs but those who attend represent only a small proportion of all teachers who are teaching Geography in Victoria. The GTAV runs about 10 professional development sessions for beginners and more advanced users each year and it also coordinates a GIS network meeting three times per year.

Western Australia – The adoption of GIS in West Australian schools has not been widespread except for within a few independent schools. This is despite the fact that computers have become more accessible and new courses in geography and earth and environmental science make direct reference and mandate the use of GIS. Meaningful professional development, easy access to data as well as software, are among the factors identified by leading WA GIS educator, Phil Houweling, as being necessary and important for GIS to be more widely integrated. Some positive steps have been taken to address these areas, particularly the issue of data accessibility. Landgate (formerly Department of Land Information) in Western Australia has developed SLIP (Shared Land Information Platform), which is a revolutionary portal that enables teachers to search, view, and access over 350 spatial datasets for use within their classroom. Western Australian Land Information System (WALIS) has been a strong supporter and partner in advancing the educational opportunities of students within Western Australia. WALIS forums have seen the inclusion of a teacher's stream, which provides teachers with an excellent professional development opportunity and enables them to connect with industry to expand their knowledge of spatial technologies. Curtin University also offers introductory and advanced GIS workshops to teachers as well as free introductory GIS workshops to students. Schools have also participated in the Spatial Technologies in Schools competitions with WA students performing very well, including those from John Calvin Christian College, arguably the state's leading GIS school.

New South Wales – In New South Wales, there have been a number of schools that have attempted to integrate GIS in their teaching of geography, with varying degrees of success. These schools are primarily independent and include The King's School, St. Ignatius' College Riverview, Barker College, and Kincoppal Rose Bay. The independent schools have been able to embrace the technology due to the fact that they are well resourced in terms of computer infrastructure and capital is available to purchase GIS data and software. It also is likely the case that in each of these schools there is an individual or small number of people solely responsible for implementing GIS because they are both interested and motivated by its potential.

Despite the positive achievement of these schools and its enthusiastic individuals, there is little evidence to support the wider adoption of GIS in New South Wales, particularly in government schools. A recent survey by Kinniburgh (2008) attempted to identify the impediments to using GIS to enhance learning and teaching in mandatory Stage 5 Geography (Years 9 and 10) in New South Wales. Whilst the study reinforced the impediments previously identified by other authors in their research, additional factors also emerged. First, it was found that 42 % of respondents indicated that their school owned some form of GIS software but it was not being used at all. A further 37% indicated that one or more teachers were investigating GIS but it was not being used in the classroom. Of particular concern was the fact that not one respondent indicated that GIS was being used on a regular basis (i.e., three times per term).

A key finding of the study was the impact of the externally set School Certificate Examination (externally set state-wide tests in English literacy; mathematics, science, Australian history, geography, civics and citizenship; and computing skills) upon the adoption of GIS in Year 5 geography classrooms. Seventy-four percent of respondents agreed or strongly agreed that the School Certificate test acted as a disincentive for teachers to use GIS to develop a student's conceptual understanding of geographical phenomenon. Another 58% agreed or strongly agreed that the emphasis placed on the School Certificate test limited the scope for utilizing GIS in Year 10 Geography. The overcrowding of the Stage 5 (Years 9 and 10) syllabus was also highlighted as being important with teachers indicating that they were unwilling to learn and utilize demanding technologies when vast amounts of syllabus content needed to be taught in preparation for the School Certificate Examination. This is in distinct contrast to South Australia and Queensland where revised curriculums have mandated the use of GIS in the teaching of geography.

2.4 The Australian Geography Curriculum

A number of detailed reviews (e.g., Erebus International, 2008) have highlighted the benefits of geography as a "core" subject. As indicated by the National Committee for Geography (2007, p. 3), the study of geography in Australian schools is essential to the development of all young people, and to the economic, environmental, and cultural future of Australia. Whilst this may be the case, a number of challenges lie ahead for each state and territory as the transition occurs from state-based syllabuses to the new Australian curriculum.

In the recently released framing paper "Draft Shape of the Australian Curriculum: Geography" (2010), it is acknowledged that geography provides many opportunities to learn and use information and communications technology (ICT) skills. It goes further to specifically acknowledge specialized spatial technologies, digital and electronic maps, 2D and 3D electronic maps, global positioning system (GPS), remote sensing, and GIS as being "rapidly growing areas of ICT" (p. 9). The growing body of research suggests that GIS-enhanced inquiry can be used to achieve meaningful alignment with curriculum standards (Coulter, 2003, p. 4) and what is needed therefore is a relevant curriculum that embraces the potential of GIS. Within the proposed geography curriculum, it is likely that spatial technologies such as GIS will be incorporated within the scope and sequence for Years 7–10 as a geographical inquiry skill. It is still unknown, however, as to whether the proposed Australian geography curriculum will fully enable GIS to be used as a resource to support inquiry based learning activities.

2.5 An Example of a GIS-Based Student Activity

Alternatives to teacher-centered instructional frameworks have emerged and include a movement toward methodologies based upon constructivist learning theory. The underlying premise of these strategies is that learning is an active process in which learners are effective sense makers who seek to build coherent and organized knowledge. GIS has been highlighted as a technology that can be an invaluable resource for extending student learning when a proper instructional framework is provided, along with data analysis and spatial reasoning concepts (Baker & White, 2003; van der Schee, 2003). GIS was not developed for education and with the constructivist philosophy in mind, however, it is well suited to learning activities that are student-centered, active, and driven by a process of investigation.

The King's School is a comprehensive boys' boarding school and the oldest independent school in Australia (founded in 1831). Since 2000, the school's geography faculty has integrated GIS-based teaching and learning strategies in all years from 7 to 12. Using the syllabus guidelines provided by Stages 4, 5 and 6 (Years 7 to 12) geography curriculum, students engage in GIS-based learning activities that range from step-by-step instruction style lessons from the Esri Press book *Mapping Our World* (Malone, Palmer, Voigt, Napolean, & Feaster, 2005) to those produced by the author, which are developed using inquiry or problem-based frameworks. Some of these are highlighted in Table 2.1.

	Stage	Year	Topic
Table 2.1 Examples of GIS-based learning activities used in the teaching of geography at The King's School in Years 7–12	4	7	• Global biomes • World heritage sites
		8	• Plate tectonics • Global climate change
	5	9	• Communities • Land use suitability assessment
		10	• Development geography • Spatial inequality in Sydney
	6	11	• Impacts of storm inundation on coastal areas • Urban growth and decline
		12	• Global dairying • Intertidal wetlands

GIS has also been used in cross-curricular applications. These include investigations within agriculture, science, and history. The school's army cadet unit has also utilized GIS technology to develop students' navigation skills in preparation for their annual camp.

The most advanced GIS-based investigations, however, are undertaken by students in their senior years. Year 11 geography students successfully partner geographic inquiry and GIS whilst completing their Senior Geography Project (SGP), a mandatory requirement of the Preliminary course. The project is unique as it relies upon geographic inquiry, the application of knowledge and research skills to produce a practical and independent student research project. In this example, students learn about the coastal biophysical environment through primary fieldwork and secondary research (Fig. 2.1).

To complete this activity, students initially learn about the functioning of coastal environments including its processes and the formation of landforms. They then design a series of research questions to investigate one issue affecting the coastal area, and which can be investigated locally. Using primary and secondary resources, students then conduct a geographical investigation of the issue.

The study area for the investigation is the Narrabeen/Collaroy embayment located on Sydney's northern beaches approximately 16 km north of the CBD. The area has been well documented and surveyed ever since coastal erosion problems were first recognized as early as 1925. Severe storms in the 1960s and 1970s (particularly 1974) focused attention on a greater need for appropriate coastal management strategies. Many of the coastal processes along this stretch of coastline do not operate in isolation but interact in complex ways. Many of these interactions create coastal hazards including inundation and coastal erosion. These problems are further exacerbated by the improper foredune development that occurs along the coastline. These issues provide the context for the student investigations in the study area.

Fig. 2.1 Year 11 geography
students from The King's
School conducting fieldwork
at Collaroy Beach along
Sydney's northern beaches

Students are required to complete fieldwork to investigate the area and examine
the competing uses along the coastal area. GPS technology is used to ground truth
and map sensitive locations where human activities and natural processes interact.
This information is collated and validated through secondary research.

When back in the classroom, students import their primary data into Esri's
ArcGIS and create a series of maps that highlight areas of concern along the coast.
Students then perform more advanced spatial analysis using Spatial Analyst and 3D
Analyst extensions. Aerial photography and a Digital Elevation Model are used by
the students to create a series of hazard maps that highlight the potential threat of
inundation and storm erosion. The reclass function within Spatial Analyst enables
the DEM to be reclassified with only two ranges displayed; data that is above and
below an identified inundation height, for example 2.5 m. The layer properties of
the resultant grid are adjusted and when overlaid onto the aerial photograph of the
area, it is possible to observe the potential impact of inundation such as storm surge
(Fig. 2.2). Students then repeat the procedure to generate other inundation scenarios
informed by research and literature review.

The adoption of GIS by Year 11 students at The King's School completing their
SGP has generated significant interest and enthusiasm in both coastal management
issues as well as GIS technology. Students clearly recognize the value of GIS technology when they realize the potential that it holds. These include being able to
generate "what-if" scenarios, enabling them to consider future problems that may
occur and the management strategies that may need to be considered in the future.

Fig. 2.2 Simulated storm inundation height of 2.5 m

The GIS not only enhances students' conceptual understanding of issues in the coastal environment but it also exposes them to "real-world" problems that are regularly generated and considered at local council and government level. There is no question that this research activity provides students with an engaging and active learning experience, enhanced by integrating primary and secondary research with GIS technology.

2.6 Conclusion

The study of geography in Australia is considered vital to the education of secondary school students, however, it is no longer appropriate for geography to be taught using traditional methods that focus upon the direct dissemination of knowledge and facts. Students should be able to explore their topic areas independently so that their learning experience is valid and worthwhile. This can be supported through the adoption of learning technologies such as GIS that facilitate research-based investigation and inquiry learning.

Whilst there are numerous examples of effective and innovative GIS use in Australian classrooms, its adoption is not widespread nor is it consistent. South Australia and Queensland (and Western Australia to a lesser degree) have provided leadership by way of revised curriculum documents that mandate the use of GIS. These initiatives have led to a range of positive initiatives that include coordinated staff professional development programs as well as the development of relationships between industry experts and teachers. In other states such as Western Australia,

Tasmania, Victoria, and New South Wales, the use of GIS in geography classrooms has fallen to a handful of individuals who are motivated by its affordances and who are interested in its potential as a learning technology. Many of these also appear to be in the independent sector where schools are well resourced and have very good computing infrastructure. There are other issues also and these include inadequate training of education students at the tertiary level.

There is no question that for GIS to be widely adopted, a more formal approach is needed. With a new national curriculum to be implemented from 2011, the opportunity to broadly integrate GIS in geography classrooms now exists. For this to be successful, a number of things are needed. First, explicit reference is required for the use of GIS as a tool to enhance learning and skills in geography. Enough flexibility is also required in the curriculum to allow the full breadth of GIS-based investigations to be undertaken. In addition to these, there needs to be a formalized program of professional development that supports teachers in schools and this could be supported by the spatial technology industry. The training of young teachers in the use of GIS must be more effective at the university level also as the current situation is not providing effective training for future geography educators. A final consideration must also be given to the nature of GIS-based pedagogies and the design of effective instructional frameworks that realize the potential of GIS in the classroom.

A number of challenges lie ahead for geography educators in Australia, however, as has been shown, a great deal has been achieved already with the integration of GIS in the classroom. It would be a great disappointment if the achievements of the past were not able to be developed further and the potential of GIS realized.

References

Baker, T. R., & White, S. H. (2003). The effects of GIS on students' attitudes, self-efficacy, and achievement in middle-school science classrooms. *Journal of Geography, 102*(6), 243–254.

Bednarz, S. W. (2004). Geographic Information Systems: A tool to support geography and environmental education? *GeoJournal, 60*, 191–199.

Coulter, B. (2003, July). *Maximising the potential for GIS to enhance education.* Paper presented at the ESRI education user conference, San Diego, CA.

Erebus International. (2008). A Study into the Teaching of Geography in Years 3–10. Australian Government Department of Education, Employment and Workplace Relations.

Ferrari, J. (2006a). Naive syllabus neglects basics. *The Australian,* 25 September, 2006, p. 1.

Ferrari, J. (2006b). The geography wars. *The Australian,* 28 September 2006, http://www.theaustralian.news.com.au/story/0,20867,20487109-28737,00.html

Houtsonen, L. (2003). Maximising the use of communication technologies in geographical education. In R. Gerber (Ed.), *International handbook on geographical education* (pp. 47–63). Dordrecht: Kluwer Academic Publishers.

Hutchinson, N., & Pritchard, B. (2006). 'True Blue' geography. *Geography Bulletin, 38*(4), 16–18.

Kerski, J. J. (2000). *The implementation and effectiveness of Geographic Information Systems Technology and methods in secondary education.* Ph.D. Dissertation, University of Colorado.

Kinniburgh, J. C. (2008). An investigation of the impediments to using Geographical Information Systems (GIS) to enhance teaching and learning in mandatory Stage 5 geography in New South Wales. *Geographical Education, 21*, 20–38.

Malone, L., Palmer, A., Voigt, C., Napolean, E., & Feaster, L. (2005). *Mapping our world: GIS lessons for educators – ArcGIS desktop edition*. Redlands, CA: Esri Press.

National Committee for Geography. (2007). *Australian's need geography*. Canberra: Australian Academy of Science.

National Geography Curriculum Steering Committee. (2009). *Towards a national geography curriculum for Australia – Background report*, June 2009. National Geography Committee. www.ngc.org.au/report/index.htm

van der Schee, J. (2003). New media will accelerate the renewal of geographical education. In R. Gerber (Ed.), *International handbook on geographical education* (pp. 205–214). Dordrecht: Kluwer.

West, B. A. (2006). Towards an understanding of conceptions of GIS. In K. Purnell, J. Lidstone, & S. Hodgson (Eds.), *The proceedings of the international geographical union commission on geographical education symposium, changes in geographical education: Past, present and future* (pp. 467–471). Brisbane: International Geographical Union Commission on Geographical Education and Royal Geographical Society of Queensland.

Chapter 3
Austria: Links Between Research Institutions and Secondary Schools for Geoinformation Research and Practice

Thomas Jekel, Alfons Koller, and Josef Strobl

3.1 Introduction

Austrian school legislation is provided at the federal level and may be amended only by a two-thirds majority of the National Council. It is this level that determines curricula in general, while the individual school has some say in the distribution of hours allotted to each subject, allowing for some specialization. At the same time, most organizational matters are dealt with at the provincial level, leading to a reasonably complex system of competencies in regard to education.

Secondary education is compulsory and free up to the age of 15 and optional and free thereafter. Thus, secondary education is divided in 'secondary education I' (ages 10–14) and 'secondary education II' (ages 14–18), which allows free access to postsecondary education. Formally, students or their parents are free to decide which school they attend, but both location and family traditions play their part in the actual decision. Secondary education I is divided into two types, which adhere to the same curriculum but employ teachers educated at different institutions (universities and university colleges of education). Secondary education II includes a variety of technical and vocational schools that provide specific possibilities of GI inclusion according to specific subjects. Even within the same type of school, quality of education varies widely – mainly across an urban–rural divide, as well as according to various family and/or migration backgrounds.

Teachers currently are trained at five universities (for secondary education I and II, 9 semesters + one year training in school, academic qualification MSc.) and 14 university colleges of education (for primary education and secondary education I, 6 semesters, bachelor of education). Due to this fragmentation of teacher training, a common qualification of teacher trainers concerning geomedia and GIS has been difficult to develop. The implementation of innovations such as GI-based learning is typically a slow process and therefore probably best achieved through in-service training for teachers. Teachers are encouraged to make use of various

T. Jekel (✉)
Institute of GIScience, Austrian Academy of Sciences, Salzburg, Austria
e-mail: thomas.jekel@oeaw.ac.at

A.J. Milson et al. (eds.), *International Perspectives on Teaching and Learning with GIS in Secondary Schools*, DOI 10.1007/978-94-007-2120-3_3,
© Springer Science+Business Media B.V. 2012

offers of in-service training coordinated by the university colleges in education. However, these support systems for professional development are not compulsory across all types of schools. Moreover, they do not adhere to a common curriculum for professional development. This makes seminars supporting GI use in school highly elective and leads to the introduction of GIS through a handful of innovative teachers.

Computer-based learning in geography education in Austria has a rather long tradition, going back to the mid-1980s. It has been greatly supported by Wolfgang Sitte, a respected leader of Austrian geography teaching, who supported a series of seminars for teachers that were later conducted by Alfons Koller and Josef Strobl. Early publications on this topic included a distinct emphasis on the Web as means to access geoinformation for geography education (Dehmer & Koller, 2000; Strobl & Koller, 1995) and argued for an approach centered on (geo-) media competencies. This work led to the first Herodot.net initiative for geography education via the Web, which later developed into the European Herodot Network. HERODOT has been a thematic network for geography in higher education with more than 220 members from around the world, funded by the EU from 2003 to 2009. Activities organized under four thematic pillars covered innovative teaching and lifelong learning, both of which supported the inclusion of GI in education (http://www.herodot.net/pillars.html). The University of Salzburg was awarded a follow-up network (www.digitalearth.eu) in summer 2010, which will focus more directly on GI at school level as well as on the development of competence centers for teacher training across Europe. Besides Web-based approaches, Wolfgang Dehmer developed a suite of materials on topics such as demography, climatology, and training for map reading including cartographic applications for the PC. However, these early settings – still dependent on available technology – relied on a mainly sender-to-recipient conception of learning. Coordinated by Kurt Trinko and later on by Alfons Koller, there has also been a regular section of the Austrian journal for geography education – GW-Unterricht – devoted to software products developed for school and openly available geomedia to be used in secondary education.

This approach changed with the wider development of the GeoWeb and has – since 2006 – been reflected in a series of conferences termed 'Learning with GI' (Jekel, Koller, Strobl, Donert, & Vogler, 2006–2010) held in conjunction with the AGIT and GI-Forum conferences at Salzburg. Finally, in 2008, digital:earth:at, a center of excellence in GI-based learning was founded by the University of Salzburg, the University Colleges of Education in Linz and Salzburg and the GIScience Institute of the Austrian Academy of Sciences (Lindner-Fally, 2009). This network targets all groups involved in secondary education, supporting pupil-oriented events such as the GIS-Day activities and students research projects (see below), providing in-service training to teachers, and organizing the scientific learning with GI conferences for teacher trainers at different institutions.

Research activities in the field of GI in secondary education have recently concentrated on the development of competence models for GI-based education, namely, the concept of spatial thinking and, more recently, the concept of spatial citizenship (Jekel & Gryl, 2010). Here, competencies are focused on everyday

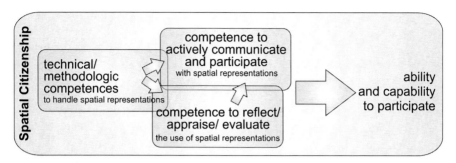

Fig. 3.1 Competencies for spatial citizenship (Gryl, Jekel, & Donert, 2010)

orientation, communication, and participation tasks in a wide variety of contexts rather than specialized scientific problem solving or software proficiency (Strobl, 2008, 2010).

Spatial citizenship may be defined as a set of competencies in addition to those mentioned in the National Research Council (2006) report on spatial thinking that emphasizes technical competencies. These may be termed competencies (a) to critically reflect and deconstruct the spatial information available from various sources and (b) to further one's own visions of social space by (c) being able to translate and actively communicate with them with the help of GI within participatory, democratic processes (Fig. 3.1). Recent technological developments actively promote these geo-communication skills, while also posing new questions pertaining to the interests involved in the production and dissemination of voluntary geographic content.

Geoinformation here is intended as a support system for participation in society rather than as a technology that should be learned. Education is bound to be oriented at basic principles and competencies to be acquired, and that is why the authors suggest that a broad spatial approach – as opposed to an approach via GI Systems – will best support these competencies.

This approach is also reflected in Austrian curricula. The Austrian ministry of Education and Culture has edited curricula that contain GIS and geoinformation in a rather indirect way (BM:UKK, 2009). They are not directly stated, but are mentioned in open phrasing in both the geography curriculum and general principles in education. Within the geography curriculum, most themes can be directly related to a spatial approach. Teachers are also urged to use new media and modern technology as much as possible in learning environments. This is supported by several general principles of education enacted by the ministry of education and culture. These include media education, citizenship, and environmental education and are applicable to all subjects. The use of online mapping, virtual globes, still images, and movies on the Web is therefore justified across a broad range of subjects. From the general principles it may be concluded that teachers have to include the use of digital media like GI-Services and GIS-software. In contrast to the situation in Germany, where GIS-skills are explicitly mentioned in several curricula, Austria concentrates on competencies in the geomedia domain and for an active spatial

citizenship at different stages of the school system. By geomedia, a wide definition is used – referring to all media that can distribute spatial information across
a variety of spatialities. This includes classical georeferenced data as well as text,
images, and film that can be linked to a location.

Students shall be able to use geoinformation in their everyday lives, ask and
answer spatial questions, use Web-GIS-services and Geo-browsers, digital maps,
and virtual globes. They shall become citizens who are aware of the personal benefits of using geoinformation and who participate in public discussion on spatial
issues. Training for special GIS software skills is not an aim of general education in
Austria, but may be conducted in secondary schools during elective courses and in
secondary vocational schools.

Although there are compulsory courses on basic concepts of geography and
spatialities for students aiming to become teachers in geography and economics,
education in the field of GIS is not compulsory in teacher training. There are
numerous seminars available for in-service teachers willing to use geoinformation
as support for instruction. Topics of these seminars include geocaching, collaborative gaming and planning, and use of remote sensing imagery in school or spatial
information for hypothesis development.

3.2 Cases: GIS in Student Research Cooperations

GI(S) has been used with a considerable range of schools in recent years. However,
apart from a few motivated teachers, only virtual globes really made it to the classroom, and even these are mainly used for illustrative purposes. A coherent body of
didactics using GI(S) is noticeable by its absence in Austrian curricula, textbooks,
and curriculum materials. This leads to 'island solutions' instead of continuously
supporting education with GI. The following case studies try to make amendments
to this state of affairs by linking research institutions and schools into common
research projects that can be structurally transferred to everyday school use.

An initiative that is probably unique in European countries is the organizational
and financial support of students' research projects. These research cooperations
are supported by a number of ministries, in order to increase students' interest in
the sciences. Apart from the primary aim of science communication, these research-
education cooperations also allow for the development of contributions to a didactics
of GI(S)-supported learning in secondary education.

GIS-based projects have been rather successful in this respect. We review here
three different projects that are funded by the SparklingScience program of the
Austrian Ministry of Sciences (BM WF, 2009). The SparklingScience program
aims at promoting science between students and invites innovative pedagogical
approaches that are meant to be transferred into everyday school once tested. While
the projects are rather different thematically as well as in the extent to which GIS is
used, they share a few common denominators: (1) all the projects are developed and
controlled by the students themselves; (2) they contain real-world research projects

based on students interests; and (3) the projects are supported by scientific staff from academic institutions.

3.2.1 Schools on Ice (2007–2009)

The Schools on Ice project included four partner schools from across Austria (both urban and rural) and generally focused on student-based research and communication of local effects of climate change. All field work concentrated on the glaciated and karstic Dachstein Mountains, which are heavily visited by tourists and therefore provided insight into both the physical and the economic effects of climate change. Students, ages 16–18, participated in 2–3 preliminary meetings at their respective schools followed by a week working in the field, analyzing data, and publishing results.

The project strongly advocated the inclusion of learners' perspectives in the conceptualization of learning materials. Based on these foundations, the project first collected and analyzed data on learners' preconceptions and interests by various methods (e.g. qualitative interviews, commented newspaper portfolios, and moderated discussions). Preconceptions were put in relation to scientific perspectives on global change to generate a pedagogical concept as a guideline for specific learning objects and learning environments (Wallentin, Jekel, Rattensberger & Binder, 2008). Students were then asked to develop their research agendas and communication designs based on their preconceptions and interests with guidance from researchers of Technological University, Vienna, and Institute of GIScience Austrian Academy of Sciences, Salzburg. In both classroom and field teaching projects, various data collecting techniques were employed, combining mobile devices, online mapping from remotely sensed data, georadar measurements of ice thickness, and interview-based analysis of local coping with climate change. Subsequent collaborative mapping was realized on digital globes. Within the project, GI was used in three different ways:

– To locate earlier measurements for a comparison of the retreat of the glacier and to provide the location of new profiles (Fig. 3.2)
– To map the recession of glaciers from 1850 to 2008 using simple overlay techniques for the years 1850–1991 and to map the 2008 ice edge directly from airborne imagery
– To provide maps of the effects of tourism in the area that provided an alternative, oppositional view to the dominant conception of local tourism agencies.

According to teachers' perceptions, motivation of students has been extraordinary and included a higher participation of students that in regular school environments was considered low key. This could be seen in project participants' motivation to work long hours in the evenings of the project weeks and also resulted in students writing a scientific article that was accepted and printed in the Austrian

Fig. 3.2 Combining
Georadar & GPS. Fieldwork
within the Schools on Ice
project. Schladminger
Gletscher, Dachstein

journal for teachers in geography and economics (Auer et al., 2008). All project groups were asked to present their results to the public, including the Austrian Minister of Science and the local mayor. This positively affected commitment and also instilled a sense of pride and a feeling of competence to be able to contribute their research results to public debate. The overall experience of the academic project staff supporting students' research projects was that students could easily fit into research teams and greatly contribute not only by their questions and ideas but also by doing the hard yards collecting and analyzing data.

3.2.2 Applications on the Move (2008–2010)

The Carinthia University of Applied Sciences, Department of Geoinformation, and the Höhere Technische Bundes-Lehr- und Versuchsanstalt Villach (HTL), as a partner school, are developing a framework for location-based games. Through the support of location-based games, interest should be aroused in handling location-based data and the students in secondary education will get insights into the functionalities of GIS. The main aim of "Applications on the Move" is the development of a modular, generic location-based service (LBS) application, which should form the basis for future applications (Anders, 2008). This development will complement the idea of GeoGames that have been held in conjunction with the annual

GIS Day events in the state of Carinthia to popularize GIS. GeoGames provided a playful access to GI technologies for high school students, allowing them easy first contact with the technology.

Differing from the Schools on Ice project described above, the 'Applications on the Move' project was distributed across a number of school subjects, therefore generating ample time and the possibility for full participation in the research and development process. The subjects involved included programming, business planning, and education in foreign languages. Students started by developing ideas for 6 different location-based games (Anders et al., 2009) they would like to play. These ideas were subsequently implemented with the help of both students and research staff from the Technical University. The project benefits both partners – students at schools, through their close contact with students and researchers at university, and academics, as they have direct and ongoing communication with their target groups.

3.2.3 Geovisualization in Participatory Decision Making Processes (GEOKOM-PEP) (2009–2011)

The final project GEOKOM-PEP probably has the closest cooperation between students and academics. The central idea is to find out about the effects of geovisualization in decision making processes (Fig. 3.3).

To this end, the project developed a collaborative and discourse-oriented spatial planning environment based on virtual globes. It does so by using lab situations that include 17-year-old students as developers and later as evaluators of the platform they have been conceiving. Students therefore are participating in all aspects of the R&D process in a continuing and ongoing cooperation between researchers and education partners.

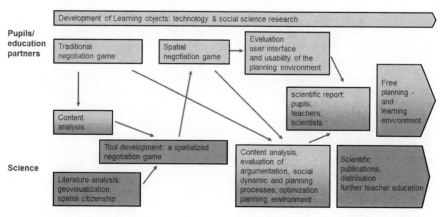

Fig. 3.3 Cooperation between education and science partners in GEOKOM-PEP (Vogler, Ahamer, & Jekel, 2010)

School partners are from the city of Salzburg to both guarantee similar experiences between groups doing spatial planning and to assure comparable experiments. Students are also encouraged to take part in the research process with paid internships offered as well as support for their research papers. Teachers included are paid a small incentive to control the experiments and compare with their everyday experience of teaching in the classroom to find out about effects on participation and learning processes.

3.3 Prospects

While the projects described above are innovative due to the partnerships between academic institutions and schools, there are several lessons to be learned related to the wider context of science education, the structural components necessary to introduce this type of project into the norm of secondary education, and the role of GI(S) in these learning environments. These refer directly to the prospects of geoinformation in secondary education.

At the moment, a project-based approach may be a highlight within the school career rather than the norm depending very much on the availability of motivated teachers in the field. For a wider implementation of science education, and, therefore, GIS in school, structural change in a variety of dimensions is needed. Various studies have called for more flexible class schedules to accommodate research projects (e.g. Specht, 2009). In addition, teacher training contains few courses to support the cooperative development of realistic research questions for school research projects. Further, with the standardization of education, few teachers are willing to grapple with the uncertainties of pupil-led project work and prefer to educate according to standards and curricula given. One of the main challenges therefore rests in the combination of the overall structure of secondary education including the number of classes to be taught by the individual teacher, the allocation of resources, a missing culture of problem-based learning, and the teacher training required.

Within all of the above projects, GIS is not the central aim of the learning process, but is embedded in real-world research and development and/or political processes. This is a considerable difference to recent attempts to 'learn GIS at school' (i.e., some form of hardware and software training). Geoinformation here is used to a thematic end. First, digital globes are used as tool to integrate and communicate students' research, ideas, and perceptions into various themes, as in the Schools on Ice project. Second, GIS is embedded in a wider context of technological development of the LBS game development spanning a range of subjects in school. Third, GI is put to use as a graphic backdrop stimulating and structuring students' discussions in local decision making processes on change. Again, communication is at the center of this learning environment that corresponds directly to the concept of spatial citizenship. In all three approaches, collaborative and participative mapping comes to the fore. All approaches however put a high emphasis on visualization rather than analysis. The technical knowledge needed to use geoinformation in this way is minimal and compatible with teachers' competencies, which have to be related to the overall responsibilities (educational, social, knowledge of two full subjects) teachers have at hand in their everyday jobs. One of the main questions over the use of geoinformatics in schools concerns the level of teacher competence needed for wider application. Any wider application may directly depend on the development of framework case studies. These should provide teachers with a conceptual framework for students' projects to be developed according to their interests, providing help with operationalization and support with public presentations.

Strobl (2010, p. 103) states that geospatial media are both ubiquitous and integrated with other media, making them a cross-curricular theme and support system. The potential of geospatial media may only be fulfilled if it is considered from a transdisciplinary perspective. With regard to teacher training, digital:earth:at plans to provide a transdisciplinary and trans-institutional module on learning with GI for both initial teacher training and in-service training to reach out to as many education partners as possible. We will also continue to develop didactics for GI-based learning across subjects and therefore hope that GI is commonplace in secondary schools a few years from now.

References

Anders, K. H. (2008). *Applications on the move. Development of a mobile gaming application for young people*. Location based games from students for students. Accessed January 25, 2010, http://www.sparklingscience.at/en/projects/14-applications-on-the-move.

Anders, K.-H., Andrae, S., Erlacher, C., Eder, K.-H., Lenzhofer, B., & Wehr, F. (2009). Applications on the move: Ortsbezogene Spiele von Schülern für Schüler. In: T. Jekel, A. Koller, & K. Donert (Eds.), *Learning with GI IV* (pp. 97–102). Heidelberg: Wichmann.

Auer, B., et al. (2008). Perspektiven österreichischer Gletscherskigebiete. *GW-Unterricht, 112*, 51–56.

BM:UKK. (Austrian Federal Ministry for Education, Arts and Culture). (2009). *Schools and education*. Accessed December 1, 2010, http://www.bmukk.gv.at/enfr/school/index.xml

BM: WF. (Austrian Federal Ministry of Sciences and Research). (2009). *SparklingScience. Science linking with school. School linking with science*. Accessed January 7, 2010, http://www.sparklingscience.at/en/

Dehmer, W., & Koller, A. (2000). Computer und Internet in der Lehrerfortbildung Österreichs, Seminarmodule und Zugänge für Lehrer. In M. Flath & G. Fuchs (Eds.), *Lernen mit Neuen Medien im Geographieunterricht* (pp. 61–75). Gotha und Stuttgart: Klett-Perthes.

Gryl, I., Jekel, T., & Donert, K. (2010). GI & spatial citizenship. In T. Jekel, A. Koller, K. Donert, & R. Vogler (Eds.), *Learning with GI V* (pp. 2–11). Berlin & Offenbach: Wichmann.

Jekel, T., & Gryl, I. (2010). Spatial citizenship. Beiträge von Geoinformation zu einer mündigen Raumaneignung. *Geographie und Schule, 186*, 39–45.

Jekel, T., Koller, A., Strobl, J., Donert, K., & Vogler, R. (Eds.). (2006–2010). *Lernen mit Geoinformation/Learning with GI I-V*. Heidelberg: Wichmann.

Lindner-Fally, M. (2009). Digital:earth:at – centre for teaching and learning geography and geoinformatics. In K. Donert, Y. Ari, M. Attard, G. O'Reilly, & D. Schmeinck (Eds.), *Geographical diversity. Proceedings of the Herodot conference in Ayvalik, Turkey* (pp. 332–338). Berlin: Mensch und Buch.

National Research Council. (2006). *Learning to think spatially: GIS as a support system in the K-12 curriculum*. Washington, DC: The National Academies Press.

Specht, W. (Ed.). (2009). *Nationaler Bildungsbericht Österreich 2009, Band 1: Das Schulsystem im Spiegel von Daten und Indikatoren*. Graz: Leykam.

Strobl, J. (2008). Digital earth brainware. A framework for education and qualification requirements. In J. Schiewe & U. Michel (Eds.), *Geoinformatics paves the highway to digital earth* (pp. 134–138). Osnabrück: Universität Osnabrück (Hrsg.).

Strobl, J. (2010). Towards a Geoinformation Society. *GIS-Development, 14*(1), 102–104.

Strobl, J., & Koller, A. (1995). Das internet und Materialien für GW. *GW-Unterricht, 59*. http://www.ph-linz.at/ZIP/didaktik/gw/strobl/strobl.htm

Vogler, R., Ahamer, G., & Jekel, T. (2010). GEOKOM-PEP – Pupil led research into the effects of geovizualisation. In T. Jekel, A. Koller, & K. Donert (Eds.), *Learning with GI V* (pp. 51–60). Heidelberg: Wichmann.

Wallentin, G., Jekel, T., Rattensberger, M., & Binder, D. (2008). 'Schools on Ice' – Einbindung von Lernendenperspektiven in GI-basiertes Lernen. In T. Jekel, A. Koller, & K. Donert (Eds.), *Lernen mit Geoinformation III* (pp. 87–95). Heidelberg: Wichmann.

Chapter 4
Canada: Teaching Geography Through Geotechnology Across a Decentralized Curriculum Landscape

Niem Tu Huynh, Bob Sharpe, Chris Charman, Jean Tong, and Iain Greensmith

4.1 Introduction

Canada is a federal state in which the governance of education is a constitutional right given to the ten provinces and three territories. Curriculum at the elementary and secondary levels is overseen by provincial and territorial ministries of education, while educational programs are administered and funded by local school boards. Secondary education in Canada is universal and compulsory; students are generally expected to stay in school until age 16. Students self-select into vocational/technical schools or university by taking designated preparatory courses (Bednarz et al. 2006; Vajoczki, 2009; Warkentin & Simpson-Housley, 2001 for a history of geography education in Canada).

Within Canada, Geographic Information System (GIS) is typically connected with the discipline of geography, although it is branching into other subject areas. As a whole, the teaching and learning of geography and GIS vary widely, each province guided by its own curriculum developed and overseen by their respective Ministry of Education. Despite differences in the curriculum, the education system is equitable in student access to schooling, quality of teaching, and general topics across the curriculum. In addition to provincial and local variations in the geography curriculum, the depth and quality of GIS education in Canada is also a function of several other influences: teacher education and comfort with technology, national and provincial teacher associations, government programs, the outreach activities of universities and professional geographers, and the initiatives of the geomatics industry.

During the elementary years, geographic concepts such as mapping, community, and land use are introduced under the umbrella of social studies. In Grade 7 and 8, geography is introduced as a stand-alone subject for the first time, with students receiving half a year of geography in each Grade. In the case of Ontario, which was the last province to abandon Grade 13, students are required to take only one compulsory credit in geography (CGC1DI or CGC1PI, geography of Canada). In

N.T. Huynh (✉)
Association of American Geographers, Washington, DC, USA
e-mail: nhuynh@aag.org

A.J. Milson et al. (eds.), *International Perspectives on Teaching and Learning with GIS in Secondary Schools*, DOI 10.1007/978-94-007-2120-3_4,
© Springer Science+Business Media B.V. 2012

Grade 10 they are required to take a mandatory history course, and at present no ministry-produced geography course exists at the Grade 10 level. Several boards have received approval for Locally Developed Courses (LDC's) at Grade 10 level to fill what is perceived as a continuity gap in students' geographic opportunities. In Grade 11, students choose between four courses, the Americas, physical geography, geographics, and travel and tourism, each of which is tailored for a specific academic level. Retention of students becomes a major issue in geography departments as students generally receive no exposure to the subject in Grade 10 when they are asked to select Grade 11 courses. As a result, few schools are able to offer more than one or two of the available courses, due to lack of student numbers. In Grade 12 the course offerings are even wider, yet the retention levels translate into a program that is generally smaller in the final year. Six courses are available at various levels in Grade 12, but in many schools only one or two of these actually run at a time.

Teachers in Canada may receive their certification in two ways. The first method is known as the Consecutive Program where students must have completed an Honours Bachelor or Bachelor degree before acceptance into a Faculty or School of Education. At this stage, teacher candidates learn educational theory and gain practical teaching experience in the classroom. When they complete the year requirements, they receive a Bachelor of Education degree and are licensed to teach in the province after successful registration with the College of Teachers in the province. The second method is to register in the Concurrent Education program where students complete their undergraduate degree (Honours Bachelor/Bachelor) simultaneously with their Bachelor of Education degree.

Teacher candidates in most programs spend three practicum sessions in the schools. A teacher qualifying to teach at the Junior/Intermediate level may choose a placement in a secondary setting in their subject area. Many will choose to do this as a stepping stone toward senior qualifications. Intermediate senior candidates are required to have two teachable subject areas (major and a minor) and will generally spend all three practicums in a secondary school setting. Normally they will spend the bulk of their time in their first teachable area (i.e., two out of three sessions) or they may find situations which allow them to work in both areas simultaneously, teaching a mixed timetable. The initial in-school sessions usually begin with observation of the environment, the students, and the teacher, leading to the candidate taking over a progressively larger portion of the associate teacher's duties. As the second and third practicums progress, the student teacher is expected to take on more and more of the host teacher's duties, until by the end they should be filling the role completely.

The major difficulty with integrating GIS technology into the geography classroom has been teacher training. One or two day workshops (even when offered) are not sufficient to bring a teacher's expertise to the required level. There are few incentives for teachers to take additional GIS courses on their own time (although the idea of a GIS Specialists certification has been explored). For many years, the hope has been that as GIS becomes a more integral part of undergraduate programs, teacher candidates would arrive fully versed in its use. Such has not proven to be the case. In general it has been noted that student teachers have had a single course (if that) and have had very little opportunity to remain current. Although they are generally quite willing to learn enough to teach a lesson or two in a computer lab, the anticipated level of expertise has not materialized. Once they secure teaching positions, they are very much in the same position as the veteran teachers they are joining.

In this chapter we survey the context of geography and GIS education across Canada. The authors broadly define GIS education to include the instruction of geography concepts and skills with a suite of digital geotechnologies, including satellite navigation systems (GPS), virtual globes, online mapping services, and GIS. The chapter provides three case studies of GIS instruction in the classroom, and then finishes with some prospects for GIS education in Canada.

4.2 Geography and GIS Education in Canada

Geography is under the larger curricular umbrella of social studies/social sciences. Although geography is taught at all levels of secondary school, provinces and territories differ in the number of geography courses offered and required for high school graduation, content and skill expectations, and in courses that provide formal instruction of GIS. For example, Ontario offers the largest selection of geography courses in Canada with twelve different courses. In Nova Scotia one geography course is offered (Grade 12 global geography) while in Newfoundland and Labrador and in Prince Edward Island, geographic subjects are integrated into other courses.

A typical geography education begins at the elementary years, where geographic concepts such as mapping, community, and land use are introduced in social studies. This trend continues in later years with geographic skills imbedded in historical/citizenship themes. Ontario is the only province that provides a geography curriculum at the middle years level although the depth of discussion, as well as how they divide their social studies time between history and geography, is determined by the teachers. At the high school level, students may learn geography as a part of a social studies course or as a stand-alone subject, with options to further their knowledge in Grade 11 or 12. In practice, however, geography is disadvantaged by its nonsequential program. In Ontario, Grade 9 students complete one compulsory credit in geography, "geography of Canada" followed by a mandatory history course in Grade 10. Geography courses are offered in Grades 11 and 12, but student enrollment is low, partly attributed to the gap in Grade 10. Geography is further disadvantaged by reduced course offerings compared to the sciences, mathematics, and English.

The actual instruction of GIS, and its depth and quality are related to the passion and practices of individual educators. The integration of GIS into the high school curriculum began in the 1990s with the efforts of pioneering and self-taught educators. The interests of these "lone-wolves" in geography and technology were the impetus for developing resources, facilitating training sessions, and sharing teaching models. To date, a small number of provinces explicitly introduce GIS as a skill within the sequence of geography study (e.g., Alberta, Manitoba, and Ontario). For example, two GIS courses have been added to the Ontario curriculum; one course is tailored to students entering the workforce or vocational education after high school while the second course is geared toward students entering university. Elsewhere in Canada, Nova Scotia has a pilot course and British Columbia has a locally developed course which was created and approved at the school board level. Others, such as Newfoundland & Labrador and Nunavut, have shown interest in bringing GIS into their curriculum. Some of the authors have worked with Nunavut during their curriculum rewriting phase to help make this a reality.

The effective delivery of GIS instruction is also related to teacher training and preparedness. Teachers receive their Bachelor of Education certification from faculties or schools of education having first completed an Honours Bachelor or Bachelor degree, or concurrently with their undergraduate degree. Geography Teacher Candidates (TCs) in high school are required to have two teachable subject areas (major and a minor) and will spend the bulk of their time in their first teachable area (i.e., two out of three sessions) with some opportunities to teach their minor. Most TCs receive their geography and GIS training in their bachelor programs. Many Faculties of Education offer little formal GIS instruction which may be limited to a lecture, guest presentation, or a few hands-on exercises. Other programs, such as York University's concurrent education program, require that TCs in geography take introductory GIS, a course cross-listed in the geography department. The University of Ontario Institute of Technology (UOIT) offers a GIS course specifically designed for teachers. Opportunities for in-service teachers are further constrained by time, and generally limited to one professional day, weekend courses, or summer training institutes.

Fortunately, teaching GIS in the classroom is supported by various associations of geography educators. At the national level the Canadian Council for Geographic Education (CCGE) publishes resources for geographic education as well as a Web site with hundreds of lesson plans, many of which include GIS activities (CCGE, 2010).

CCGE is a national organization that promotes geographic education and geographic literacy in the classroom and in the public. It was established in 1993 as a joint initiative of The Royal Canadian Geographical Society and the National Geographic Society of Washington, D.C. A nationwide initiative was the introduction of the "Canadian National Standards for Geography: A Standards-Based Guide to K-12 Geography" (CCGE, 2001). Since its adoption is voluntary, there is little data to suggest the extent of its implementation in the classroom.

At the provincial level, the Ontario Association of Geography and Environmental Educators (OAGEE), for example, has championed the integration of GIS into the Ontario curriculum as well as putting geotechnologies at the forefront of their annual fall conference (OAGEE, 2010). In Manitoba, the Manitoba Social Studies Teachers' Association (MSST, 2010) has had a similar influence.

In Canada, geography has been the "champion" subject for GIS to integrate as a tool for learning. There is a stigma that GIS is "just maps" and therefore many teachers do not see the value of using it in other disciplines. Despite this perception, there are a handful of educators that use GIS as an analytical tool and understand the power of maps to teach ideas beyond the geography classroom. We have seen the use of GIS to teach concepts in mathematics, history, and science. Moving forward, as GIS becomes more prominent in the professional community, we see its adoption in a wider variety of disciplines.

Federal agencies also offer significant support to GIS education in Canada. Natural Resources Canada maintains the Atlas of Canada, which provides, at no cost, a comprehensive collection of maps, geospatial data, and lesson plans (NRC, 2007). Similarly, Statistics Canada provides high quality data sets along with lesson plans (Statistics Canada, 2010). Both these resources have been helpful in the dissemination and integration of GIS into classrooms across Canada. Another federal program, Skills Canada, sponsors annual GIS competitions for secondary school students in several districts across Canada (Skills Canada, 2010).

Operating at the national and local levels is the Canadian Association of Geographers (CAG), which represents professional geographers in universities, colleges, and schools, as well as in governments and business (CAG, 2009a). The CAG and its Geographic Education study group, established in 2003, co-sponsored with the CCGE a symposium on "Projecting Geography in the Public Domain in Canada" (CCGE, 2005). In 2006 the CAG established a program to promote and provide resources for Geographic Awareness Week and Geographic Information Systems Day (CAG, 2009b). CAG members at local universities are also active in geographic education (CAG, 2008). Several geography departments across Canada offer a variety of geography outreach programs, including GIS workshops and training institutes for teachers (e.g., in British Columbia, Manitoba, and Ontario). There are a few geography faculty who undertake research and supervise graduate study in geographic and GIS education (some recent graduate thesis/dissertations include Crechiolo, 1997; Huynh, 2009; Storie, 2000).

Lastly, the geomatics industry has also had a major role in both teacher training and the provision of resources. Several different vendors have been involved, but of particular note is Esri Canada's formation of a dedicated education team in 1997 to support educators that included training (face-to-face and online), development of teaching materials, data repository (ArcCanada), and technical support. In particular, GIS resources have played a pivotal role in helping teachers successfully integrate GIS into schools. In 2000 Ontario adopted a provincial license for Esri software followed by British Columbia, Manitoba, Nova Scotia, Newfoundland, and Labrador.

4.3 Case Studies from Across Canada

4.3.1 Case Study 1 – Teresa Kewachuk, Hants East Rural High School, Milford Station (Shubenacadie), Nova Scotia

Teresa Kewachuk is the department head for social studies and teaches the Grade 12 students global geography and geomatics courses at Hants East Rural High School

(Grades 9–12). Students in Nova Scotia must pass at least one of the global courses to graduate (i.e., economics, history, or geography). Teresa was part of the writing team (2007) that developed the geomatics course, which can be applied as either a technology credit or a social studies credit. Students in her geomatics class experience learning at different levels ranging from enhancing their critical thinking and problem-solving skills to independent work on tutorials. Students learn new skills in various assessment activities. For example, during the process of retrieving data from Statistics Canada and other sources, students can be heard commenting about the facts they are learning, which range from topics on earthquake patterns, comparisons between rich and poor countries, to historic settlement patterns (Fig. 4.1).

In an ongoing project, Teresa takes her students to a local cemetery where they collect data using a GPS unit and record historic information from tombstones. Students create maps with the data to gain an understanding of various societal issues such as soldiers who died during the war, ethnic groups in the area, as well as the symbols and the statement implied by tombstones. On another field trip to the local municipal building, students see how GIS is applied by the local government.

Despite these successes, a key challenge to any GIS implementation is access to data. The filtering systems used by school boards block noneducational sites including those that are pre-approved by the Department of Education for downloading. GIS may be slowly making its way into the classroom through software like Esri's ArcGIS Desktop or via online geo-browsers like Google Earth, but students who leave Teresa's classroom have a defined sense of spatial awareness and a better understanding of how geotechnologies can be used and applied beyond the walls of the classroom.

In the province of Nova Scotia, Teresa's students' experiences seem to be unique. Although there may be few high schools with computer labs with access to GIS software, Hants East has a lab dedicated to geomatics. The GPS units and GIS software used in Teresa's global geography 12 and geomatics classes are supported by a grant from the provincial teachers union.

Fig. 4.1 Students working on GIS

4.3.2 Case Study 2 – Rob Langston, Neelin High School, Brandon, Manitoba

Rob Langston is a Social Science teacher at Neelin High School (NHS) (Grades 9–12). NHS is Brandon's only French immersion program and is one of only a small number of schools worldwide approved to offer the International Baccalaureate curriculum.

Students in Rob's classes are exposed to GIS through a variety of teaching methods. Learning typically begins with lecture and demonstration, and as students' confidence and abilities increase, they are encouraged to drive their own learning. Class field trips and guest speakers increase students' appreciation of the real world applications of GPS and GIS. NHS administration is supportive of Rob's activities and he has recently partnered with a local community college to develop two courses offered to NHS students in the fall of 2010.

Like many cities in Canada, Brandon is experiencing a change in demographics, where recent immigrants and their families are moving into the area. An interdisciplinary exercise conducted between the Grade 11 geography students and English as an Additional Language (EAL) students supported EAL students' learning of the provinces, territories, and capitals in Canada by creating a map. As a result, EAL students responded positively to the collaborative task where GIS was successfully used as a tool for learning about their new country.

Success at NHS is evident in the increasing enrollment in Grade 11 and 12 geography courses, both elective courses in Manitoba. Students are encouraged and eager to share their GIS experiences. For example, Grade 10 geography students were paired with Grade 2 students from a local elementary school to conduct a community study. Grade 2 students collected GPS waypoints of significant features in their community (e.g., churches, playgrounds, and green areas), and then worked with Grade 10 students to map the data using a GIS.

NHS is arguably a leader in K-12 GIS education in Manitoba. All Grade 9 social science teachers use GIS in their courses, which ensures that students are exposed to GIS during their first year of high school. Rob believes that the use of GIS in Manitoba schools will continue to grow.

4.3.3 Case Study 3 – Kirsten Davel and Cheryl Murtland, SMUS, Victoria, British Columbia

The Senior School at St. Michaels University School (SMUS) is located in Victoria, British Columbia (BC) with a population of approximately 550 students aged 13–18. Kirsten Davel (Head of Geography) and Cheryl Murtland (Assistant Director) have shown incredible leadership with GIS at SMUS.

Although geography is not a distinct discipline until Grade 12 in the BC social studies curriculum, geography courses are offered from Grades 9 through 12 at SMUS. Grade 9 students are introduced to the concept and functionality of GIS. Geography 10 offers students an opportunity to apply their skills in desktop and

Fig. 4.2 GPS field use Grade
10 GIS experiential program

Web-based assignments. For example, twenty-four Grade 10 students participated in an experiential program where they spent two days in Mount Tolmie Park (located behind SMUS) collecting field data and using GIS to map and analyze native and invasive flora (Fig. 4.2).

The true power of GIS becomes evident to students in geography 11 when they create a map to support an independent study research paper on a global issue. By the end of the course, students have developed sufficient skills to mine the Web for data, import, and use it to create a map for analysis. In addition, advanced placement human geography and geography 12 students utilize GIS to support their learning of the curriculum. The skills continuum ensures a sound foundation in GIS as an invaluable tool for geographic inquiry, analysis, and application to real world issues.

Kirsten and Cheryl's success is apparent in their students' level of engagement; students soon become excited about the endless possibilities of GIS. For example, some students moved from being passive learners to connecting actively through GIS-based assignments. A student who was not previously engaged with the subject noticed a syntax error and suggested a solution which moved the class forward. From that point, geography came alive for him and he participated with enthusiasm.

Unfortunately, GIS is not recognized in the BC social studies curriculum. There is no mandate to teach GIS but teachers such as Kirsten and Cheryl, who believe strongly in its power, have made a commitment to its implementation in the classroom. Kirsten and Cheryl's continuing outreach efforts in BC encourage teachers to explore the possibilities of GIS as a necessary tool in geographic education.

4.4 Prospects

Canadian educators are faced with many challenges in the course of integrating GIS into their classrooms. These range from the lack of availability of school computers to insufficient teacher expertise in GIS. These are briefly described below.

4.4.1 Access to Technology

GIS software is often installed in word-processing labs where computers cannot handle the software – thus low adoption. Partnerships between school boards and companies may be one solution as exemplified by an Ontario school board, partnering with a computer company to install 1,500 computers ($1,000,000). A related issue is the rate at which new computers are installed; this varies across the country and is decided upon anywhere from the principal level to the provincial Ministry of Education level. The second obstacle is a lack of sufficient access to computer labs for students to develop needed skills.

4.4.2 IT Conflicts

Educators often face barriers from IT departments who will not install the required software and data. Some reasons include the space the software takes on a network, the IT department's lack of knowledge of GIS, and the low priority given to GIS in the school.

4.4.3 Time/Training

Teachers receive limited release time to learn geotechnology that is needed to develop sufficient skills and expertise to impart to students. Even when weekend and day workshops are offered, some obstacles include a lack of incentives to attract teacher interest as well as insufficient training time to raise a teacher's GIS skills and confidence level.

4.4.4 Being the Expert

GIS is an excellent teaching tool for educators to be facilitators of learning rather than sole transmitters of knowledge, however, many educators have difficulty including technology if they are not experts themselves. Teachers who have successfully engaged students with GIS affirm that students quickly surpass their knowledge and students advance their own learning through exploration.

4.4.5 Education Policy

A fundamental barrier to geography education in Canada is rooted in the value that society places on it. The value and contributions of geography are not always recognized or understood. Thus, decision makers in education have not explicitly recognized its value and consequently diminished course offerings and weakened a sequential curriculum compared to that seen in the mathematics and sciences.

Despite these challenges, GIS has established a niche in the Canadian education system that will likely expand, although its future direction will be subject to much change. Geographic and spatial literacy along with numeracy, graphicacy, and visuality will continue to have an important role in the curriculum. As the curriculum is periodically reviewed, geography educators will be asked to further integrate the concepts (Sharpe & Huynh, 2005) and skills of GIS. Likewise, teachers will prepare and revise their lesson plans so that their objectives and methods incorporate the evolving technologies.

In the future, geotechnologies will continue to diffuse throughout the school system and into the classroom. These geotechnologies will also further evolve accompanied by the increasing availability of high quality geospatial data suitable for GIS analysis. Rapid developments in IT and a growing assemblage of digital imaging technologies and visualizing practices, such as CCTV systems, distributed sensor networks, and innumerable mobile camera phones, will provide rich and exciting prospects for educators. In this environment, the role of GIS in secondary education is likely to change. Even now, educators find the functionality and low cost of virtual globes, like Google Earth, free online GIS (ArcGIS online - http://www.esri.com/software/arcgis/arcgisonline/index.html) and map making (ChartBins - http://chartsbin.com/) sites to be sufficient for geographic visualization as well as to provide an introductory awareness of GIS. Even so, GIS will continue to have a role in engaging critical thinking (Bahbahani & Huynh, 2008) and problem solving that involve geographic information and spatial analysis.

References

Bahbahani, K., & Huynh, N. T. (2008). *Teaching about geographical thinking*. R. Case & B. Sharpe (Series Eds.).Vancouver, BC: The Critical Thinking Consortium and The Royal Canadian Geographical Society.

Bednarz, S., Bednarz, R. S., Mansfield, T. D., Semple, S., Dorn, R., & Libbee, M. (2006). Geographical education in North America (Canada and the United States of America). In J. Lidstone & M. Williams (Eds.), *Geographical education in a changing world: Past experience, current trends, and future challenges* (pp. 107–126). Dordrecht, The Netherlands: Springer.

Canadian Association of Geographers (CAG). (2008). *Canadian Association of Geographers Geographic Education Study Group*. Accessed September 12, 2010, http://info.wlu.ca/%7Ewwwgeog/CAGEDU.htm

Canadian Association of Geographers (CAG). (2009a). *Canadian Association of Geographers Homepage*. Accessed September 12, 2010, http://www.cag-acg.ca/en/index.html

Canadian Association of Geographers (CAG). (2009b). *Geographic awareness week and geographic information systems day*. Accessed September 12, 2010, http://www.cag-acg.ca/en/geography_week.html.

Canadian Council for Geographic Education (CCGE). (2001). *Canadian national standards for geography: A standards-based guide to K-12 geography*. Accessed September 12, 2010, http://www.ccge.org/programs/geoliteracy/geography_standards.asp

Canadian Council for Geographic Education (CCGE). (2005). *Projecting geography in the Public Domain in Canada.* Accessed September 12, 2010, http://www.ccge.org/programs/geoliteracy/geolit_symposium.asp

Canadian Council for Geographic Education (CCGE). (2010). Accessed September 15, 2010, http://www.ccge.org/

Crechiolo, A. (1997). *Teaching secondary school geography with the use of a Geographical Information System (GIS).* Unpublished master's thesis. Wilfrid Laurier University, Waterloo, ON.

Huynh, N. T. (2009). *The role of geospatial knowledge and geographic skills when reasoning in geography using Geographic Information System (GIS).* Unpublished doctoral dissertation. Wilfrid Laurier University, Waterloo, ON.

Manitoba Social Science Teachers' Association. (2010). Accessed September 15, 2010, http://www.mssta.mb.ca/

Natural Resources Canada (NRC). (2007). *Atlas of Canada.* Accessed September 12, 2010, http://atlas.nrcan.gc.ca/site/english/index.html

Ontario Association for Geographic and Environmental Education (OAGEE). (2010). Accessed September 15, 2010, http://www.oagee.org/

Sharpe, B., & Huynh, N. T. (2005). *Geospatial knowledge areas and concepts across the Ontario Curriculum.* Technical report for the GeoSkills Program, GeoConnections. Natural Resources Canada, 22 p.

Skills Canada. (2010). *Skills Canada.* Accessed September 12, 2010, http://www.skillscanada.com/

Statistics Canada. (2010). *Statistics Canada learning resources.* Accessed September 12, 2010, http://www.statcan.gc.ca/edu/index-eng.htm

Storie, C. (2000). *Assessing the role of Geographical Information System (GIS) in the geography classroom.* Unpublished Master's thesis. Wilfrid Laurier University, Waterloo, ON.

Vajoczki, S. (2009). Geography education in Canada. In O. Muniz-Solari & R. G. Boehm (Eds.), *Geography education: Pan American perspectives* (pp. 139–155). Austin, TX: The Grosvenor Center for Geographic Education.

Warkentin, J., & Simpson-Housley, P. (2001). The development of geographical study in Canada, 1870–2000. In G. S. Dunbar (Ed.), *Geography: Discipline, profession and subject since 1870* (pp. 281–316). Dordrecht, The Netherlands: Kluwer.

Chapter 5
Chile: GIS and the Reduction of the Digital Divide in the Pan-American World

Osvaldo Muñiz-Solari and Gustavo Moreira-Riveros

5.1 Introduction

The impact of information and communication technologies (ICT) and the subsequent challenge to learn geography with the assistance of new technologies could not be more evident in Chile. Among several universities which have geography and pedagogy as joint fields of professional development in Chile, ICT and GIS go together. This new trend has to do with the evident reduction of the digital divide in the Pan-American world (Muñiz-Solari, 2009). Chile, among other countries, looks for quality in education through innovative projects and programs. One major innovation is the expansion of infrastructure and management to improve ICT. The Open Access Initiative, created in Europe in 2002, has also had an effect on South American society in general and particularly the way teachers and students access information. GIS has been one of the most manifest examples of this technological transformation, and education represents the path to channel this hidden but increasing revolution. The aim of this chapter is threefold: (1) to analyze the educational trends in Chile and their consequences in new ways of teaching and learning based on ICT; (2) to identify problems and solutions related to the use of ICT and GIS; and (3) to present innovations in GIS applications developed by teachers and students in secondary schools.

5.2 A Holistic Curriculum for a 'Knowledge Society'

The government of Chile has considered primary and secondary school mandatory for all the people since 1965. Free and compulsory secondary education was established for all the inhabitants of Chile up to 18 years of age in 2003. This translates as practically universal education, a unique condition among Latin American countries. The secondary school system across the country is divided between

O. Muñiz-Solari (✉)
Texas State University, San Marcos, TX, USA
e-mail: o.muniz@txstate.edu

A.J. Milson et al. (eds.), *International Perspectives on Teaching and Learning with GIS in Secondary Schools*, DOI 10.1007/978-94-007-2120-3_5, scientific-humanist and technical-professional schools. The first is the regular option while the second is vocational. While these options are taken freely by any student, the second type of education is more oriented to public schools or Liceos.

The new challenges raised by the information society around the world created a transformation in education at all levels during the early 1990s in Chile. Following the return to democracy in 1991, new measures to improve education were developed by changing the organization of the educational system. The educational reform process that has been conducted since the mid-1990s is based on some important pillars, two of which are focused on improving teaching methods and transforming the educational curriculum (Arellano-Marín, 2001). The new curriculum framework encourages the importance of competencies from a holistic perspective to develop a "knowledge society" through the use of ICT. The components intended to promote the mastery of higher-order skills are critical thinking, abstract reasoning, problem solving, information processing, communication, and interaction (Delannoy, 2000). Intersecting objectives focus the learning process on human–environment relationships, social aspects such as tolerance and solidarity, as well as national and international cooperation.

Secondary teacher training is performed as a university level career of five years. After several practices in the selected schools, they earn a certificate and pass a national exam to be ready for professional activity. Chile has had over a hundred years of experience in which a system of teacher preparation offered subject specialization together with pedagogical learning. However, education is promoted as a holistic curriculum in the schools where teachers are still inclined to perform traditional teaching styles. An important transformation is about to occur in order to practice critical thinking, to persuade self-guided learning and encourage innovation.

The new educational reforms place teacher–student interaction as an essential process for citizen formation. The 1990s were years for improving teacher education quality through the creation of a special program across the country known as "Strengthening Initial Teacher Training" (Fortalecimiento de la Formación Inicial Docente, FFID). Several improvements have been carried out since that time, such as infrastructure and teaching resources (libraries and ICT), collaboration between teachers and institutions, and allowing teachers to engage in study tours to international centers and universities (Avalos, 2002). More recently, the Ministry of Education has developed standards for initial teacher education following international recommendations. They also include teacher performance evaluation and self-evaluation as well as accreditation of undergraduate university teaching programs (Avalos, 2005).

Some specific programs to improve and modernize teaching methods concentrate on suitable training for teachers to use basic information technology as a teaching tool. In 1992, the Ministry of Education created a program called ENLACES (LINKS) to introduce digital technologies in the public schools as a means to reduce the digital divide and as new learning and teaching resources. By the year 2008, 87% of the student population had access to information technologies through the school (ENLACES, 2009). Each school in the ENLACES network received computers to implement laboratories, local networks, educational and productivity software, teacher training in the basic use of ICT, and technical and pedagogical assistance provided by a network of 24 universities all over the country. More than 80% of teachers have been trained to use ICT (Hepp, Hinostroza, & Laval, 2004). About 75% of Chilean schools are currently connected to the Internet and have access to specially created educational content relevant to the Chilean curriculum (EDUCAR CHILE, 2005). However, ENLACES is still concentrated on the adoption of computers and laboratories to use the Internet and basic software to learn science and mathematics. GIS is not included in the formal curriculum in the Chilean schools. Consequently, teachers are not required to teach about GIS and only those who are good innovators in social sciences teach geography with GIS.

5.3 Classrooms as Laboratories: A Problem-Based Learning Environment

The transition from traditional teaching methods to pedagogical innovations required a new perspective to deal with a more student-centered learning process. The educational reform brought about an important change toward a more applied perspective inside and outside the classroom in secondary schools. Nature and the environment started to play crucial roles and, as a consequence, spatial dimensions and map skills are developed by observing not only local areas, such as nearby towns, but also regions and territories (Muñiz, 2004).

The applied perspective creates a sense of community and teamwork among students not only working in the classroom but also in the field to learn about spatial problems through data collection. This approach has been attempted to encourage deep cognitive learning to solve open-ended problems in which uncertainty is present. Even when this new learning process is not widespread in schools around the country, it is recognized in some pedagogic approaches such as problem-based learning (PBL). The Ministry of Education funded a PBL project through the Program for Improvement of Teacher Education in 1998 to disseminate this new learning and teaching environment (Iglesias, 2002).

Availability and access to ICT is clearly demonstrated by a systematic national survey conducted by the Center for Technology and Education (CTE) of the Ministry of Education in 2004. On average, secondary schools have 37 computers, with a ratio of 26 students per computer. Most of the computers in secondary schools are located in computer labs (62%). Only 6% are in school classrooms. Computer labs are used 37 hours per week, and more than 60% of the schools report that they have an ICT-related strategy, a key factor that promotes information and communication technologies (Hinostroza, Labbé, & Claro, 2005). It is noticeable that the number of hours spent in the labs and the use of computers as free time available to students are strategies for ICT insertion into every school's culture.

CTE develops the Technology Plan for Quality Education that reaches schools in the form of plans to improve educational outcomes through the use of technology. These plans articulate the three pillars of the program ENLACES (i.e., infrastructure, teaching and ICT skills, and content) available to the schools (ENLACES, 2005).

Chile currently records a ratio of 16 students per computer and is expected to reach the rate of 10 students per computer in 2010 with the Technology Plan for Quality Education. Be it in a traditional classroom, laboratory, or the field, the process of active learning is not without important risks and setbacks. There still are structural bottlenecks in the educational system affecting the future expansion and development of ICT.

5.4 ICT and the Use of GIS: Problems and Solutions

One of the important challenges of the national program coordinated by ENLACES is the integration of ICT with the curriculum. The national program's long-term objective is to accomplish an extensive use of ICT in the educational structure. At the center of this program is the main strategy to combine the use of ICT with new teaching procedures to facilitate a more advanced learning process.

Moreira-Riveros points out the need for an "infusing learning" and a "platform" approach based on the Web to improve knowledge comprehension for the student (Moreira-Riveros, 2004). According to his experience, learning abilities, cognitive operations, and personal values and attitudes are critical steps in the new learning system in the Chilean schools. A possible solution is to create a good e-learning environment.

Solutions are not without problems when ICT have to be implemented in secondary schools. A new educational perspective that deals with student-centered learning requires the availability of extensive hours for training and practice. However, school schedules are insufficient to conduct workshops on a continuous basis and teachers are more inclined to traditional teaching than innovative approaches with high risks and uncertainty in the final outcomes.

Geography teachers and particularly those using GIS tend to focus on critical thinking and problem-solving procedures when using computers. Their students tend to pursue abstract reasoning through visual interpretation of complex components (e.g., scales, symbols, profiles, layers, buffers) often immersed in a typical PBL environment to resolve real problems. Some of these teachers and their students do not recognize themselves as PBL fans. However, their work represents a transition between a traditional class and an innovative laboratory. Others are in fact working in problem-resolution environments that bring new generations into a straight student-centered learning environment that in turn promotes increasing innovation.

Problem resolution is seen as a new step in the learning process practiced by innovative teachers in some of the most advanced secondary schools in Chile. The same innovators are those who tend to interact with other actors of the ICT–GIS world. Three of the most critical actors encouraging change with the use of ICT and GIS in Chile are the government, industries, and universities.

Esri is the primary GIS source of assistance to public offices, private companies, and universities in Chile. Several ministries and regional offices rely on software and technical assistance provided by Esri-Chile. The Chilean government as a whole facilitates access to data sources. Digital map data produced by professionals are available to be disseminated to schools and teachers in order to contextualize their learning activities. Two examples are the Territorial Information System (TIS) of the Ministry of Education (SITE, 2009) and Geographic System of the Ministry of National Goods (GEOPORTAL, 2010).

One of the most common means to facilitate access to GIS is based on the open-source initiative (Silva, 2008). Although Esri is a closed-source software vendor, it has taken an open systems approach since the early 1980s. The ArcGIS platform is one example of using open-source tools, languages, integrated development environments (IDEs), libraries, and Web server technologies (Esri, 2010). The open-source initiative has been fundamental to inspire innovation and encourage GIS development in university curriculum.

Influenced by Esri, some universities have developed GIS research centers. Geography has been one of the fields where GIS has a special position. However, the pedagogical preparation of geography teachers does not include GIS as a critical technology for learning. Consequently, teaching about GIS and with GIS is not widespread among secondary schools.

5.5 Examples of GIS Innovations and Applications

Teaching with GIS tools in secondary schools in Chile entails the right environment and lab setting to deal with abstractions, visualization, and spatial thinking. The two selected examples represent excellent innovations as pedagogical practices and special training for students.

The first example is a case study that involves e-learning using an open-source platform and the ArcReader application by high school students. Through CLAROLINE, a learning management system (LMS) with a general public license (GPL) based on the Web, information is collected and organized (Moreira-Riveros, 2004). Procedures are developed by steps based on geographic data (e.g., demography, climate, environment) from the National Service of Territorial Information (SNIT) that have been collected at local settings where schools are located. Use of local data stored in the database MySQL of the LMS allows the students to generate layers. The students enter the learning activities by getting access to the tool offered by CLAROLINE, which is supported by ArcReader (Fig. 5.1).

The case study is the Pascua Lama Transnational Corporation Gold Mining Project in the Andes. This project will eventually affect rural population in Chile and Argentina. In Chile, the case study is presented to secondary school students to analyze the rural valleys of the Atacama Region. The activity is focused on water contamination of two valleys, and the students can simulate water quality monitoring using Excel tables. Geographic location of population affected by contamination could be determined by using the layer for Poblados or rural settlements (Fig. 5.2).

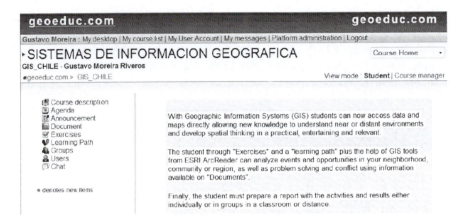

Fig. 5.1 E-Learning platform based on CLAROLINE (Moreira-Riveros, 2006)

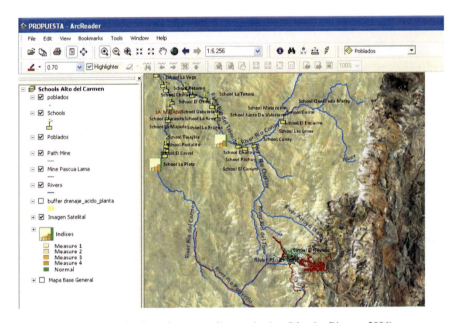

Fig. 5.2 Settlements, schools, and water quality monitoring (Moreira-Riveros, 2006)

The second example is a project carried out by Liceo Carmela Carvajal de Prat (CCP) located in Santiago. A group of female students, ages 16–17 years, were taught GPS and GIS principles and procedures and applied their new knowledge and skills in order to encourage the use of bike trails in the local community of Providencia, an administrative subdivision in the capital that is home to a large upper-middle class population. GPS and GIS knowledge and practice also allowed the students to propose that their school be considered a magnet place since it attracts students who reside in Providencia

Fig. 5.3 Students working in their project at CCP and attending the XV Conference of Esri Users in Santiago, Chile (Moreira-Riveros, 2006)

as well as those who reside outside of this community and far away from Providencia. The result of this second GPS and GIS experience was a total success thanks in part to a well-organized training for students. This activity took place in the school and was guided by their teacher who also received international assistance to conduct lab exercises and activities in the field. In addition to face-to-face guidance conducted by the 4-H[3] Groups from the University of North Carolina and the University of Oregon, Esri documents prepared by Napoleon and Brook were critical to complement the training sessions (Napoleon & Brook, 2010). The students presented their findings at the XV Conference of Esri Users held in Santiago in 2008 (Fig. 5.3).

5.6 Conclusions

The educational reform of the 1990s in Chile developed a holistic curriculum to encourage a new tendency in teaching and learning. While critical thinking, abstract reasoning, problem solving, and information processing are the main components of the new trend in the schools, the holistic perspective endorsed competencies to achieve a "knowledge society." High quality education via technology and the

Internet has turned out to be one of the most important goals to enhance the educational curriculum in the classrooms of both public and private schools in Chile. The evident reduction of the digital divide by new communication systems and widely available technological tools is a great opportunity to initiate the process of GIS learning and GIS adoption in secondary schools. However, GIS is not viewed as one of the main technologies that should be incorporated in schools today.

The transition from traditional teaching methods to pedagogical innovations requires a new perspective to deal with a more student-centered learning process. A new pedagogy is in place once the government considered the ICT as important strategy to reach a knowledge society. ENLACES has paved the way to promote ICT in the schools, but GIS is not considered a key technology, in part because of a weak pedagogical strategy to teach with new technologies and a lack of regulation in the preparation of teachers at the university level to teach with GIS in the secondary schools. The school system has to overcome a structural bottleneck related to the still strong influence of traditional teaching. Traditional pedagogy in numerous secondary schools, especially those controlled by public education, and reduced school hours to conduct extracurricular activities are two of the main factors that inhibit innovation among teachers.

The first important challenge to promote and make GIS available has to be focused on government officials in education. They have to be guided in order to instill a strong public awareness of the need for GIS education. A second challenge is to prepare teachers who should learn that GIS is a requirement with the new pedagogical approach based on self-learning activities. They need to be prepared to teach not only about GIS but also with GIS. In contrast, innovation in several private and semiprivate secondary schools is clearly present where new technologies are always included. Geography teachers and particularly those who are interested in ICT are also interested in GIS development. Esri has been the link to allow and perform innovation based on student-centered activities in some private and semiprivate secondary schools. Examples of ICT innovations and GIS applications, such as those already described, must be multiplied in the secondary schools with the assistance of the Ministry of Education and other public offices that support the Territorial Information System. Finally, open-source software must be widely available in the schools by increasing Internet coverage, expanding ICT networks to connect schools in the rural areas and creating efficient e-learning systems.

References

Arellano-Marín, J. P. (2001). Educational reform in Chile. *Cepal Review, 73*, 81–91.
Avalos, B. (2002). *Profesores para Chile. Historia de un proyecto.* Santiago: Ministerio de Educación.
Avalos, B. (2005). *Secondary teacher education in Chile: An assessment in the light of demands of the knowledge society.* Paper submitted to the Ministry of Education, Santiago.
Delannoy, F. (2000). Education reforms in Chile, 1980–1998: A lesson in pragmatism. Country studies. *Education reform and management publication series* (Vol. 1(1), pp. 1–80). Washington, DC: The World Bank.

EDUCAR CHILE. (2005). *The education portal.* Accessed November 14, 2009, http://www. educarchile.cl/Portal.herramientas/quienessomosingles/index.html

ENLACES. (2005). *Centro de Educación y Tecnología.* Ministerio de Educación, Santiago, Chile.

ENLACES. (2009). *Centro de Educación y Tecnología.* Ministerio de Educación, Santiago, Chile. Accessed November 2, 2009, http://www.enlaces.cl/index.php?t=44%i=2%cc=800%tm=2

Esri. (2010). *Open source.* Accessed January 12, 2010, http://www.esri.com/technology-topics/ open-source/index.html.

GEOPORTAL. (2010). *Ministry of national goods.* Accessed February 8, 2010, http://www. geoportal.cl/Portal/ptk

Hepp, P., Hinostroza, J. E., & Laval, E. (2004). A systematic approach to educational renewal with new technologies: Empowering learning communities in Chile. In A. Brown & N. Davis (Eds.), *World yearbook of education 2004: Digital technologies, communities and education* (pp. 299–311). London: Routledge Falmer.

Hinostroza, J. E., Labbé, C., & Claro, M. (2005). ICT in Chilean schools: Student's and teacher's access to and use of ICT. *Human Technology, 1*(2), 246–264.

Iglesias, J. (2002). Problem-based learning in initial teacher education. *Prospects, 32*(2), 319–332.

Moreira-Riveros, G. (2004). Sistema de aprendizaje basado en la web para facilitar el desarrollo del pensamiento. In J. Sánchez-Ilabaca (Ed.), *Memorias, IX taller internacional de software educativo, TISE* (pp. 29–32). Santiago: Universidad de Chile.

Moreira-Riveros, G. (2006). *Activities of learning of geography using open source and ArcReader.* Paper 1951, Proceedings of the Conference of Esri Users, San Diego, California. Accessed October 26, 2009, http://proceedings.esri.com/library/userconf/educ06/abstracts/a1951.html

Muñiz, O. (2004). School geography in Chile. In A. Kent, L. Rawling, & A. Robinson (Eds.), *Geographical education. Expanding horizons in a shrinking world. Geographical Education Commission of the International Geographical Union* (pp. 177–180). Glasgow: Scottish Association of Geography Teachers.

Muñiz-Solari, O. (2009). Geography education: 'The North' and 'The South'. In O. Muñiz-Solari & R. G. Boehm (Eds.), *Geography education. Pan American perspectives, a volume in the International Geography Education Series (IGES)* (pp. 5–33). The Gilbert M. Grosvenor Center for Geographic Education, Austin, TX: Allen Griffith.

Napoleon, E. J., & Brook, E. A. (2010). *Thinking spatially using GIS: Our world GIS education, Level 1.* Student Workbook, Esri. ISBN 9781589481848.

Silva, J. E. (2008). *Tendencias y futuro de los GIS en gobierno.* PowerPoint presentation, XV Conferencia de Usuarios Latinoamericanos de Esri, Santiago, Chile. Accessed October 30, 2009, http://www.esri-chile.com/lauc2008/13.proceeding_4h-2.htm

SITE. (2009). *Sistema de información territorial de educación.* Ministerio de Educación, Chile. Accessed December 10, 2009, http://atlas.mineduc.cl/pmgt/

Chapter 6
China: Teacher Preparation for GIS in the National Geography Curriculum

Pinliang Dong and Peiying Lin

6.1 Introduction

In China, universal secondary education has been part of the *Law on Nine-Year Compulsory Education* since 1986. There are two levels of secondary education: compulsory education for junior students aged 13–15, and non-compulsory education for senior students aged 16–18. Senior students can choose general high schools to prepare for college entrance examinations, or vocational high schools to obtain job skills. The curriculum of secondary schools has three levels: national curriculum, local curriculum, and school curriculum. The national curriculum is administered by the Ministry of Education, while the local curriculum and school curriculum are made by the Department of Education at the provincial level and by individual schools.

Teachers of secondary schools are mainly produced by teachers' universities or colleges. Certificates are required for secondary school teachers. Some graduates from teachers' universities or colleges can receive certificates automatically, whereas some may pass extra tests to obtain teacher certificates depending on the policy of the local government. Graduates of general universities can also obtain teacher certificates and become secondary school teachers. GIS courses are included in both required and optional courses in teacher preparation programs. In some cities (such as Beijing), GIS is part of the exam to become certified to teach geography. However, a national standard for the teacher certification exam is still under discussion. Secondary school teachers are required to participate in continuing education to earn certain credits every year, which will be used as part of the criteria for annual evaluation or promotion. Various forms of continuing education for secondary school teachers are available, such as courses for advanced degrees, online training, and workshops. GIS is part of the continuing education courses in some programs. For example, GIS courses are included in the summer courses of the Capital Normal University for secondary school geography teachers.

P. Dong (✉)
University of North Texas, Denton, TX, USA
e-mail: Pinliang.Dong@unt.edu

A.J. Milson et al. (eds.), *International Perspectives on Teaching and Learning with GIS in Secondary Schools*, DOI 10.1007/978-94-007-2120-3_6, © Springer Science+Business Media B.V. 2012

Research and education in GIS in China started in the early 1980s. Although Chinese secondary schools have a long history of teaching geography at both junior and senior levels, GIS was not part of the curriculum until 2004, when revised curriculum standards were put into effect for senior high school students (Ministry of Education, 2001). The Geography Curriculum Standards Working Group organized by the Ministry of Education played a key role in putting GIS in the geography curriculum for secondary schools. At the junior level, the concept of GIS is only briefly introduced along with applications of digital maps and satellite images (Ministry of Education, 2001). At the senior level, more details of GIS are introduced. In required course, the School Geography Standards have 47 items, and one of them is specifically for GIS with focus on applications in urban management (Ministry of Education, 2003). In addition, an elective course on "Applications of Geographic Information Technology" is available for senior students. Students are expected to understand the following after completion of the course: (1) General GIS functionality and application examples; (2) Methods of map digitizing; (3) Creating attribute tables using GIS software; (4) Changing symbology of map layers; (5) Basic skills of querying features; and (6) Map making using GIS software (Ministry of Education, 2003). Although statistics are not available from different regions, it is believed that many high schools would cover GIS, global positioning systems (GPS), and remote sensing in one class. Since many high school GIS classes are taught by the instructor through lectures and demonstrations, students do not have many opportunities to do hands-on exercises. GIS implementation in high schools is primarily driven by increasing use of the technology, not necessarily by the high demand of GIS at the university level.

6.2 Cases

The following two cases of teaching GIS applications were developed by faculty members of the College of Resources, Environment and Tourism at the Capital Normal University and geography teachers of secondary schools.

6.2.1 Case 1: Geography Class for Second Year High School Students (Arts Majors)

This class was taught in May 2009 in an ordinary classroom with a multimedia projector at the Experimental High School affiliated with Beijing Normal University. The teacher graduated from the department of geography, Beijing Teachers' College. She had over 20 years of teaching experience, but was relatively new to GIS. The class started with a lecture session on the concept of GIS, followed by a group project session, a discussion session, and a demonstration session. The topic of the group project was GIS for agricultural planning. The students were asked to identify suitable areas for wheat and fruit plantation in Beijing Municipality based on four paper maps: municipality planning map, topographic map, soil map, and land use map. The students were then divided into two groups, one for wheat plantation, and the other for fruit plantation. Each group was further divided into several subgroups for discussion. About eight minutes later, representatives from the two groups started to show their results using the projector, and explain their methods. At this point, the teacher summarized student activities using one sentence: "What you have done can be conveniently implemented in GIS." She then demonstrated GIS methods for the same tasks using a PowerPoint presentation.

After the project and discussion session, the teacher demonstrated GIS modeling capabilities using the relocation of the *Capital Steel and Iron* company as example. Air quality plume modeling was conducted in GIS to show the extent of air pollution caused by the company. Finally, the teacher demonstrated the application of digital maps.

In this case study, students experienced traditional map overlay process and learned that the same process can be carried out in GIS, which helped the students' understanding of the functionality and benefits of GIS. Since the class was in an ordinary classroom instead of a computer laboratory, students did not have a hands-on experience with GIS operations (Fig. 6.1). Student feedback suggests that a hands-on exercise should be introduced for better understanding of the concept and functionality of GIS.

6.2.2 Case 2: Geography Class for Second Year High School Students (Science Majors)

This class was taught in May 2009 in a computer laboratory at the Experimental High School affiliated with Beijing Normal University. The teacher graduated from

Fig. 6.1 Students experiencing GIS in an ordinary classroom

Fig. 6.2 Students experiencing GIS individually in a geography class

the Department of Geography, Capital Normal University. She studied GIS courses in college, and had 8 years of teaching experience. In this class, each student had access to a networked computer in the computer laboratory (Fig. 6.2). The topic of the class was "How to use digital maps?" The teacher briefly demonstrated the use of Google Maps, and the students were then divided into two groups to work on two individual tasks using Google Maps. The students in the first group were asked to (1) find the best route for taking buses or driving from a suburban location to the school, and (2) find an apartment near the school within a certain rental price range. The task of the students in the second group was to (1) find restaurants and supermarkets with international foods that can meet the needs of an international high school student and her parents, and (2) find road sections with high traffic volume on the way from the home of the student to the workplace of her parents. After completion of the tasks, selected students were asked to demonstrate their results using the teacher's computer and answer the teacher's questions. Based on the demonstration and hands-on practice, the teacher summarized the functionality and applications of digital maps and GIS, and introduced the second question: Where are the suitable areas for rice plantation in Beijing Municipality? Unlike the first case study in which students worked on similar questions using paper maps, the answer to this question was demonstrated by the teacher using GIS data layers for soil, topography, and land use.

The two teachers in the above two cases had discussed the contents before the classes, and decided to use different contents based on the different preparations of the teachers and arts and science students. Although a formal questionnaire was not designed to get student feedback on the classes, discussions with individual students indicated that students were satisfied with the structure of class sessions, level of difficulty, and relevance to daily life. It should be noted that the two cases were initially designed for research on GIS education in secondary schools. The purpose was to identify gaps in teachers' preparation in GIS and students' interest level in learning GIS. The results of the project indicate that: (1) The motivation and self-learning ability of the instructor play an important role in high school GIS education. Through self-learning and preparation, teachers without formal GIS training can still deliver a successful GIS class. (2) Following the same curriculum standards, instructors can develop different teaching and learning styles depending on student interests and classroom facilities. (3) Collaboration between universities and high schools can effectively improve the quality of GIS teaching and learning in high schools. In terms of class contents and styles, most secondary schools in China can adapt contents and styles similar to the above cases. As an open laboratory for high school teachers, the GIS Laboratory of the College of Resources, Environment, and Tourism at the Capital Normal University has trained many high school geography teachers.

6.3 Prospects

GIS has been widely used in China, and GIS education has been included in the curriculum standards for compulsory and elective courses in Chinese high schools. All high schools should teach GIS contents in the compulsory courses, whereas no statistical data are available on how many high schools also teach GIS contents in elective courses. Currently, there are three major problems in GIS education in Chinese secondary schools: (1) Many geography teachers in secondary schools do not have enough GIS knowledge; (2) GIS software and data are not widely available for secondary schools; and (3) Secondary schools in some less developed areas do not have enough computing facilities to support GIS education. To meet the requirements of the curriculum standards, it is important to improve GIS training for secondary school geography teachers, and to collaborate with international and domestic GIS software vendors and universities to improve the availability of GIS software and data. Among the four major elements of GIS (hardware, software, data, and people), we believe that people are the most important element at this stage for improving GIS education in secondary schools in China. Only through well-trained teachers can hardware, software, and data be integrated into GIS in a classroom setting. We believe that the Internet and wireless communication technologies will have a major impact on the application of GIS in daily life and GIS education in Chinese secondary schools. According to the China Internet Network Information Centre (CNNIC, 2010), China had 420 million Internet users in July 2010, about one third of China's population of 1.3 billion. In July 2010, the number of mobile phone users in mainland China reached 800 million (CNTV, 2010), about 62% of the total population. Digital maps on the Internet and mobile Location Based Services (LBS) continue to attract an increasing number of users. Meanwhile, over 140 Chinese universities are producing undergraduate and graduate students in GIS to meet the demands of the public and private sectors. With the improvement in hardware, software, GIS data availability, teacher preparation, and student interest level, GIS education in Chinese secondary schools will move into a new stage (Han & Zheng, 2006).

References

CNNIC. (2010). *Statistical report on Internet development in China. China Internet Network Information Centre (CNNIC)*. Accessed July 2010, Website: http://www.cnnic.net.cn

CNTV. (2010). *China's mobile phone users top 800 mln*. Accessed September 2010, http://english.cntv.cn/20100721/101314.shtml

Han, L., & Zheng, J. (2006). Practical methods for GIS education in China – How to meet social need of high quality human resources. *IEEE international conference on geoscience and remote sensing* (pp. 999–1002). Denver, CO.

Ministry of Education. (2001). *The people's Republic of China. Geography curriculum standards for full-time compulsory education (experimental)*. Beijing: Beijing Normal University Press (in Chinese).

Ministry of Education. (2003). *The people's Republic of China. High school geography curriculum standards (trial version) [S]*. Beijing: People's Education Press (in Chinese).

Chapter 7
Colombia: Development of a Prototype Web-Based GIS Application for Teaching Geography

Luz Angela Rocha Salamanca and Natalia Andrea Diaz Vega

7.1 Status of Geography Education in Colombia

In Colombia, basic secondary education represents Grades 6–9 and is based on Law 115 (1994), which is identified as the "General Educational Law." This law has the philosophy that education performs a social function regarding the necessities and requirements of the people, family, and Colombian society and its principles are in the political constitution of the country. Law 115 of 1994 designates three levels for education in Colombia, as follows:

a. Kindergarten: one obligatory year course
b. Basic education: 9 grades in two cycles: Basic primary education (5 grades) and basic secondary education (4 grades)
c. Media education (2 grades culminating in a bachelor's degree) and higher education

According to Article 67 of the Constitution of 1991, education "is compulsory between five and fifteen years of age and includes at least one year of preschool and nine of basic education." Colombian schools can be private or public.

> The curriculum for secondary education has been changing from time to time, according to the rules or resolutions given by the Ministry of Education. Nowadays, education is based on basic standards of competence established by the Ministry of Education, with public and clear criteria that provide the guidelines on how students should be taught and establish the standard of learning in each area and level. This represents a reference guide to educational institutions, urban or rural, private or public across the country (MEN, 2004, p. 5).

L.A. Rocha Salamanca (✉)
Universidad Distrital F.J.C., Bogotá, D.C. Colombia
e-mail: lrocha@udistrital.edu.co

A.J. Milson et al. (eds.), *International Perspectives on Teaching and Learning with GIS in Secondary Schools*, DOI 10.1007/978-94-007-2120-3_7,
© Springer Science+Business Media B.V. 2012

Geography in Colombia is not taught as a single subject; rather, it is taught as a social studies lecture and involves many themes for different levels at the secondary school. Geography is taught as a subject integrated with other sciences; therefore, students learn history, geography, and economics.

Although they must respect the national curriculum, certain autonomy is given to institutions and colleges in the applications of their programs. The methodologies change depending on the college or institution (Leon, 2002). On the other hand, the decentralization of education affects the access of the students to it, because a child is more likely to complete basic education in cities or in an urban area than in the rural zones. In addition, the education of each child depends on the economic capacity of the family.

In Colombia, there are educational bachelor's degree programs for people to study social studies in order to become secondary teachers. Public and private universities supply these programs. For example, the Distrital University of Bogota offers the social studies bachelors' degree program (Licenciatura en Educación Básica con Énfasis en Ciencias Sociales) whose main objective is to prepare students to become professionals in education and research in the social studies area (Universidad Distrital, 2010). Geographic Information Systems (GIS) is not compulsory in most programs in the country, yet it can be taken as an elective course.

Currently, school teaching is mainly done using conventional means. The use of computers is a new challenge particularly at public schools. In most of the private schools, students have more access to technical resources like computers and the Internet. However, teachers of secondary school, in most cases, do not use computers in their teaching. Social studies education is based on competencies. Each competence requires knowledge, abilities, and capabilities in order to achieve the respective skills. The main characteristics of the curriculum are the following (MEN, 1994):

1. The education process is centered on the student in order to enable him to be an integrated person as a community member.
2. The education programs must maintain the balance between the theory and the practical application of the knowledge.
3. The curriculum must be a dynamic system for the personal formation and social integration of the student.
4. The education process must involve the study of the national and international problems and relevant events.

In the current context in Colombia, geography includes the spatial and environmental relationships and the economy in order to understand different ways of human organizations and their relationship with the natural landscape. The Colombian standards regarding the geography subject are as shown in Table 7.1 (MEN, 2004).

Table 7.1 Colombian standards regarding the subject of geography

VI to VII Grade	VIII to IX Grade
• The student is able to understand the earth characteristics as a living planet • The student is able to use coordinates, legends, and scales in order to work with maps • The student recognizes and uses the time zones • The student is able to understand the geographic space and the different physical aspects of the environment • The student can establish relationships between the geo-spatial location and the climate characteristics of the different cultures • The student can identify the production systems of different cultures and different time periods and show relationships between all of them • The student is able to understand the different economic organizations in Colombia and make analysis and explanations between their differences and similarities • The student is able to describe the characteristics of the natural regions of Colombia • The student is able to identify the economic, social, political, and geographic factors that have been generated in the different recruitment processes	• The student is able to describe the principal physical characteristics of the different ecosystems • The student can explain the impact of the environment in the type of social and economic organization in the different Colombian regions • The student can compare the ways of relationships between different cultures and communities in Colombia • The student can compare the causes of the migration and human recruitment in the territory in the XIX century • The student can explain the impact of migration in the politic, social, economic, and cultural life in the XIX century • The student can identify the modernization process in Colombia in the XIX and XX centuries

7.2 Geography Contents at the Secondary Basic Education in Colombia

The program of social studies integrates the subjects of geography and history and, at the same time, tries to introduce these two subjects between the Latin American and worldwide framework. In Grade 6, geography includes the basic concepts regarding evolution, adaptation, culture, civilization, transformation, changes, and differences and whether changes were produced by natural effects or by human renovation of the landscape (MEN, 1988). In Grade 7, it is focused on the following topics: space, time, sociocultural structure with emphasis on three elements: economic relationships, political relationships, and knowledge and collective expressions (MEN, 1989). Geography comprises the social economic changes in Europe, the industrial revolution, and the romantic age in Grade 8. The program also includes the imperialism, the colonial expansion, and the causes of the First World War (MEN, 1990). In Grade 9, the program is designed for the students to gain an integral knowledge about society through relevant events (MEN, 1991).

7.3 History of GIS in Schools in the Country

GIS in Colombia is related to information technology (IT). IT has been recently implemented in order to make people more competitive and also to have an education modernized to provide better opportunities to students. Colombia is improving the use of technology especially at the higher levels of education, but there is still very poor use of technology in schools. Consequently, the Ministry of Information Technology and Communications recently gave guidelines to address the need to include information technologies in education.

In Colombia, GIS is recognized mostly in public and private companies but not in secondary education; we can say that the implementation of GIS in the classroom is just starting. Regarding this, some universities have been working on how to introduce the concepts of GIS education and many projects are to be developed. GIS is not formally included in the curriculum for secondary education; however, with the recent history and new initiatives, some schools are adopting these tools for education, starting with geography subject matter. On the other hand, teachers in secondary education are still not prepared to use GIS techniques for teaching. Therefore, one of the goals and strategies is to create mechanisms that persuade them of the significance of using GIS as a tool in the classroom.

7.4 The GIS Prototype

As explained above, GIS has not yet been introduced in Colombia at the basic secondary education level. One of the problems is that in public schools, especially in rural areas, access to computers is quite difficult. Even in big cities like Bogotá, there are some schools with few computers for teaching. In various private schools, the students have more access to technical resources like computers and the Internet. However, the general problem is that the teachers in secondary schools, in most cases, do not use GIS concepts in their work in the classroom.

For these reasons, the Distrital University of Bogotá, as a superior academic institution, joined its efforts with the SELPER chapter Colombia (Association of Specialists in GIS and Remote Sensing of Colombia) in order to create a GIS application project, which supports teaching using GIS at schools. The aim of the project was to create a prototype of a GIS application for teaching the geography curriculum in schools, using the academic standards of basic secondary education in Colombia, taking into account that geography is a very important aspect of the territory and an essential element of GIS (Fig. 7.1). The second objective was to familiarize the secondary teachers with the use of GIS technology as a tool that can support their work in the classroom.

Fig. 7.1   GIS prototype produced for teaching geography at schools

The prototype of a GIS application for teaching geography handles physical and socioeconomic information of the country and simulates different scenarios (http://soportesig.procalculoprosis.com/geosig2/). The application was constructed using commercial software (ArcGIS and Flex) and incorporates the topics regarding the Colombian standards for teaching geography:

- Using maps to find geographic information.
- Understanding economic geography.
- Understanding Colombia, including relief, natural regions, and natural resources.

The creation of the GIS application required a decision about the selection of the software. For this project, the best solution was ArcGIS Server. Since the application was designed for young students, it was necessary to create an attractive interface. Thus, the ArcGIS API for Flex was used because Flex allows one to create dynamic and functional environments. The goal was to ensure that the students felt interested in learning geography with the GIS application. Creating this application also required reviewing and evaluating thematic information together with spatial information. Once the final thematic maps to be included in the application were determined, they were created in ArcGIS Desktop 9.3. The final maps were then created in ArcGIS Server services and, finally, were set in the Flex builder application in order to publish the project online. Because the GIS prototype is a Flex Shareable Application, it can be used for teachers who decide to work with this in the classroom. The only condition is to have a computer connected to the Internet. This project is really important because it is one of the first initiatives in Colombia and will be implemented not only in public schools but also in private schools using the Internet (Diaz, 2010). The development of the project involved an important group of professionals such as geographers, teachers of social studies with an emphasis on geography, computer systems engineers, GIS specialists, and students of the Cadastral and Geodesy program.

7.5 Prospects

The opportunity to include GIS techniques in basic secondary education is compatible with the goals of the Ministry of Education to achieve a modern education, which allows for a better development of skills in students (MTIC, 2008). It is also compatible with the goal of improving current teaching methodologies and fostering learning from each teacher and student as well. However, there are many challenges. The Ministry of Education in Colombia has plans and programs with policies and guidelines (MEN, 2008), but the implementation of these is slow and the inclusion of GIS will be complex due to two basic problems. First is the challenge for teachers to create new teaching methods to introduce techniques in the classroom where students can interact with GIS. In order to make this expectation come true, the teachers must be familiar with GIS tools. Therefore, new education programs regarding concepts and use of GIS techniques could be implemented. A second challenge is to include GIS in the basic secondary curriculum. This will take more time and perhaps will be more difficult.

In conclusion, the use of GIS in secondary schools in Colombia is now starting. Several projects and plans have been created in order to introduce GIS concepts and applications for social studies with an emphasis on geography teachers, and to develop new tools for teaching, such as the use of GIS applications in the classroom. One of these projects is to do a second phase of the GIS prototype. This stage will offer a GIS course for teachers of secondary school to introduce them to this technology and get better results in the use of the application. These courses are planned first at local level in Bogotá public schools as an initiative of the Distrital University. It is expected that in the future the project will be implemented in other cities of Colombia. Also in the second phase, the GIS prototype application will be improved. It is planned to include other subjects, such as history or biology, to increase the use of GIS for teachers at secondary schools, thereby creating a necessity for both students and teachers to use GIS in their academic lives.

References

Diaz, N. (2010). *Prototipo de aplicación de los sistemas de información geográfica S.I.G en la enseñanza de geografía en la educación básica secundaria en Colombia*. Tesis para optar por el título de Ingeniera Catastral y Geodesta. Bogotá D.C.: Universidad Distrital Francisco José de Caldas.

Leon G. G. (2002). *La imposición de modelo pedagógicos en Colombia – Siglo XX*. *Revista de Estudios Latinoamericanos*. Pasto: Universidad de Nariño, Centro de Estudios e Investigaciones Latinoamericanas.

Ministerio de Educación Nacional (MEN). (1988). Propuesta de programa curricular. Ciencias Sociales. Sexto Grado. Educación Básica Secundaria. Bogotá D.C.: MEN.

Ministerio de Educación Nacional (MEN). (1989). Propuesta de programa curricular. Ciencias Sociales. Séptimo Grado. Educación Básica Secundaria. Bogotá D.C.: MEN.

Ministerio de Educación Nacional (MEN). (1990). Propuesta de programa curricular. Ciencias Sociales. Octavo Grado. Educación Básica Secundaria. Bogotá D.C.: MEN.

Ministerio de Educación Nacional (MEN). (1991). Propuesta de programa curricular. Ciencias Sociales. Noveno Grado. Educación Básica Secundaria. Bogotá D.C.: MEN.

Ministerio de Educación Nacional (MEN). (1994). *Ley 115 de 1994 (8 de Febrero). Ley General de Educación*. Accessed August 10, 2010, http://www.mineducacion.gov.co/1621/w3-channel. html

Ministerio de Educación Nacional (MEN). (2004). *Estándares Básicos de Competencias en Ciencias Naturales y Ciencias Sociales*. SERIE GUÍAS No 7. Formar en Ciencias: ¡El desafió! Lo que necesitamos saber y saber hacer. Bogotá D.C.: MEN.

Ministerio de Educación Nacional (MEN). (2008). *Revolución educativa: Plan Sectorial de educación 2006–2010*. Bogotá D.C.: MEN.

Ministerio de Tecnologías de la Información y las Comunicaciones (MTIC). (2008). *Plan Nacional de Tecnologías de la Información y las Comunicaciones 2008–2019*. Bogotá D.C.: MTIC.

Universidad Distrital Francisco José de Caldas. (2010). *Licenciatura en educación básica con Énfasis en Ciencias Sociales*. Accessed August 13, 2010, http://www.udistrital.edu.co/ academia/pregrado/lsocial/

Chapter 8
Denmark: Early Adoption and Continued Progress of GIS for Education

Torben P. Jensen

8.1 Introduction

Danish geography teachers have used geography-oriented IT programs since the early 1990s, and thus have been one of the pioneer subject-areas to introduce IT to the classroom. One of the first programs, called Hefaistos, could display a computer model of plate tectonics, where the localization and depth of earthquakes and volcanoes could be studied. Other programs could construct age pyramids, make demographic prognoses, and simulate weather stations. Many of these early programs were conceived and constructed by individual enthusiasts, who created everything from the initial idea to the finished product. This meant that product development depended too much on individual initiative. If, instead, basic standard programs were readily available, enthusiastic teachers would have more time and space for teaching experiments. This was realized in 1998 when the Remote Sensing program Science Image became one of the standards in digital image processing. It was a very efficient teaching tool that could make animations of various Meteosat images and "cloud movies" of global wind systems with students controlling the playing speed of the movie. Such up-to-date, topical satellite images entered the classroom via satellite or the Internet, and this aroused curiosity and created an interest in studying such digital images in class.

In 1999, a number of geography teachers in Denmark began to use another standard program, ArcView 3.0, to process GIS data. GIS can be used to "build a bridge" between two "islands" consisting of the statistical, geo-related data and student-centered teaching methods. GIS is a good example of an IT tool that can be used to analyze and visualize huge quantities of historical and topical geo-data and invite new in-depth studies.

The increasing use of IT in primary and secondary education in Denmark has led to great changes in classroom teaching practices. Since all schools have fast

T.P. Jensen (✉)
Langkær Gymnasium and HF, Tilst, Denmark
e-mail: tpj@tpj.dk

A.J. Milson et al. (eds.), *International Perspectives on Teaching and Learning with GIS in Secondary Schools*, DOI 10.1007/978-94-007-2120-3_8,
© Springer Science+Business Media B.V. 2012

Internet connections as well as wireless Internet, Web GIS is easily available for the teachers and the students. However, Web GIS does not offer the same quality analysis facilities as more specialized GIS programs such as ArcGIS. Until recently, GIS instruction took place in special computer-equipped classrooms with ArcGIS installed on the computers, but that is changing now as many schools are dismantling their computer rooms. Esri offers a model that would give students access to ArcGIS if the school is licensed so that ArcGIS will be available on the students' own PCs/laptops. This will make it much easier to use GIS in schools in the future.

8.2 Primary and Secondary Education in Denmark

Basic education in Denmark comprises primary and lower-secondary education and lasts for nine or ten years (the tenth year is optional). Upon completion of class 9 or 10 of the Folkeskole (public community schools) pupils may go on to vocational or general upper-secondary education. The three-year Gymnasium program is the academic, general upper-secondary program. Students are admitted if they have average or above average grades.

The Ministry of Education is responsible for the national regulations of both primary and secondary education. It shares the control of the Gymnasiums with the individual school boards. However, the specific contents of the courses are decided by the schools and their boards and by the teachers and the students. Vocational education and training is also controlled by the Ministry of Education in cooperation with local school boards and the labor market organizations. The secondary schools are self-governing state schools with different histories and academic profiles. They are direct grant schools run on a democratic basis with staff and student representation on the boards. The extensive reform of upper-secondary education in Denmark in 2005 introduced in the Gymnasium a new interdisciplinary basic science course that includes topics from biology, chemistry, geography, and physics. GIS is introduced in this course and is used to strengthen project teamwork and for such tasks as categorizing data and testing hypotheses. GIS is now a compulsory part of the teaching of geography at intermediate level in the Gymnasium.

8.3 In-service Training for Upper-Secondary Teachers

In Denmark, professional upper-secondary teachers can apply for refresher courses or in-service training courses on an individual and voluntary basis. These courses may be offered by the various subject teachers' associations and may be established on the initiative of individual members or the association. The courses are sanctioned by the Ministry of Education and announced on the Ministry website. Often a new course will be offered nationwide at first and later some of the participants may assist in organizing and lecturing on regional or school-based courses. Presently, there are several local school-based GIS courses run all over Denmark. An experienced GIS teacher may organize a course locally at his school or at a neighboring school. These local courses help strengthen the teachers' understanding of GIS, and the hands-on exercises can be organized to fit specific local conditions.

The introduction of GIS in Danish schools and the succeeding in-service training was initiated in 1998 at the initiative of a small group of geography teachers, who got together in the GOGIS—now GEOGIS group—under the Geografforlaget (The Geographers' Publishing House). In close cooperation with the Informi GIS company, who provided the professional know-how of the GIS programs, national and regional courses were developed and arranged all around the country. High quality GIS data were available to the course members during the course, and every school received regional GIS data sets on CD-ROM afterwards. Each of the courses was one day in length and alternated between hands-on exercises and lectures by the Informi GIS experts and experienced high school teachers who demonstrated how to include GIS in the teaching of various subject areas. The courses were free of charge and extremely popular. In a very short time, 35% of all geography teachers in Denmark had participated. As a result, many gymnasiums bought the GIS software ArcView. After the course, the website gisgeo.dk (formerly gogis.dk) supported teachers with additional cases with download of data, suggestions, and inspiration for classroom activities. The cases had the same structure: background material, problems to be addressed, GIS data for downloading, and links for inspiration.

8.4 The History of GIS in Denmark

The use of GIS in Denmark began in the late 1970s when the Danish Institute of Plant and Soil Science programmed their own GIS system. In the 1980s, major progress was made in the GIS field when the national natural gas project demanded registration of all major natural gas pipelines. Electricity boards followed suit, as did water works, central heating companies, and sewage boards to have their networks GIS registered. A national central land registry office also started a systematic, national map production using GIS. In the 1990s, the use of GIS expanded to new areas. The first alternative uses were typically the geographical income distribution or the traffic density in various geographical areas. Today GIS is widely used by public as well as private institutions.

8.4.1 From Remote Sensing to GIS

In 1991, Peter Brøgger Sørensen and Karl-Erik Christensen completed a set of Remote Sensing (RS) booklets and related data (called DIBB for Digital Image Processing) for Geografforlaget. It did not become widespread until 1998, however, when the DIBB material was transformed, expanded and rewritten into the digital teaching material COURSE (Colleagues Using Remote Sensing in Education), which was to become very popular and widely recognized and used. The COURSE Project adopted the teaching model of high school teachers communicating

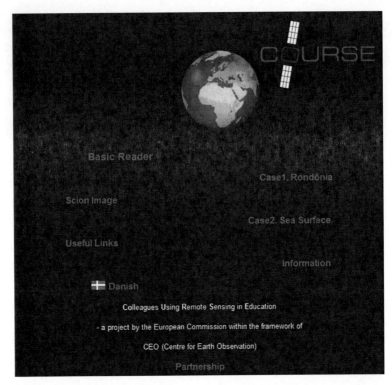

Fig. 8.1 COURSE – Digital teaching material from 1998 containing basic reader, data, and guided exercises. The teacher could use this website as a starting point to get to know Remote Sensing (www.gisgeo.dk/course)

scientific research data (Fig. 8.1). The COURSE model meant that scientists, who had no time for communicating their results, could get help from experienced high school teachers, who were given access to relevant data and who helped adapt the results to the classroom scene, helped organize courses, and set up combined course and teaching websites.

The Ministry of Education supported a study trip to the Joint Research Centre (JRC) in Northern Italy. Important data, such as sea temperatures in the Atlantic Ocean and Vegetation in Africa, were made available for the COURSE project, and thus became available for upper-secondary students and teachers. These data were later supplemented with data from Norsk Rumcenter (Norwegian Space Center) and Sarepta. It was soon evident that there was a growing interest in combining Remote Sensing and GIS. The use of GIS was introduced by a group of geographers after a study trip to the United States in 1999. They visited universities with strong GIS programs and the Esri headquarters. This study trip was also supported by the Danish Ministry of Education. As a result, members of the newly established GOGIS group participated in Esri's first educational conference in 1999 and later in the large-scale Esri users' conference in San Diego, California, USA.

The close and fruitful cooperation between Informi GIS and Esri resulted in the introduction of the GIS software ArcView 3.0 at five gymnasiums in Denmark, Ikast Gymnasium, Langkaer Gymnasium and HF, Silkeborg Gymnasium, Toftlund Studenterkursus, and Aabenraa Gymnasium. GIS was now included in the traditional collaboration between teachers of geography and the geography departments at the universities in Denmark. They had already worked together on Remote Sensing courses and teaching materials, and now the GIS experience of the universities contributed to the first teaching materials using GIS. The goodwill and patronage of data suppliers were also very important at this early stage. For example, the use of very detailed topographical maps from the TOP10DK collection by the Danish National Survey and Cadastre was made possible for a GIS case on Copenhagen. In this way the schools gained an insight into these quality data.

8.4.2 Establishment of ArcIMS in 2003

When Karl-Erik Christensen and Torben P. Jensen did postgraduate GIS studies at Aalborg University, they established an ArcIMS server – a Web GIS server- in cooperation with Geografforlaget (Christensen, Thomas, & Torben, 2003). They used the server as a Digital Atlas in connection with their studies (Fig. 8.2). Almost at the same time the Joint Research Centre (JRC) in Italy was looking to establish a Science Learning Pool to demonstrate how different sources and types of data can

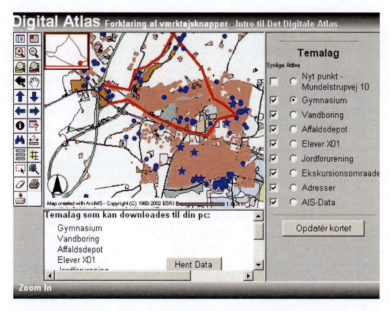

Fig. 8.2 The Digital Atlas was implemented on a standard ArcIMS server with added extract and download facilities. The figure shows a local area around a gymnasium and shows that it is possible to download data

be pooled together using an IMS server to provide a one-stop-shop for high quality eContent for the eLearning market. EduGIS supplied this in 2003. EduGIS was founded in 2002 by four upper-secondary school teachers doing work in the field of producing teaching and learning material utilizing the possibilities of modern IT-technology. EduGIS has done most of its work in the field of using remote sensing and GIS in various subject areas. EduGIS has also contributed material to Eduspace, which is a subdivision of ESA, creating learning and teaching tools for secondary schools such as LeoWorks (digital image processing and GIS).

8.5 GIS in Primary Education 2003–2004

The project "GIS in Primary Schools" was the first attempt to introduce GIS at this level. It involved four schools around Denmark. Several big Danish data suppliers made a number of local data (detailed topographical maps from the TOP10DK collection by the Danish National Survey and Cadastre and orthophotos from COWI a/s) available to the schools. With the support of the Ministry of Education, the next project was continued as "Best Practice with GIS." The main point was to use the GIS tool to render visible conditions and connections locally as well as nationally and internationally. This was later followed by the establishment of siGIS at the Geografforlaget. At the same time, in-service training courses were organized for teacher training teaching staff and some began teaching GIS afterwards.

8.5.1 NatGIS – An Interdisciplinary Project Involving Biology–Geography–Physics 2003

Geography has been a kind of dynamo in the use of GIS in the Danish Gymnasium. In 2003, other subjects joined in and a study trip to the United States was arranged, in which representatives from astronomy, biology, and physics participated together with geography representatives. After the study trip, teaching materials were developed, which could be used in the cross-disciplinary teaching projects required by the Education Reform Act of 2005.

The following NatGIS cases give inspiration to teaching at all levels of the gymnasium:

(1) "Radon," including biology, geography, and physics, gives background information and invites analysis of the connection between the spreading of Radon and the frequency of diseases; (2) "Giant Hogweed," mainly created for biology, but also offers geography aspects; (3) "Controversy," constructed as a court case, where the arguments for the extinction of the dinosaurs have been drawn up rather sharply, and parts of it can be acted out as a kind of drama where the natural science arguments are pitted against each other. GIS is used in the analysis; (4) "The lake project" is an example of the development in the cultural landscape of a local area; and (5) "Wind" is an analysis of the environmental and energy-optimal location of windmills.

8.5.2 *Annual GIS Day in Denmark*

The annual GIS day is often celebrated by specific events including schools and local business. Local companies visit the schools to demonstrate how they employ GIS, and the students show how they work with GIS in class. Often these events lead to new ways of working with GIS in education.

8.6 Challenges

Finding good quality data is one of the main challenges to using GIS in schools. Although many of the Danish data are available for free, they are scattered all over the Internet. To ease the burden of searching for GIS data, the GISGEO group has striven to get access to these data and place them on the website www.gisgeo.dk. There has been an interim period in Denmark after the administrative reform in 2007 when fourteen counties were reduced to five regions. This meant the annulment of the old agreements between the map providers, the schools, and the counties. New agreements are being created and many public service organizations have begun making new GIS data available on the Internet.

As schools dismantle their computer rooms, new arrangements must be made so that the students can access ArcGIS on their own computers at schools with a school license. It will be interesting to follow the development in this area, as an optimal solution will strengthen the daily use of GIS in the classroom. Today the students may access Web GIS servers belonging to the local municipality, the state, or sources such as Google Earth. However, Web GIS may offer only a limited scope for analysis, which is why the students should be required to work with GIS using a real GIS program.

8.7 Case Study: Using GIS for Distance Analysis in Silkeborg

The following case was a study project at Silkeborg Gymnasium made by second-year high-school science students (17-year-old students) and their teacher Karl-Erik Christensen. The project was a geography GIS project. One of the scenarios for the location of a new freeway running through the town of Silkeborg has given rise to a heated debate. The scenario shows the freeway running through a built-up area. With GIS it is possible to find out who and how many will be living close to the freeway. A buffer zone of 300 meters has been designated along the proposed freeway. By combining the buffer zone and the theme with the buildings, GIS can select the buildings situated in the buffer zone (Fig. 8.3). The data base contains information about the buildings whether they are housing or business. A simple query selects the buildings that are housing. They are shown in red. Other buildings in the buffer zone (business) are shown in blue. All buildings outside the buffer zone are shown in orange. If the classification has an equal number of municipalities in

Fig. 8.3 Project Freeway. Using GIS for distance analysis in Silkeborg

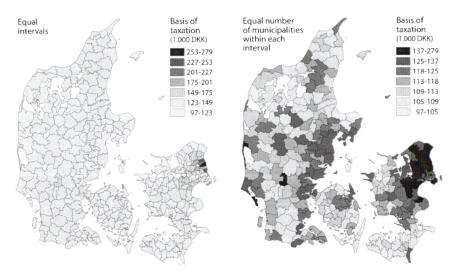

Fig. 8.4 Welfare Denmark and unequal Denmark

each class, an impression of Welfare Denmark is created, where many earn well and few have too little. If the municipalities are classified in equal income intervals, the high income areas north of Copenhagen are highlighted. This creates an impression of an unequal Denmark (Fig. 8.4). So even though the same data are used, changing the visualization method delivers a different message.

8.8 The Future of GIS in Schools

In the future there will be a growing demand for a more open GIS technology storing data in new standardized formats, so that data can be reused across all systems and thus be used by more people than before. A growing number of public services will be available on the Internet GIS servers, so that the citizens can search, find, and comment directly on the desired information. It may relate to changes in the local area where various GIS departments in the region or the municipality make more and more information available to the citizen. If the citizens can help themselves and master GIS, the authorities will eventually be spared a large number of time-consuming enquiries. The active citizen will gain a better insight into the decision-making processes behind changes, for example in local development plans.

The most significant increase in the use of GIS is in education. Today students are encouraged to find information on the Internet and help themselves. They have access to ArcGIS on their laptops and iPhones, which makes it readily available for integration in their day-to-day studies. This is also a part of their democratic training, which is also an important part of school life. The Danish development shows that most subjects can use GIS as part of their teaching. This applies not only to science subjects, but also to social studies and second language teaching. It turns out that GIS is eminently suited to strengthen cross-curricular collaboration. The successful introduction and use of GIS in education are dependent upon an easy access to quality data on the Internet. This can only be achieved through extensive collaboration between data providers, schools, and the Ministry of Education to ensure attractive websites and relevant teacher training.

Reference

Christensen, K. E., Thomas, J., & Torben, P. J. (2003). *Det digitale atlas*. MTM-Geoinformatik, Aalborg Universitet.

Internet Sources

http://www.gisgeo.dk – GIS Portal – Danish
http://www.edugis.dk – IMS-server – Danish. EduGIS has done most of its work in the field of using remote sensing and GIS (Geographical Information Systems) in various subject areas
http://www.geoforum.dk/ – the Danish Association for Geographic Information

http://www.esa.int/SPECIALS/Education/index.html – Eduspace is the website for Earth Observation (EO) enabling secondary schools to bring EO into the classroom for teaching and learning. Eduspace is a subdivision of ESA, creating learning and teaching tools for secondary schools such as LeoWorks, a tool for digital image processing and GIS

http://gisdag.dk/ – Informi GIS about education – Danish

http://www.dmu.dk/en/ – National Environmental Research Institute (DMU)

http://www.dmu.dk/udgivelser/kortoggeodata/ais/aisdownload/esriformat/ – Region maps of Danmark, ArcView format

http://www.dst.dk/HomeUK.aspx – Statistics Denmark

http://www.kms.dk/English/ – National Survey and Cadastre

http://www.COWI.dk – with maps and geo-data products

http://www.hiskis.net/ – History GIS Maps – Danish

Chapter 9
Dominican Republic: Prospects for the Incorporation of GIS into the School Curriculum

Quinta Ana Pérez Sierra and Victoria Castro De la Rosa

9.1 Introduction

The education system in the Dominican Republic is experiencing important changes with regard to the school curriculum. As the curriculum for school subjects such as geography, history, and natural science is developed, the geographical situation and geological composition of the Dominican Republic are being considered along with the effects of climate change on the island. The updates in the curriculum provide an opportunity to integrate GIS technology into schools. Presently, GIS is not as directly applied as it should be given its importance in the teaching–learning process. The aim of this chapter is to analyze the integration of the GIS technology in the school curriculum of the Dominican Republic, discuss a pilot GIS project for schools, and make recommendations for further application of GIS in the schools of the Dominican Republic.

The Information and Knowledge Society has been characterized by the use of Information and Communication Technologies (ICT) in all human activity. The introduction of ICT into society requires citizens to develop new social, personal, and professional contexts in order to confront the ongoing changes present in our society. Educational institutions are not isolated from these changes. On the contrary, they remain in full connection with the surrounding world through computer systems. ICT provides a way to access information that can be helpful in the completion of objectives and the application of educational curriculum. In fact, ICT has been incorporated in multiple curriculum areas, proving that it can be a crucial resource. In this chapter, we focus on the integration of the GIS technologies into the school curriculum in the Dominican Republic.

Q.A. Pérez Sierra (✉)
Instituto Tecnólogico de las Américas, Santo Domingo, Dominican Republic
e-mail: qperez@gmail.com

A.J. Milson et al. (eds.), *International Perspectives on Teaching and Learning with GIS in Secondary Schools*, DOI 10.1007/978-94-007-2120-3_9,
© Springer Science+Business Media B.V. 2012

9.2 Incorporating GIS in the School Curriculum in the Dominican Republic

The structure of secondary school education in the Dominican Republic consists of four years divided into two cycles of two years each. Each grade consists of two 4-month periods called *cuatrimestres*. The average age of secondary school students is between 14 and 18 years. Secondary school includes three modalities: the general modality, the technical professional modality, and the modality of art. The general modality includes all the specifications for the first and second cycle. The first cycle is general and common for all the students that attend secondary school. It allows students to develop abilities that teach them life lessons that will allow them to continue humanistic studies. The second cycle consists of three modalities: general, technical professional, and art. The general modality provides students with a foundation in the different areas of the scientific–humanistic and technological knowledge. The technical professional modality prepares students for qualified technical careers in the following sectors: industrial, services, and farming. The modality of art provides a general artistic foundation for the student to develop competencies and skills in different manifestations of art, such as scenic arts, music, visual, and applied arts. The transformation of the Dominican secondary school curriculum focuses on connecting the students' knowledge/learning with economic, cultural, and social life (Table 9.1).

According to the content recommended in the new curriculum map, GIS technologies could play a crucial role in the incorporation of the ICT in the process of teaching/learning in subjects like history, geography, and natural science. These subjects are present in all four years, including the first and second cycle. In this chapter, we will take the subject of geology as an example of the content and knowledge of natural science of the second cycle and second year. For example, the content and knowledge related to "seismic zones of greater frequency in Santo Domingo (Island)" include the expectation that students "interpret, value, and participate in the application and diffusion of procedures or measures on earthquakes occurrences in order to reduce human vulnerability." GIS technology resources recommended for this goal could include (1) digital cartography editors from the superposition of layers of remote perception images, (2) superposition of layers of soil and land use, and (3) layers of seismic risk. With all these resources, the students will have the opportunity to examine the entire island and critically analyze the problem using the seismicity map of the island. With these resources they should be able to design their own risk maps from the population layer and layers of seismic risk (Fig. 9.1).

The study of type of soils, components, modified soils, and soil destruction provides another content of great potential to be developed with the resources of GIS and remote sensing–based analysis. The expected knowledge/learning for this topic includes, "explore soil components, the changes that it goes through, advantages and disadvantages according to the type of material present in the formation of the soil." Similarly, the "ecosystem of the Dominican Republic" topic involves teaching students to "value, conceptualize and relate the conservation and transformation of energy in the hydrological, carbon and nitrogen cycle … and its importance for the

Table 9.1   Curriculum map of education in the Dominican Republic

First cycle

First year	Weekly hours	Second year	Weekly hours
Language	5	Language	5
Mathematics	5	Mathematics	5
Geography and history (Prehistory-XV Century)	5	Geography and history (XV century- mid-XIX century)	5
Natural Science (Biology)	5	Natural science (chemistry)	5
Artistic education	2	Artistic education	2
English	3	English	3
French	2	French	2
Human formation and religion	1	Human formation and religion	1
Physical education	2	Physical education	2
Total hours	30	Total hours	30

Second cycle

Third year	Weekly hours	Fourth year	Weekly hours
Language	5	Language	5
Mathematics	5	Mathematics	5
Geography and history (Mid-XIX century – Mid-XX century)	5	Geography and history (Mid-XX century – present)	5
Natural science (physics)	5	Natural science (environment and geology)	5
Artistic education	2	Artistic education	2
English	2	English	2
French	2	French	2
Human formation and religion	2	Human formation and religion	2
Physical education	2	Physical education	2
Total hours	30	Total hours	30

living." Students are also expected to "understand and discuss the laws and norms that regulate protected areas in the Dominican Republic and its application for the living." Furthermore, students are expected to "explore/investigate, compile scientific reports about natural resources in the Dominican Republic, its classification and adequate management for its viable handling."

These topics could be very attractive and of great importance, but at the same time they may be difficult for students due to the lack of learning resources that would allow them to interact directly with this content and the reality surrounding them. This content is being taught largely without the aid of computers and software, yet we believe that the integration of GIS technologies and audiovisual equipment is necessary in order to improve our education system. After analyzing the topics and recommendations for the new curriculum, we identified several

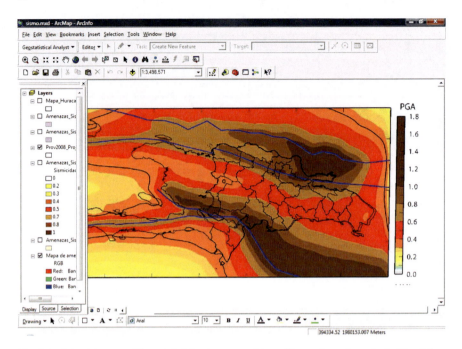

Fig. 9.1 Seismicity map, Hispaniola Island, superposition of layer of provinces in the Dominican Republic, as modeled in ArcGIS software

GIS resources that could be incorporated for the learning of this content. These resources include (1) terrestrial space image interpretation, (2) applications of spatial and geographical location, (3) applications for the analysis of remote perception images, and (4) applications of vector cartography development and edition. In addition to online sources, there are valuable places for acquiring cartography resources in the Dominican Republic to support GIS in the schools, such as (1) the Minister of Hydraulic Resources, Department of Geomatics, (2) the National Statistics Office, (3) the Minister of the Environment, Department of Environmental Cartography, and (4) the Geological Services National Office, Department of Risk Maps.

9.3 Pilot Study: Environmental Education in the Dominican Republic with GIS

We regard environmental education as the cornerstone for sustainable development in the Dominican Republic. The main objective of this project was to create awareness of environmental protection in students through the use of ICT as a tool for constructivist learning. The Dominican Republic suffers global environmental problems of deforestation, disasters, biodiversity loss, soil degradation, climate change, thinning of the ozone layer, pollution, water shortages, and population growth. Our intent was to attack the identified problems by applying GIS. We believe that starting with basic primary education (children between the ages of 7 and 11) is the first step to achieve changes in our environment by raising awareness of the issues among children who are in what Piaget (1937) called the "Period of Concrete Operations." A pilot project was developed by a research group of the Instituto Tecnólogico de las Américas (ITLA) (www.itla.edu.do) to bring students around Boca Chica, Santo Domingo, the knowledge of how to take care of their environment. With the collaboration of teachers of each course, contents about environmental care were selected. To develop this theme, the research group decided to focus on a protected area of Española's Island called *Los Haitises*. The first lesson involved the identification of the island in relation to its position on the globe using Google Earth. A second lesson involved mapping Los Haitises National Park using Esri's ArcGIS. The preliminary results of the project are encouraging the group to use GIS technology to develop other projects about specific locations in Dominican communities, including places of high waste concentration, mosquito breeding grounds, and classification of pets. There is no doubt for us that GIS technologies are a useful tool for developing educational projects in different areas of curricular content in the Dominican Republic.

9.4 Conclusion

GIS can have a positive influence on the teaching and learning of the school curriculum in the Dominican Republic. The geographic location and tectonic structure of the Dominican Republic and the effects of climate change and environmental degradation on the island provide numerous learning opportunities. For this reason, we recommend improving the Dominican education system by (1) developing learning/teaching materials for understanding environmental change processes, (2) providing graphics of the real world that will allow students to analyze relationships and interactions and to reach their own conclusions, (3) using GIS technologies for teaching history, geography, and natural science in order to inquire about the spatial nature of these subjects, and (4) facilitating access to geographic information sources through remote sensing image servers free of charge for the development of geospatial content activities.

Reference

Piaget, J. (1937). *La construction du réel chez l'enfant/The construction of reality in the child*. New York: Basic Books.

Additional Sources

Garcia, O. J. (2005). *El uso didáctico de la cartografía digital (SIG) como instrumento de análisis del paisaje y desarrollo de valores ambientales, Escuela Universitaria de Magisterio*

de Ciudad Real Universidad de Castilla-La Mancha, Ciudad Real, España. http://www.
bibliotecaspublicas.es/villarrubiadelosojos/imagenes/contenido_7642.pdf

Joyanes, L. (1997), Cibersociedad: Realidad o utopía, Universidad Pontificia de Salamanca en
Madrid, 1996, pág. 469. Madrid: McGraw-Hill.

Kerski, J. J. (1999). *A Nationwide analysis of the implementation of GIS in high school education.*
Proceedings of the 1999 Esri international user conference. http://proceedings.esri.com/library/
userconf/proc99/proceed/papers/pap202/p202.htm.

Ministerio de Educación (2010). *Currículo actualizado de educación media de la república
dominicana Revisión Agosto 2010.*

Piedrahita P. F. (2010). Rector Universidad Icelsi, Cali, Colombia. Un Modelo para integrar
TIC en el Currículo. Accessed September 25, 2010, http://www.eduteka.org/tema_mes.php3?
TemaID=0018

Sánchez, L. (2004). Tecnologías de la información y la comunicación en la enseñanza y en el
currículo de la educación secundaria obligatoria y del bachillerato de la comunidad de Madrid:
Modelo de Enseñanza Virtual. Tesis Doctoral.

Sánchez, L., Lombardo, J. M., Riesco, M., & Joyanes, L. (2004). Las TIC y la formación del profesorado en la enseñanza secundaria. Accessed September 27, 2010, http://www.cesdonbosco.
com/revista/TEMATICAS/tic.asp

Wanner, S., & Kerski, J. J. (1999). *The effectiveness of GIS in high school education.* Proceedings
of 1999 Esri International User Conference. http://proceedings.esri.com/library/userconf/
proc99/proceed/papers/pap203/p203.htm

Wiegand, P. (2001). Geographical Information Systems (GIS) in education. *International Research
in Geographical and Environmental Education, 10*(1), 68–71.

Chapter 10
Finland: Diffusion of GIS in Schools from Local Innovations to the Implementation of a National Curriculum

Tino P. Johansson

10.1 The Structure of Secondary Education in Finland

The education system in Finland is categorized into four stages, namely pre-primary, primary, secondary, and tertiary education. Local authorities provide pre-primary education for free to all 6-year-old children. Although, pre-primary education is not compulsory, nearly all eligible children participated in pre-primary education in Finland in 2008. Compulsory education begins at the year of the student's seventh birthday and is provided for free by municipalities in comprehensive schools. This basic education is comprised of primary and lower secondary education. The scope of the basic education syllabus is nine years and it is divided into nine forms, where 1–6 belong to primary education and forms 7–9 to lower secondary education. The comprehensive schools may also provide an additional but elective tenth basic education year for pupils who have not entered upper-secondary general education or vocational education and training levels. The tenth year is also available for pupils who need extra time to clarify their study plans for the future. The pupils will generally apply to upper-secondary general education and upper-secondary vocational education and training through the national joint application system at the age of 16. The students will be selected to upper-secondary general education on the basis of their previous study record. The scope of the syllabus of upper-secondary general education is three years, but students may complete it in two to four years. During the last year of study at the general upper secondary school, the students will participate in a matriculation examination which is required for graduation. The completion of either upper secondary education line gives students eligibility to apply to higher education in universities and polytechnics, representing the tertiary stage in the educational system. The Finnish National Board of Education verifies the national core curriculum for the comprehensive schools, upper-secondary general education, and upper-secondary vocational education and training in Finland. The national core curriculum includes objectives, contents, and assessment criteria

T.P. Johansson (✉)
University of Helsinki, Helsinki, Finland
e-mail: tino.johansson@helsinki.fi

A.J. Milson et al. (eds.), *International Perspectives on Teaching and Learning with GIS in Secondary Schools*, DOI 10.1007/978-94-007-2120-3_10,
© Springer Science+Business Media B.V. 2012

that provide a framework for municipalities and schools to form their own local curriculum. Teachers are able to select their own teaching methods and materials for learning. The current national core curriculum for comprehensive and upper secondary schools was approved in 2004 (European Commission, 2009).

10.2 Secondary School Teachers Possess a Master's Degree

Universities and vocational institutes of higher education provide teacher training in Finland. There are two categories of teachers who are qualified to work in the comprehensive schools. The first category is classroom teachers who obtain a master's degree in Educational Science (300 European Credit Transfer System: ECTS credit units). This degree also qualifies them to work as pre-primary school teachers. The second category of teachers working in secondary education is the subject teachers. They will study for a master's degree in the selected subject, such as geography or chemistry. The extent of the studies will be 300 ECTS units. The subject teachers start their degree studies along the same degree requirements as other master's degree students but include pedagogical studies for their degree at the later stage. One may also apply directly to subject teacher training in Finland but it is only possible in a few subjects, such as musical and physical education, home economics, and visual arts, at this time. The only category of teachers qualified to work at the upper secondary schools are the subject teachers. The vocational school teachers must also have a degree from the university or alternatively from a vocational institute of higher education. After graduation, these teachers usually work for a few years and then complete their pedagogical studies at a vocational institute of higher education.

The lifelong learning approach of the Finnish educational system ensures that teachers at the comprehensive schools and upper secondary schools, and also teachers at other educational levels, get support for their professional development through in-service teacher training organized by the National Board of Education and all Finnish universities. In-service teacher training is voluntary and usually free for the participants. The content of in-service training varies from university to university according to the profile of their degree structure and content. Some universities have established National Resource Centers to support teachers and degree students by providing learning and teaching materials and in-service training workshops on certain subjects. The University of Helsinki coordinates a National Center for supporting natural science education (LUMA), which hosts Resource Centers for biology, physics, chemistry, geography, mathematics, and pedagogy. GeoPiste is a National Resource Center of Geographical Education and it operates under the LUMA Center. GeoPiste provides technical and pedagogical support for teachers and university students by organizing training on the latest teaching methods and by borrowing equipment and literature.

10.3 The History of GIS in Finnish Schools

The introduction of GIS-assisted learning into the general upper secondary schools started to evolve gradually through a few local innovative learning experiments of pioneering teachers across the country in the late 1990s. Since the beginning of the new millennium, the Finnish Ministry of Education has funded a few projects that aimed at introducing GIS in schools. One of the funded projects was a national three-year pilot project on GIS in teacher training carried out by the Department of Geography of the University of Helsinki between 2001 and 2004. The pilot project was managed by the author and it focused on the incorporation of GIS tools into environmental education and hands-on teacher training. The project involved collaboration with the teachers of the Hausjärvi secondary and upper secondary schools and adopted a collaborative learning approach for studying the variation and spatial differences in water quality. The empirical work carried out by the teachers and students focused on understanding the causes and effects of water quality change. Teachers and their students regularly visited a local lake and a river and collected data on water quality with GPS receivers and other instruments (Fig. 10.1). The collected data was analyzed and visualized by the students with GIS software (Fig. 10.2). Another large project funded by the Ministry of Education and partly by the participating municipalities is the GIS and Cartography Education Network for Schools. The project coordinated by five upper secondary schools (2002–2006) focused on training the pioneering teachers and building a network of schools for the common purpose of developing the use of GIS in the classrooms. The project also developed virtual GIS courses and a GIS portal for the schools participating in the network. These services and in-service teacher training were later extended nationwide. From 2006 onward the GIS and Cartography Education Network continued to maintain the online portal for schools and developed good practices for using GIS in schools.

Fig. 10.1 Hausjärvi lower secondary school students collect data on water quality from Lake Mommilanjärvi with their teacher in 2003 (Photo by Terttu Talman)

Fig. 10.2 Data collected from the field was saved into a GIS database and visualized on a local topographic map of 1:20,000 by the students of Hausjärvi lower secondary school in 2003 (Photo by Tino Johansson)

The first GIS Day was organized in a few Finnish lower secondary and upper secondary schools in November 2003. The National Board of Education together with the Water of Life Project advertised the idea to the schools for organizing a GIS Day in Finland. The primary objective of organizing a GIS Day was to introduce GIS to Finnish teachers and students and to demonstrate how GIS is used in society. Another objective was to encourage schools to create contacts with municipal authorities, various organizations, and companies that work with GIS. The program of the first GIS Day included visits to regional planning authorities and to the Finnish Environmental Institute. One school was visited by a log truck demonstrating the use of GPS navigation and digital maps in locating the piles of logs for loading and protected areas where access is prohibited to big trucks.

The concept of GIS was introduced in geography textbooks in Finland from 2003 onward. However, the majority of secondary schools still remained uninformed about the potentials and possibilities of GIS as an educational tool for geography, as well as for several other subjects. The reform of the National Curriculum Standards for Upper Secondary Schools in Finland, which took effect in 2005, emphasizes interdisciplinary learning, aims at enhancing students' problem-solving skills and collaborative learning, and simultaneously paves the way for the incorporation of GIS into upper-secondary school classrooms. Currently, the Finnish upper secondary schools are obliged to provide an elective course in geography (GE4) containing basic GIS theory and principles as well as the introduction of some available GIS applications. The learning objectives of the GE4 course include the comprehension of the basics of cartography as well as the basic principles and applications of GIS in society. Moreover, the student should learn to collect regional data with different methods (observations, interviews, and questionnaires) and from various sources (atlases, statistics, and maps). The student should be able to use information networks in data collection and learn how to visualize regional information with maps, diagrams, and photographs. In addition, the student should know how to analyze and interpret the collected data and comprehend the basics of scientific writing.

These learning objects described in the National Curriculum provide a framework for the municipalities and upper secondary schools for selecting the contents of their own locally applied curricula. The fluency of teachers in GIS has generally determined the scope of implementation of these tools into the GE4 course. The students' regional studies in the GE4 course have traditionally been carried out without any computer-assisted cartography or GIS software. The National Board of Education and the Finnish universities, especially the University of Turku and University of Helsinki, have organized in-service teacher training for geography and biology teachers before and after the curriculum reform. However, the majority of geography teachers have not studied any GIS use during their degree studies at the universities. Therefore, the younger generation of teachers is more fluent in GIS than their older colleagues, trying to introduce these learning tools into the classrooms.

The University of Helsinki coordinated a three-year project funded through the MInisterial NEtwoRk for Valorising Activities in digitization (MINERVA) of the European Commission in 2004–2006. The aim of this Geographical Informations Systems Applications for Schools (GISAS) project was to introduce GIS into European secondary and upper secondary schools. The GISAS project utilized the lessons learned from the previous teacher training project carried out in Finland. One of the objectives was to create a model on how to incorporate GIS into geography and environmental education at these educational levels. In total, 35 in-service teachers and over 220 students from 9 European countries participated in the project. A few teachers from the Viikki Teacher Training School of the University of Helsinki also participated in the GISAS project, learned to use desktop GIS software and used different educational materials created by the project consortium (Johansson, 2006).

10.4 Case Study: Jaakkola Lower Secondary School in Kerava, Finland

The national curriculum does not oblige lower secondary schools to introduce GIS in Finland. Geography teacher Merja Tuurala from the Jaakkola lower secondary school had earlier participated in in-service teacher training on GIS and she invited the author to their school to organize a brief GIS-based activity involving both lab and field work during the GIS Day in 2003. Tino Johansson and Merja Tuurala decided in a preparatory meeting that the activity should be both entertaining and educational. It was decided that a geocaching exercise would fulfill both requirements. The five geocaches hidden in the school surroundings by the teachers a day before the actual exercise were located with GPS receivers. The geocaches were established by placing small pieces of paper with questions on map symbols on one side and questions on forest types and vegetation on the other side into plastic photographic film containers. The questions on basic map symbols were for the pupils of the 7th Grade (13-years old) and the questions on forest types were for the pupils of the 8th Grade (14-years old). In total, 42 pupils, both girls and boys, participated in the exercise, using GPS receivers to locate the hidden containers and then answering the questions inside. The pupils formed groups of four to five pupils each for a total of ten groups. Before the field exercise, the pupils were given a piece of paper with the X and Y coordinates of each geocache. They also received one-hour training on the use of the GPS receiver and instruction on how to move and find the right direction on the basis of the changing coordinate values on the screen of the GPS. The idea was not only to teach them about map symbols and forest types but also to teach them to communicate with the exact cardinal points when navigating in the forest. The groups were told to start from different locations and to place each geocache exactly in the same place after they had opened it, read the question, and written the collaboratively decided answer on the reply paper. They were given a maximum of one hour to find the geocaches and answer the questions. It was evident that the pupil groups began to compete with each other, so we instructed them that the quality of answers was more important than which group found all the hidden containers first. After all groups returned to the classroom equipped with GIS software, the teachers showed how the locations of the geocaches could be visualized on a digital map. Each group presented its answers to the teachers and the locations of the geocaches were discussed with the pupils by using the GIS software and digital maps reflected on the screen. The feedback session took one hour and it aimed at introducing a few basic GIS tools to the pupils and teaching them how to interpret map symbols and locations of forest types on a digital map. In the end, the groups received prizes, which were GIS-related supplies, such as pencils and caps.

The implementation of the geocaching exercise took one and a half work days at school. One hour for locating five geocaches was too little time for some groups. The accuracy of the GPS receivers varied a lot from one group to another. During the morning, one GPS receiver gave a group a location that was 50 meters away from the actual geocache site and the pupils could not find it without the help of the teacher. For the other groups, the GPS receivers guided the pupils within one meter from the geocache. The pupils liked the exercise primarily because it provided them a different way to learn the contents of the curriculum. The exercise contained both adventurous and competitive elements, which encouraged the pupils to actively participate in the group work. Despite the technical obstacles, the teacher Merja Tuurala considered the geocaching exercise to be a useful and fun learning method, especially once carefully planned and implemented with more time. Nowadays, many adults around the world participate in geocaching as a hobby, but as a learning method it still remains an example of how innovations in lab and field work can be integrated while using modern technology such as GIS in the lower and upper secondary schools. The geocaching exercise may be utilized in different subjects and by combining learning objectives of several subjects. The interdisciplinary nature of the exercise, however, was not yet fully utilized during the case study due to the lack of time at the school.

10.5 Prospects for Learning with and About GIS in Finland

An online questionnaire survey on the use of GIS in the upper secondary schools of Finland, carried out by Tino Johansson in 2006, showed that over 62% of the respondents had, at least to some extent, used GIS technologies in education (Yuda, Satori, & Johansson, 2009). The results show that the reforms of the national curriculum have encouraged more and more geography teachers to experiment and introduce either desktop or WebGIS applications. The locally-adapted national curriculum framework also allows more GIS-fluent teachers to use these technologies in other subjects and courses available at the upper secondary level. The rather high percentage of teachers who mentioned in the questionnaire survey that they have used GIS in the classroom, however, does not reveal anything about the ways of using the technology in the classroom or how intensively and for how many hours they actually used it. Despite the promising first steps taken toward a large-scale incorporation of GIS-assisted teaching and learning at upper secondary schools, there are still several obstacles to reaching nationwide adoption and use. Lack of software, databases, and pedagogical instructions have partially limited the use of GIS in secondary schools, but recently developed Web-based learning environments for schools, such as PaikkaOppi in 2008–2011, may overcome these barriers in Finland. PaikkaOppi (http://www.paikkaoppi.fi/) is a project to develop a Web-based learning environment for GIS in the Finnish language funded by the National Board of Education. The learning environment contains ready-to-use materials and exercises for teachers and students. There is also an online mapping interface which allows the users to view existing maps and to create and edit their own maps. The PaikkaOppi project is currently looking for new schools and teachers for the piloting stage. The Web-based learning environment does not require any previous GIS skills so the project has great potential for encouraging more teachers to experiment with its tools. The project tries to overcome obstacles experienced in previous efforts to implement GIS in schools in Finland. For example, it does not rely on expensive software licenses or complicated data sets. The PaikkaOppi project utilizes up-to-date GIS databases from various Finnish organizations, such as the National Land Survey, Geological Survey, and the Environmental Institute. The approach of the PaikkaOppi project will potentially help Finnish schools to integrate GIS into the teaching and learning of various subjects in the next few years. Currently, using GIS in schools is easier than it ever has been in Finland, but the greatest challenge will be to convince teachers that GIS will bring added value to learning geography and other subjects. The value of GIS for spatial thinking and problem-solving has to be scientifically proven at schools and demonstrated for the teachers in order to find more followers for the GIS-enlightened group of teachers in Finland.

References

European Commission. (2009). *EURYDICE. National summary sheets on education system in Europe and ongoing reforms*. Finland, August 2009. Education, Audiovisual & Culture Executive Agency. Accessed August 20, 2010, http://eacea.ec.europa.eu/education/eurydice/documents/eurybase/national_summary_sheets/047_FI_EN.pdf

Johansson, T. (2006). GISAS project in a nutshell. In T. Johansson (Ed.), *Geographical information systems applications for schools–GISAS* (pp. 7–21). Helsinki: Dark Oy.

Yuda, M., Satori, I., & Johansson, T. (2009). Geographical information systems in upper-secondary school education in Japan and Finland: A comparative study. *The Shin-Chiri (The New Geography), 57,* 156–165.

Chapter 11
France: Dogmatic Innovations, Innovative Teachers, and Parallel Experimentations

Eric Sanchez, Sylvain Genevois, and Thierry Joliveau

11.1 Introduction

This chapter aims at drawing a portrait of the use of geotechnologies in secondary schools in France. We first describe the context of the integration of these technologies into the curriculum. This context depends on the organization of the educational system and on the history of the integration of remote sensing and digital cartography into French secondary schools. In this first part of the paper, we address the question of the influence of this context on GIS integration. Second, we give a short overview of the uses of geotechnologies by secondary teachers based on the results of a survey and a case description. This leads us to address the question of the potential of geotechnologies for teaching and learning. Third, we discuss the prospects for the development of the use of these technologies for educational purposes in France and formulate recommendations for policy makers to foster this development.

11.2 The French Context: Innovative Teachers and National Initiatives

The French educational system can be defined by the decisive role of the central government in the area of educational policy (European Commission, 2008). This role encompasses both the writing of a detailed curriculum and guidelines for teachers and the administration of the recruitment, training, and management of the teaching staff. The French territory is divided into 30 academies, which are responsible for the implementation of national educational policy.

Education is compulsory between the ages of 6 and 16, and 79% attend secondary education. The school that a student attends (general, technological, or vocational

E. Sanchez (✉)
Université de Sherbrooke, Sherbrooke, Québec, Canada
e-mail: eric.sanchez@usherbrooke.ca

A.J. Milson et al. (eds.), *International Perspectives on Teaching and Learning with GIS in Secondary Schools*, DOI 10.1007/978-94-007-2120-3_11,
© Springer Science+Business Media B.V. 2012

school) is determined by the plan of the student (and of his/her parents) and the decisions made by the head of the school and the *conseil de classe* (committee of teachers).

Until reform was adopted in 2009, teacher training was provided at *Instituts Universitaires de Formation des Maîtres* (IUFM) after 3 years of university studies. The first year was devoted to preparation for a competitive examination, and the second year consisted of a part-time placement of the preservice teacher in a school and training in IUFM. This reform abolishes the previous IUFM system of one year preservice teacher training so that teacher education is now at the master degree level.

The core categories of teachers who use GIS in secondary education in France are geography/history teachers and geology/biology teachers. Some projects related to the integration of geotechnologies in secondary education have been fostered by national or regional educational authorities, and they appear as a dogmatic innovation (Alter, 2002), that is, a rigid top–down process that aims at encouraging teachers to introduce these technologies into their practices. It is also possible to identify many individual initiatives among geography and geology teachers. At the same time, "lateral" or "parallel" initiatives, based on partnerships with researchers or companies, have appeared.

In France, the history of the integration of geotechnologies in education is rooted in the development of the uses of satellite imagery after the first SPOT satellite was launched. At the beginning of the 1990s, the Ministry of Education (MEN) decided to supply SPOT data and support the development of TITUS (*Traitement d'Images de Télédétection à Usage Scolaire*: http://histoire-geographie.ac-dijon.fr/SIG/Carto/satellite/Somsat.htm), a software package applicable for educational purposes (Fig. 11.1). At the beginning of the 2000s, some teachers began to consider using digital mapping software, but the use of GIS was still uncommon because of the cost and the availability of software applications and data. Innovative teachers developed some new tools, mostly for digital mapping and educational GIS software applications (Genevois, 2008). The Ministry of Education set up a new working group in order to develop the use of GIS in schools. Some of these "pioneers" were involved in this group, while others joined other initiatives based on local partnerships and funded by research credits. These experimentation groups designed new tools more adapted for educational purposes (e.g., Géoanalyste: http://dossier.univ-st-etienne.fr/crenam/www/recherche/geoanalyste.php). Their aim was to understand the impact of these new tools on teaching practices as well (Collicard, Trisson, Genevois, & Joliveau, 2005; Genevois, Carlot, Joliveau, & Collicard, 2003). Another initiative was the Dakini project, an Anglo-French educational program cofunded by the European Regional Development Fund, which displayed, in collaboration with local authorities, software and data with the objective to foster the use of geotechnologies in schools (http://www.thenrgroup.net/theme/dakini.htm and http://hgc.ac-creteil.fr/spip/rubrique.php3?id_rubrique=78%20).

Lately, the initiatives to develop the use of geotechnologies in schools have become merely a top–down process again. Edugéo (http://www.edugeo.fr), the

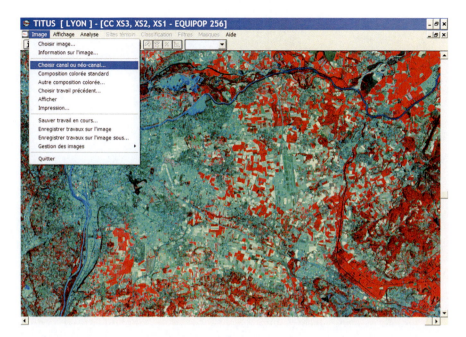

Fig. 11.1 SPOT image analysis with ©TITUS

educational side of the French virtual globe, Géoportail (http://www.geoportail.fr), has been designed by the IGN for geography teachers. Geology teachers can access geoinformation with Infoterre (http://infoterre.brgm.fr), a Web mapping service designed by the *Bureau de Recherche Géologiques et Minières* (BRGM), a French leading public institution involved in the Earth Science field for the sustainable management of natural resources and surface and subsurface risks. A national *Lithothèque* provides many examples of fieldtrips for geosciences teaching (http://www.educnet.education.fr/svt/ressources-numeriques/banques-de-donnees/litho). These initiatives mainly concern the development of Web mapping or digital globes. Recently, some initiatives have arisen to directly promote GIS in the curriculum. For instance, the introduction of the curriculum for geography teachers for lower secondary school published in 2008 underlines the importance of the use of GIS as a part of citizenship. Pairform@nce, a blended in-service teacher training program, includes four courses that relate to the uses of virtual globes for secondary education (http://edu-fc.pairformance.education.fr).

A French study about the uses of GIS in France (Fontanieu, Genevois, & Sanchez, 2007) revealed that teachers express a high interest for the uses of GIS material for educational purposes and identified many teachers who have developed interesting and innovative uses of GIS. Despite the fact that 80% of the teachers expressed willingness to use geotechnologies for geography or geology teaching, they mainly used virtual globes rather than GIS. This can be explained by the transfer of personal practices into their professional activity and by the fact that these

Fig. 11.2 Students prepare a geology field trip with ©Geonote

technologies are recent. Most teachers have not benefited from adequate training for the use of GIS (Sanchez, 2009).

Nevertheless, many teachers are engaged in teaching with GIS and GPS using innovative teaching methods such as fieldwork (Fig. 11.2), project-based learning, inquiry-based learning, game-based learning, or collaborative learning. Sometimes, these examples are based on specific new tools such as online geoinformation platforms designed for educators (Genevois & Joliveau, 2009). For example, the geowebexplorer platform is a Web-based learning environment which has been developed by the University of Saint-Etienne to enable teachers and learners to easily deal with geoinformation (Joliveau, Calcagni, & Mayoud, 2005). This educational platform provides free basic geomatics tools, data sets, and geography lesson plans. This application was tested in 12 classrooms by nearly 400 children in French high schools from 2005 to 2008. Beyond the benefits of an online GIS platform, the teaching sequences designed by the teachers aim at developing problem solving, complex thinking, and systemic analysis (Genevois, 2008). However, few initiatives provide for the dissemination of such practices within the educational system.

11.3 Teaching/Learning with GIS in France

Within the context of the development of geotechnologies for personal and professional uses (Longley, Goodchild, Maguire, & Rhind, 2005), it has been advocated that GIS offers the opportunity to change methods of teaching and learning (Kerski, 2003). Geotechnologies make it possible to deal with embedded knowledge by taking into account the complexity of the world, and they are adapted for education in sustainable development. These opportunities also relate to the type of teaching that is practiced, since a student-centered approach is fostered by the uses of geotechnologies. Nevertheless, the French survey points out that these tools are used in

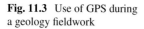

Fig. 11.3 Use of GPS during a geology fieldwork

different pedagogical contexts for the teaching of a wide range of subjects, and that the teachers use geotechnologies mainly to allow students to access geographical/geological information and to visualize geographical/geological features rather than involving students in the process of data gathering and data processing during fieldwork (Fig. 11.3).

One aspect that seems important for explaining the success of geotechnologies is their capacity to allow the user, teacher or student, to create his/her own content. This can be illustrated by cases of the use of geotechnologies to implement game-based learning in the classroom, which have been developed for research projects carried out by the *Institut National de Recherche Pédagogique* (INRP).

One project, begun in the 2007/2008 school year (Sanchez & Jouneau-Sion, 2009), involved using Google Earth to design a pretend game based on a real land use management plan: the design of a high-speed railway loop for the test of a high-speed train (TGV). During 4 hours (2 weeks), 20 students from a lower secondary school (ISCED level 2A-13/14 years old) were asked to play the role of different characters: citizens living close to the project, citizens who do not want their environment to be partially destroyed, a citizen looking for a job, local political authorities, and a factory director. The students' task consists of convincing the other characters that this project should or should not be set up and finding a location for the implementation of the project (Fig. 11.4). The students are given access to a set of data designed by the teacher to help them to decide and to convince the others.

The results of the research emphasize several benefits of the uses of the virtual globe to design the game. The students can access information about their role and the rules of the game, visualize and gather different data about the project, create their own set of data in order to argue their point of view, and use the virtual globe to communicate with other participants. The fact that virtual globes are virtual worlds based on Web 2.0 technologies and partly based on user-generated content facilitates their use by teachers and students.

Fig. 11.4 Students' presentation using geotechnologies during a pretend game

A second result relates to the involvement of the students in the game. They managed to fulfill their task by themselves and to use most of the ideas that they had learned during previous geography lessons. The game offers them the opportunity to use embedded knowledge, give meaning to the tasks that they have completed, and develop situated knowledge and competencies. This involvement is fostered by the collaborative work allowed by the use of the virtual globe. The students are asked to form interest groups, and they develop the awareness that cooperation is important to completing the task.

Another project (school year 2009/2010) about the use of geotechnologies in order to implement game-based learning in secondary education is now in progress (Sanchez, Delorme, Jouneau-Sion, & Prat, 2010). In this pretend game, 30 students (ISCED level 3A-15/16 years old) play the role of companies' staff specializing in different sustainable energies, opponent citizens, or policy makers. They have to design or discuss a project that aims at renewing the way of producing energy for their city. The design of the game aims at providing a rich, complex, and realistic environment with geotechnologies (including augmented reality and fieldwork) in which students gain knowledge and competencies about sustainable development and citizenship. An experimentation of geolocated games (http://geoeduc3d.scg.ulaval.ca/) was started in 2009 in cooperation with the GéoEduc3D international Project (Daniel & Badard, 2008).

11.4 The Development of the Uses of GIS for Educational Purposes, a Matter of Professional Development

The use of GIS in the French educational system should be considered as one side of the introduction of Information and Communication Technologies (ICT) for secondary education in France. It is expected that the use of these technologies will foster both student-centered methods and the development of student ICT literacy.

On one hand, teachers should implement student-centered pedagogies based on problem solving and collaborative learning. On the other hand, students should improve their capacity to handle geographic information and develop relevant uses of geotechnologies as an integral part of their digital culture. For example, students should be able to assess the relevance of geographical data. They should have the capacity to identify information that results from empirical evidence. They should be able to identify the consequences of the choices made by the information provider and the models implemented in the software. Students need to have basic knowledge about the use of GIS. As a result, these objectives are accomplished when teachers allow students to be involved in the process of learning, carry out their own investigations, and have the opportunity to engage in the process of data gathering and data processing with GIS. Furthermore, the potential of the use of GIS for model-centered approaches to teaching should be considered. GIS allows students to be engaged in modeling activities in order to investigate complex problems and to identify the factors that should be taken in account.

In addition to the improvement of "information literacy," the use of GIS allows for the development of spatial competencies (National Research Council, 2005). Through the use of digital images and maps, especially 3D images or georeferenced maps, students can develop their "spatial literacy" when using GIS. Using GIS to improve the learners' "spatial literacy" for various disciplines is a promising idea that is not yet at the top of the agenda in France where secondary teachers have typically taught both history and geography since the end of the nineteenth century. Furthermore, the university background of 90% of new teachers is history, and the portion of the effective curriculum dedicated to geography in high school programs represents less than 30%. Historians are typically less concerned with space and spatial thinking than geographers and are less interested in tools that are designed to address spatial questions. Since GIS was not traditionally designed to allow diachronic studies, GIS diffusion among history teachers remains slow. However, if the history departments at the university level will start using GIS as part of their teaching, some changes might take place in the future regarding skills of teachers. Spatiotemporal literacy concerns both geology and history, but space conceptualization is different between the natural and social sciences. The development of qualitative GIS applications must help to bridge this gap (Cope & Elwood, 2009). To develop the uses of geotechnologies in secondary education, teachers should be provided with GIS designed for educational purposes (e.g., which allow using the main format of available data and permit the main types of data processing). Teachers should be provided with relevant data that fit with their expectations to choose subjects related to both their surrounding environment and worldwide topics. Regardless, the main problem to solve is to improve the capacity of teachers to use GIS and to integrate them in their teaching practices. Most French geography and geology teachers did not benefit from GIS training during university education, and they lack confidence to use such tools. As a result, their training needs to focus on confidence building as much as on competence building (Welzel-Breuer et al., 2010).

To enhance the confidence and competence of French teachers for the use of GIS in the classroom, it is necessary that their professional development takes into account the need for basic technical skills. They need to understand the core concepts implemented in these tools and the methods of data processing with GIS. They need also to feel free to experiment with new ways of teaching in order to develop student-centered pedagogies. One way to reach this goal could be to give the teachers the opportunity to share experiences and design lessons that take into account the latest results of research into the uses of GIS for educational purposes.

11.5 Conclusion

What are the prospects for the use of GIS in French secondary schools? The time when educational authorities, innovative teachers, and parallel experiments coexisted separately is probably over. The uses of geomatics for everyday life purposes and the democratization of GIS are now well established and should enhance the dynamic of the adoption of geospatial tools by teachers. This dynamic will depend on the capacity of the French educational system to adopt a more flexible model than dogmatic innovation. This dynamic needs support rather than control, orientations rather than centralized decisions. The policy should be to offer the teachers more space for experimentation, establish a systematic training policy, encourage innovation among communities of practice, and set up an interoperability framework in order to facilitate the diffusion and the convergence of rich and various local initiatives.

References

Alter, N. (2002). *L'innovation ordinaire*. Paris: Presses Universitaires de france.
Collicard, J. P., Trisson, C. M., Genevois, S., & Joliveau, T. (2005). *L'utilisation d'un Système d'Information Géographique en classe*. Bilan d'une expérimentation Paper presented at the Actes de Géoforum Lille 2005 "Savoir penser et partager l'information géographique: les SIG", Lille.
Cope, M., & Elwood, S. (2009). *Qualitatife GIS. A mixed methods approach*. Thousand Oaks, CA: SAGE Publications Ltd.
Daniel, S., & Badard, T. (2008). *Mobile geospatial augmented reality, games and education: The Geoeduc3D project*. Paper presented at the Second International Workshop on Mobile Géospatial Augmented Reality, August 28–29, Quebec.
European Commission. (2008). The Education system in France: Directorate-General for Education and Culture.
Fontanieu, V., Genevois, S., & Sanchez, E. (2007). Les pratiques géomatiques en collège-lycée. D'après les résultats d'une enquête nationale sur les usages des outils géomatiques dans l'enseignement de l'Histoire-Géographie et des sciences de la vie et de la Terre. *Géomatique Expert*, 49–55.
Genevois, S. (2008). Quand la géomatique rentre en classe. Usages cartographiques et nouvelle éducation géographique dans l'enseignement secondaire. Thèse de doctorat, Université de Saint-Etienne, UMR 5600.

Genevois, S., Carlot, Y., Joliveau, T., & Collicard, J. P. (2003). Le SIG: un outil didactique innovant pour la géographie scolaire. *Dossiers de l'ingéniérie éducative* (n°44), 10–13.

Genevois, S., & Joliveau, T. (2009). Using a geoinformation-based learning environment in geography education: The GeoWebExplorer platform. In T. Jekel, A. Koller, & K. Donert (Eds.), *Learning with Geoinformation IV* (pp. 113–120). Heidelberg: Wichmann.

Joliveau, T., Calcagni, Y., & Mayoud, R. (2005). *Geowebexplorer, un outil géomatique collaboratif au service des enseignants et des élèves.* Paper presented at the Actes de Géoforum Lille 2005 "Savoir penser et partager l'information géographique: les SIG", Lille.

Kerski, J. (2003). The implementation and effectiveness of Geographic Information Systems Technology and methods in secondary education. *Journal of Geography, 102*(3), 128–137.

Longley, P., Goodchild, M., Maguire, D., & Rhind, D. (2005). *Geographic Information Systems and science* (2nd ed.). New York: Wiley.

National Research Council. (2005). *Learning to think spatially.* Washington, DC: National Academic Press.

Sanchez, E. (2009). Innovative teaching/learning with geotechnologies in secondary education. In A. Tatnall & T. Jones (Eds.), *Education and technology for a better World* (pp. 65–74). Berlin: Springer.

Sanchez, E., Delorme, L., Jouneau-Sion, C., & Prat, A. (2010). Designing a pretend game with geotechnologies: Toward active citizenship. In T. Jekel, A. Koller, K. Donert, & R. Vogler (Eds.), *Learning with geoinformation V* (pp. 31–40). Heidelberg: Wichman.

Sanchez, E., & Jouneau-Sion, C. (2009). Playing in the classroom with a virtual globe for geography learning. In T. Jekel, A. Koller, & K. Donert (Eds.), *Learning with Geoinformation IV* (pp. 78–86). Heidelberg: Wichmann.

Welzel-Breuer, M., Graf, S., Sanchez, E., Fontanieu, V., Stadler, H., Raykova, Z., et al., (2010). Application of computer aided learning environments in schools of six European countries. In G. Cakmakci & M.F. Taşar (Eds.), *Contemporary science education research: scientific literacy and social aspects of science* (pp. 317–326). Ankara, Turkey: Pegem Akademi.

Chapter 12
Germany: Diverse GIS Implementations within a Diverse Educational Landscape

Kathrin Viehrig and Alexander Siegmund

12.1 Introduction

Despite being a relatively small country, Germany has a very diverse educational landscape. Each of the 16 federal states has the primary responsibility for education, and thus much freedom in how to shape its educational system, the curricula, and the way teacher training is organized. However, educational issues of supraregional importance and student mobility between the federal states are managed by a "Standing Conference of the Ministers of Education and Cultural Affairs of the Federal States in the Federal Republic of Germany"; the *Kultusministerkonferenz* (KMK). State schools and many private schools belong to one of the following basic categories of secondary schools:

- *Hauptschule* (low stream; certificate generally at the end of Grade 9, ≈ 15 years old)
- *Realschule* (middle stream; certificate generally at the end of Grade 10, ≈ 16 years old)
- *Gymnasium* (high stream; certificate generally at the end of Grade 12 (rarely still 13), ≈ 18–19 years old)
- Comprehensive schools (varying organizational forms; e.g., division in different ability groups on a subject by subject basis or streaming in most or all subjects but within the same school; different certificates)
- Special needs schools.

These categories cover only the basic structure of the German school system – the variety is made even more complex by special forms in individual federal states as well as a growing number of private schools that often do not follow

K. Viehrig (✉)
University of Education Heidelberg, Heidelberg, Germany
e-mail: viehrig@ph-heidelberg.de

A.J. Milson et al. (eds.), *International Perspectives on Teaching and Learning with GIS in Secondary Schools*, DOI 10.1007/978-94-007-2120-3_12,
© Springer Science+Business Media B.V. 2012

the state system. These include for instance schools offering the International Baccalaureate or Walldorf schools.

Notable is that in the state system, except for some comprehensive schools, students are streamed at a very early point in their school career (mostly after Grade 4, in a few states after Grade 6). Streaming occurs largely based on the students' marks and perceived ability. The Programme for International Student Assessment (PISA) showed that when students reach the age of about 15, there are extreme differences in achievement between the school types. In the federal state of Hamburg, for instance, the average score for Gymnasium students in reading literacy was 573, for Realschule students 479, and for Hauptschule students only 345 (Prenzel et al., 2008, p. 190). The boundaries between the different school streams are set up to be permeable in both directions. In practice, while many students having difficulties are simply sent to a "lower" stream, it is not easy to switch to and succeed in a "higher" stream. Some students choose to get a "higher" school certificate later on through special (evening) classes.

The type of school leaving certificate determines the options students have. Students with a Hauptschule certificate have only limited options, for instance to apply for some professions to full-time vocational schools or apprenticeships (comprising vocational school and on the job training). In contrast, students with a Gymnasium certificate have earned the "*allgemeine Hochschulreife*" (literally, general university maturity) and thus, in principle, have full choice between full-time vocational schools, apprenticeships, dual studies (college level study combined with on-the-job training), and studying at colleges and universities.

Similar to the student level, teacher training varies widely between the states. In general, aspiring teachers have to choose which school type(s) they want to teach later on and study in a teachers' course for several years at a university (includes a number of teaching internships), earning their 1st State Exam. Teachers then spend between one to two years teaching in a school while taking classes at a teachers' seminary. If they pass, they earn their 2nd State Exam. Only then can they apply for normal full-time state teaching positions. Private schools do not have the same restrictions, but often follow the state requirements to some degree. Furthermore, many federal states offer special application schemes in subjects such as physics or Latin where there are not enough teachers to fill open positions even in state schools.

In-service teachers have a variety of professional development course options. Moreover, teachers fulfilling certain conditions can apply to special programs helping them to earn qualifications such as a PhD.

The educational system, as diverse as it is, is currently undergoing various reforms, resulting for instance from the "PISA shock" 2000 – which not only indicated lower levels of achievement for students in Germany than desirable, but also considerable differences in achievement between the different federal states. The reforms include the development of national educational standards for some defined core subjects (e.g., mathematics, German, English/ French, biology, chemistry, and physics), based on a KMK decision (KMK, 2010). Representatives of some of the subjects not included in that list – such as geography – decided to develop national standards themselves, starting with those for the 10th Grade certificate (Deutsche

Gesellschaft für Geographie, 2008). These national standards thus far have only partly been integrated into the curricula of the individual federal states.

12.2 GIS in Secondary Schools

In 1996–1997, supported by Esri, a working group for GIS in schools formed at a teacher professional development facility in Bavaria, called ALP Dillingen. Yet, the number of schools working with GIS in the late 1990s was still small (Cremer, Richter, & Schäfer, 2004). Since then, the number of schools using GIS increased slowly.

Not only have individual schools been sharing their GIS projects (e.g., Engelhardt, 2009; Heyden, 2009; Schäfer & Ortmann, 2004; and case below), but over the years, the body of publications and teaching materials for GIS implementation has also increased substantially. These include for instance articles and sample materials in teachers' journals (e.g., *Praxis Geographie, Geographie und Schule; Geographie heute*), a limited number of books dealing with GIS in schools (e.g., Falk & Nöthen, 2005), the inclusion of the topic into introductory textbooks for geography teaching (e.g., Haubrich, 2006), lessons on online platforms (e.g., www.lehrer-online.de), and efforts to develop GIS-specific competency models (e.g., Volz, Viehrig, & Siegmund, 2010). Esri provides project examples, school licenses for ArcGIS ArcView, a regular section "schools and universities" in the ArcAktuell, GIS summer camps, and support through their GIS education program. Moreover, Desktop-GIS versions have been developed specifically for classroom use (DierckeGIS first version, 1998; SchulGIS first version, 2001) as well as accompanying materials (Cremer et al., 2004). Especially in recent years, these have become supplemented by a growing number of online GIS-based applications designed for use in schools, developed by publishers (e.g., http://diercke.webgis-server.de; www.klett.de), researchers (e.g., www.webgis-schule.de), teachers (e.g., www.sn.schule.de/~gis/), or state institutions (e.g., http://www.gis.bildung-lsa.de/; http://webgis.bildung-rp.de/).

Furthermore, the number of GIS professional development courses offered to in-service teachers has increased in recent years. They are provided by a wide variety of stakeholders (e.g., companies, teachers, universities) and range from a few hours to several months in duration. With regard to the content, many focus on GIS skills and methodology. Courses that focus primarily on a content topic and only use GIS as a tool to explore it seem to be more rare.

In the past, students preparing to become teachers only in some courses had to participate in a full "introduction to GIS" class or in classes such as "using new media in the geography classroom" that could include GIS. The newly published framework of requirements for geography teacher education at universities in Germany now explicitly demands GIS and GIS education skills (Deutsche Gesellschaft für Geographie, 2009). While it will take considerable time until these demands are fully integrated into all universities' course requirement structures, it can be expected that in the future teachers will be quite well prepared to integrate GIS into their classrooms. The current extent of in-service-teacher preparedness, in contrast, has been hardly studied and is thus difficult to gauge. Studies-in-progress in this area include for instance those by Volz, Viehrig, and Siegmund (2008) or Höhnle, Schubert, and Uphues (2010).

Moreover, GIS has been explicitly included into an increasing number of (geography related) curricula and schoolbooks, mostly for Gymnasium (e.g., Siegmund & Naumann, 2009). It is also explicitly included in the national educational standards for the 10th Grade certificate for geography (Deutsche Gesellschaft für Geographie, 2008). The demands vary considerably between the different documents. At the end of Grade 10, it ranges from students not explicitly needing any GIS skills, over students using GIS as an information source, to students using GIS to create maps or as an analysis tool (e.g., Niedersächsisches Kultusministerium, 2008, p. 17; Deutsche Gesellschaft für Geographie, 2008, p. 18; Ministerium für Kultus Jugend und Sport Baden-Württemberg, 2004, p. 242). Even in those states where GIS is not explicitly mentioned, the curriculum would allow for integration, under categories such as using electronic sources of information or methods competency (e.g., Ministerium für Bildung Wissenschaft und Kultur Mecklenburg-Vorpommern, 2002). Nationally, there is a general focus on computer literacy and improvements in school computer infrastructure.

In recent years, a variety of programs and competency centers seeking to increase the use of GIS in schools have been established, either specifically for that purpose (e.g., www.gis-station.info) or as part of a more general GIS program (e.g., www.gis.uni-kiel.de). Thus, there currently seems to exist at least one outreach program in many of the federal states (e.g., www. gi-at-school.de, www.lis.bremen.de/sixcms/detail.php?gsid=bremen56.c.14799.de, http://gis-im-unterricht.de). Moreover, international and national conferences such as Learning with Geoinformation (Jekel, Koller, & Donert, 2008, 2009; Jekel, Koller, Donert, & Vogler, 2010; Jekel, Koller, & Strobl, 2006, 2007) or the GIS *Ausbildungstagung* have become important venues to network and share progress reports and ideas.

Despite all these initiatives, data collected in the federal state of Schleswig-Holstein from 44 teachers showed that GIS was still among the least used media of those included in the study, with 81.4% stating to never use GIS and 88.4% stating to never use WebGIS in the classroom (Klein, 2007, p. 154). A more recent study by Höhnle et al. (2010, p. 151) with 410 teachers nationwide gives a lower number of those stating to never use GIS in the classroom (WebGIS: 44.7%; Desktop GIS: 71%).

12.3 Cases

One common approach for implementing GIS in classrooms is to acquire prefabricated, curriculum-relevant materials and let the students work through them with an accompanying Web-based GIS application or a Desktop GIS. In the German educational community, this approach is especially recommended for beginning users

(Püschel, Hofmann, & Hermann, 2007; Schäfer, 2003) and included in schoolbooks (e.g., Gaffga et al., 2008). As few studies regarding the effectiveness of this approach exist in the German context, it has been chosen as part of an ongoing study in three federal states. Within the study design, students had a maximum of four lessons in order to work with a partner through a set of worksheets with fairly closed-ended tasks, which aimed at helping them to explore the topic "Tourism in Kenya." The GIS group either used the Web-based GIS application (http://diercke.webgis-server. de) or a corresponding data set for ArcGIS 9.3. Classes working on the same topic with paper maps were chosen as a comparison group.

Another approach for GIS implementation is to generate one's own GIS projects for use with one's students. In the German educational discourse, this is usually recommended for more advanced users (Püschel et al., 2007; Schäfer, 2003). This approach is less common, but individual teachers have used it in recent years. Some of these teachers are well connected with each other, active in the larger GIS education community, serve as trainers of other teachers, have experience with the implementation of a wide variety of GIS, Web-based applications, and GPS in their own classrooms, and share those experiences in journals and online. One of the teachers who has been active in this way for a long time is Matthias Schenkel, a geography teacher at the private Thadden Gymnasium in Heidelberg. One of the projects he conducted in 2003 with his students was a tree assessment in the schoolyard by type and age. The 11th Graders made a map (by hand) of the schoolyard, based on a ground plan and tree data provided by the environmental protection officer. The geometry and attribute data was then generated at the PC in SchulGIS. In the four lessons allocated for the project, the students could cover only part of the schoolyard. Stressing the authentic learning aspect, the teacher opted consciously to not finish the map himself. Teachers participating in an in-house training led by a staff member of Esri in 2009 conducted further work on the project using ArcGIS. The teachers used ArcPad and a geo-referenced orthophoto to map the trees. The new technique helps in making both the generation of geometry data and updates easier and faster.

12.4 Prospects

Although GIS implementation in secondary schools has increased in the last decade, it is still very much a "work-in-progress" in Germany. Challenges include issues such as competition for computer room access and getting the support of the school's computer administrator for installing and updating desktop GIS, insufficient teacher training, difficult-to-change traditions, and a lack of research regarding the effectiveness of GIS for reaching content competency objectives as well as educational concepts and teaching materials that are deduced from that research. Furthermore, the existing initiatives aimed at fostering the implementation of GIS in schools still seem to be somewhat separate.

Opportunities include the continuation and expansion of those factors that have led to an increase in the use of GIS in schools in recent years. Moreover, recently published frameworks, such as the explicit inclusion of GIS in the national educational standards and the framework of requirements for teacher education, will increasingly be integrated into the curricula of the individual federal states. This will likely also lead to an increased awareness and implementation of GIS in the future, helped by a noticeable increase in interest in GIS in the German educational research community. At the moment, GIS is mainly a topic within the framework of geography education. This leaves much room for expansion into other subjects such as history or science, as well as for interdisciplinary projects. Thus, the chances for the development of GIS as a component of secondary education in Germany look good.

References

Cremer, P., Richter, B., & Schäfer, D. (2004). GIS im Geographieunterricht – Einführung und Überblick. *Praxis Geographie, 34*(2), 4–7.

Deutsche Gesellschaft für Geographie. (2008). *Bildungsstandards im Fach Geographie für den mittleren Schulabschluss – mit Aufgabenbeispielen*. 5th Edition. Kiel, Germany: Author.

Deutsche Gesellschaft für Geographie. (Ed.). (2009). *Rahmenvorgaben für die Lehrerausbildung im Fach Geographie an deutschen Universitäten und Hochschulen*. Bonn: Selbstverlag Deutsche Gesellschaft für Geographie.

Engelhardt, R. (2009). *OHG-Schüler erzielen ersten Preis*. Accessed January 15, 2010, http://www.ohg-ka.de/

Falk, G., & Nöthen, E. (2005). *GIS in der Schule Potenziale und Grenzen*. Berlin: Mensch & Buch-Verlag.

Gaffga, P., Barheier, K., Bloching, K., Engelmann, D., Felsch, M., Harms, E., et al. (2008). *Diercke GWG für Gymnasien in Baden-Württemberg 5*. Braunschweig: Westermann.

Haubrich, H. (Ed.). (2006). *Geographie unterrichten lernen, Die neue Didaktik der Geographie konkret* (2 ed.). München: Oldenbourg.

Heyden, K. (2009). *Einsatz von geographischen Informationssystemen an der LG*. Accessed January 15, 2010, http://www.lg-ratzeburg.de/index.php?id=172

Höhnle, S., Schubert, J. C., & Uphues, R. (2010). The frequency of GI(S) use in the geography classroom – Results of an empirical study in German secondary schools. In T. Jekel, A. Koller, K. Donert, & R. Vogler (Eds.), *Learning with Geoinformation V – Lernen mit Geoinformation V* (pp. 148–158). Berlin: Wichmann.

Jekel, T., Koller, A., & Donert, K. (Eds.). (2008). *Learning with Geoinformation III – Lernen mit Geoinformation III*. Heidelberg: Wichmann.

Jekel, T., Koller, A., & Donert, K. (Eds.). (2009). *Learning with Geoinformation IV – Lernen mit Geoinformation IV*. Heidelberg: Wichmann.

Jekel, T., Koller, A., Donert, K., & Vogler, R. (Eds.). (2010). *Learning with Geoinformation V – Lernen mit Geoinformation V*. Berlin: Wichmann.

Jekel, T., Koller, A., & Strobl, J. (Eds.). (2006). *Lernen mit Geoinformation*. Heidelberg: Wichmann.

Jekel, T., Koller, A., & Strobl, J. (Eds.). (2007). *Lernen mit Geoinformation II*. Heidelberg: Wichmann.

Klein, U. (2007). *Geomedienkompetenz. Untersuchung zur Akzeptanz und Anwendung von Geomedien im Geographieunterricht unter besonderer Berücksichtigung moderner Informations- und Kommunikationstechniken*. Dissertation, Christian-Albrechts-Universität, Kiel.

KMK. (2010). *Bundesweit geltende Bildungsstandards.* Accessed September 15, 2010, http://www.kmk.org/bildung-schule/qualitaetssicherung-in-schulen/bildungsstandards/ueberblick.html

Ministerium für Bildung Wissenschaft und Kultur Mecklenburg-Vorpommern. (2002). *Rahmenplan Geographie - Regionale Schule, Verbundene Haupt- und Realschule, Hauptschule, Realschule, Integrierte Gesamtschule.* Accessed September 10, 2007, http://www.bildungsserver-mv.de/download/rahmenplaene/rp-geografie-7-10-reg.pdf

Ministerium für Kultus Jugend und Sport Baden-Württemberg. (2004). *Bildungsplan allgemeinbildendes Gymnasium.* Accessed July 10, 2007, www.bildung-staerkt-menschen.de/service/downloads/Bildungsplaene/Gymnasium/Gymnasium_Bildungsplan_Gesamt.pdf

Niedersächsisches Kultusministerium. (2008). *Kerncurriculum für die Realschule Schuljahrgänge 5-10 Erdkunde.* Hannover: Niedersächsischer Bildungsserver (NIBIS).

Prenzel, M., Artelt, C., Baumert, J., Blum, W., Hammann, M., Klieme, E., et al. (2008). *PISA 2006 in Deutschland - Die Kompetenzen der Jugendlichen im dritten Ländervergleich.* Accessed January 5, 2010, http://www.wir-wollen-lernen.de/resources/PISA-E_2006_vollst_Bericht_HH.pdf

Püschel, L., Hofmann, K., & Hermann, N. (2007). *Blickpunkt WebGIS. Der Einstieg für Schulen in Geographische Informationssysteme (GIS).* Koblenz: Landesmedienzentrum Rheinland-Pfalz.

Schäfer, D. (2003). *Scalable GIS applications for schools – examples from Germany.* Paper presented at the Esri 2003, 18th European User Conference, 10. Deutschsprachige Anwenderkonferenz. Conference CD. Accessed January 2, 2010, http://www.staff.uni-mainz.de/dschaefe/pdf/14_30_Fr_Gr_schaefer.pdf

Schäfer, D., & Ortmann, G. (2004). Biotopkartierung in Rheda-Wiedenbrück. Gemeinsames Projekt zwischen Stadtverwaltung, Schule und Universität Mainz. ArcAktuell(1/2004), 44.

Siegmund, A., & Naumann, S. (2009). GIS in der Schule. Potenziale für den Geographieunterricht von heute. *Praxis Geographie, 39*(2), 4–8.

Volz, D., Viehrig, K., & Siegmund, A. (2008). GIS as a means for competence development. In: T. Jekel, A. Koller, & K. Donert (Eds.), *Learning with Geoinformation III – Lernen mit Geoinformation III* (pp. 42–48). Wichmann: Heidelberg.

Volz, D., Viehrig, K., & Siegmund, A. (2010). Informationsgewinnung mit Hilfe Geographischer Informationssysteme – Schlüsselkompetenz einer modernen Geokommunikation. *Geographie und ihre Didaktik, 38*(2), 102–108.

Chapter 13
Ghana: Prospects for Secondary School GIS Education in a Developing Country

Joseph R. Oppong and Benjamin Ofori-Amoah

13.1 Introduction

Over the past three decades or so, Geographic Information Systems (GIS) has become widely accepted as an essential decision-making tool in both the public and private sectors in developed countries. From natural resources management, disease mapping and monitoring, crime analysis, and city and regional planning to planning and management of transportation and distribution services, as well as emergency response, GIS has become the central tool. Consequently, employment in geospatial technologies is booming and has become one of the most rapidly growing sectors (US Department of Labor, 2007, 2010). The consequent demand for geographic education centering on GIS has led to the proliferation of GIS certificate and degree courses in universities and colleges. In fact, GIS education has become so pervasive that it is emerging in secondary schools in a number of countries (see for example Kerski, 2003; Meyer, Butterick, Olkin, & Zack, 1999).

In developing countries, however, the story is entirely different. While cell phone use has increased dramatically, individual access to computers remains rare. Due to the high cost of computer technology, access is restricted to urban centers with the communication infrastructure, uninterrupted electricity supply, and skilled personnel. Even in these places, besides the few computers in major universities and government offices, public access is through Internet cafes that charge by the hour for use. For the rest of the population living on less than $2.00 a day, Internet use (at sixty cents an hour) remains an unaffordable luxury. In fact, although Internet use in Africa has grown three times faster than the rest of the world since 2000, only 5.3% of Africa's population used it in 2008 (Oppong, 2010). Moreover, due to cost, the most popular services remain electronic mail services using free Web-based services such as Yahoo, Hotmail, and Gmail. The most rapid growth in computer-based

J.R. Oppong (✉)
University of North Texas, Denton, TX, USA
e-mail: Joseph.Oppong@unt.edu

A.J. Milson et al. (eds.), *International Perspectives on Teaching and Learning with GIS in Secondary Schools*, DOI 10.1007/978-94-007-2120-3_13,
© Springer Science+Business Media B.V. 2012

services in African countries today remain centered on cybercafés and telecenters – a term that encompasses virtual village halls, telelearning centers, telecottages, electronic cottages, community technology centers, networked learning centers, or digital clubhouses (Share, 1997).

Ghana's case typifies the gap between developed and developing countries regarding the use of GIS. In both the public and the private sectors, awareness of the value of GIS as a decision-making tool is matched with very little application. In government departments, there is unceasing talk about using GIS, but the resources to make this work do not exist. GIS education is just beginning and currently limited to a few universities – Kwame Nkrumah University of Science & Technology (KNUST), University of Ghana, the University of Cape Coast, and the Western University College at Tarkwa. Even here, computer-based GIS and remote sensing education is only available to graduate students, arguably due to shortage of computing facilities. Undergraduate GIS education mostly entails students learning only the theory about GIS without actually touching a computer. Not surprisingly, GIS is yet to reach Ghanaian secondary schools due to a variety of problems.

In this chapter we acknowledge the need for GIS in Ghanaian secondary school education and set out to evaluate it. However, since GIS education currently does not exist in secondary schools in Ghana, we devote our attention to discussing the challenges and prospects of introducing GIS education in Ghana's secondary schools. The chapter is divided into four sections. We begin with an overview of Ghana's secondary education, followed by an examination of the state of GIS in Ghana. The challenges and prospects of introducing GIS into secondary schools are then discussed, and we conclude with recommendations for tapping this potential.

13.2 Secondary Education in Ghana – An Overview

Secondary education began in Ghana in 1876 when the Methodist Missionary Society of London supported the establishment of the Wesleyan High School, later on called Mfantsipim, in Cape Coast. From that period on, secondary education grew through the 1920s. The postcolonial period has seen two major explosions in the growth of secondary schools. The first occurred in the early postcolonial period of the 1960s and 1970s, as the new country of Ghana placed a high priority on education to train the much-needed human resources. The second wave came in the 1990s following the implementation of the 1987 education reforms.

As a former British colony, Ghana patterned its secondary education after the British system. Thus, a five-year secondary education system was established followed by a two-year sixth form (lower sixth and upper sixth forms). Students were accepted into secondary school after successfully completing a 6-year primary education followed by at least a two-year middle school education and passing a Common Entrance Examination. At the end of Form 5, students wrote the General Certificate of Education Ordinary Level (GCE O-Level) examination. After the sixth form, students wrote the GCE Advanced Level (GCE A-Level) examinations. Both examinations were standardized tests administered by the West African Examinations Council (WAEC). At both the O- and A-levels, the subject that offered the closest exposure to GIS was, of course, geography, which was a stand-alone subject. At both O-and A-levels students studying geography were required to study physical, human, and regional geography as well as map work. The A-level map work included map projections, dot mapping, simple choropleth mapping, and surveying.

In 1987, however, secondary education in Ghana underwent a dramatic reform. The old system was replaced by a six-year primary education followed by a three-year junior secondary school (JSS) and a four-year senior secondary school (SSS). Students may study in any of eleven local languages for much of the first three years, after which English becomes the medium of instruction. Students continue to study a local language and French as classroom subjects through at least the ninth Grade. All textbooks and materials are otherwise in English.

At the junior secondary school level, English, mathematics, social studies, and integrated science, including agricultural science, a Ghanaian language, technical, vocational, information, and communication, are the major subjects. After passing the Basic Education Certificate Examination (BECE) at the end of JSS 3 (ninth Grade) in nine or ten subjects, students are admitted to senior secondary school. Here the core curriculum comprises English language, integrated science, mathematics, and social studies. Each student also takes three or four elective subjects, chosen from one of seven groups: sciences, "Arts" (social sciences and humanities), vocational (visual arts or home economics), technical, business, or agriculture.

The standardized examinations were still maintained, but there was a restructuring of the curriculum, particularly with the subject of geography. Rather than being a stand-alone subject, geography became part of social studies at both the JSS and SSS levels. At the SSS level, however, it became possible for student electing in the general arts area to study geography as their option.

While secondary school curriculum is pretty much standardized, educational infrastructure has always been spatially unequal since the introduction of formal education in Ghana. Postcolonial educational reforms, policies and practices have merely perpetuated, or arguably, compounded it. Premier schools such as Mfantsipim, Achimota, Prempeh College, and Wesley Girls, which date back to colonial times, were established as elite British secondary schools. In contrast, most rural secondary schools were established in the 1970s with poorer infrastructure. In the 1990s, implementation of the senior secondary school (SSS) concept led to a proliferation of even more poorly equipped rural secondary schools exacerbating the quality gap between rural and urban schools (Ministry of Education, 1974, 1999a, 1999b). Thus, compared to urban schools, rural schools have always been poorly equipped (Mfum-Mensah, 2003) and poorly staffed and serviced (Asiedu-Akrofi, 1982; Graham, 1971). These differences in turn create differences in student performance.

13.3 GIS and Secondary Education in Ghana

There is strong awareness of GIS technology in Ghana; however, its use has been more limited to the work of expatriates whose projects remain unsustainable after their departure (Karikari, Stillwell, & Carver, 2003). Due to poor commitment of host agencies, usually due to poor or inadequate maintenance of equipment, such pilot projects usually fail shortly after they are completed. The absence of trained locals inhibits the successful deployment of GIS tools in such contexts (Taylor, 1991). Apart from these, the other forms of applications are the client-based GIS services provided by the Center for Remote Sensing and GIS (CERGIS) at the University of Ghana.

GIS education in Ghana is offered through formal institutions and short-term training. With formal institutions, GIS education is only available at the tertiary level, particularly at the country's three premier universities: the University of Ghana, Kwame Nkrumah University of Science and Technology, and the University of Cape Coast. Beside these the only other institution that offers GIS courses is the Western University College, formerly the Tarkwa School of Mines. Even here only graduate students receive computer-based GIS and remote sensing education.

The short-term GIS courses are run by different organizations for specific projects. These tend to be short term and project specific and thus lack the long-term vision and continuity needed to support a national GIS expansion effort. For example, Esri has run short-term courses to introduce Esri products to the Ghanaian market, but access remains limited to urban residents. Thus, GIS education is currently nonexistent in Ghanaian secondary education. The question then is, What are the main challenges facing introduction of GIS into secondary education in Ghana and what are the prospects that it might succeed?

13.4 Challenges and Prospects of Introducing GIS into Secondary Education in Ghana

13.4.1 The Challenge of Recognition

The curriculum of secondary education in Ghana is highly standardized, without much room for change by teachers in terms of content and topics. This is largely due to the fact that students have to pass standardized tests in the end. Thus, teachers are expected to teach certain specific topics to their students and students are expected to learn what they are taught if they want to pass the standardized tests. This means that for any subject or topic to be seriously taught in the system, it must become part of the curriculum. GIS is currently not part of the secondary school curriculum. Thus, although geography is in both the JSS and SSS curricula, GIS is not part of the topics to be covered under geography and as result cannot be taught.

Unfortunately, for all the talk about ICT and the government policies aimed at using ICT for education, the place of GIS appears to be missing. Instead most of the interest in ICT is about the use and application of Internet technology in such things as e-learning, e-commerce, and e-trading. There is no specific reference to the promotion and use of GIS and other geospatial technologies. This lack of recognition makes it difficult to make GIS a priority. In addition, Ghana's leading universities are struggling to make GIS acceptable to both public and private sectors. Despite the much proclaimed GIS awareness within the government and private sector, there currently is neither a demonstrable demand for GIS-trained personnel nor a commitment to provide the necessary environment to hire them. The challenge will be to sell GIS as a necessary subject to become part of the high school curriculum.

13.4.2 The Challenge of Weak National Infrastructure in ICT

Even if GIS education becomes part of the secondary school curriculum, the next challenge will be the current weak national ICT infrastructure. First and foremost is the lack of uninterrupted access to electricity. The rural areas of Ghana, where about 70% of the population live, lack uninterrupted access to electricity. Even in the urban areas, including Accra, the national capital, frequent interruptions in electricity supply is the norm. Such unstable electricity supply precludes effective use of computers. Thus, the government's one laptop per child policy and the promise to extend computers and the Internet services to every secondary school still remains a dream. Second, although Ghana has an ICT policy, the translation of the policy into reality has been hampered by both limited physical and human resources. Through NGO activities, many urban secondary schools now have computer labs to provide basic computer literacy. Some of these, usually premier secondary schools, have Internet capabilities (Dankwa, 1997; Hawkins, 2002; Parthemore, 2003). The schools that have taken part in the Ghana Education Service–sponsored scheme receive one computer system for every hundred textbooks purchased from a private firm. Yet these unlinked computers suffer severely from frequent power interruptions.

Many rural communities are yet to be connected to the electricity grid. Most rural communities that have secondary schools do not currently have access to electricity and telephone services. In such localities, the idea of promoting computers in classrooms will require more financial backing, and a considerable amount of time, considering the pace of development in Ghana. In a recent Ghanaian case study (Ismail, 2002), it became apparent that the high costs for providing electricity (where there is none) and connectivity to telephone services are major setbacks to providing ICT in rural areas in Ghana.

13.4.3 The Cost of GIS Computer Hardware and Software

GIS cannot be learned theoretically. To be effective, there needs to be computer labs equipped with GIS software to provide the much-needed hands-on experience, and the cost of these laboratory facilities poses another challenge. It is true that the prices of computers have dropped dramatically over the past decade and will continue to drop in the future. It is also true that NGOs and other donor organizations are making efforts to extend information technology to secondary schools, but these efforts remain confined to urban areas (Dankwa, 1997; Parthemore, 2003). Efforts by rural schools to acquire computers are usually hampered by cost and also adequate facilities to house these computers given that some of the schools may actually be located in thatched-roof mud buildings, with no access to electricity.

Beyond this, the real challenge in supporting GIS education is not the one-time start-up cost but the ability to sustain the education once it is started. This is because hardware wears down and GIS software technology is evolving all the time. Thus, frequent upgrades in both hardware and software are required. Unfortunately, as experience with GIS education in universities in Ghana and in most African countries have shown, there is often a lack of continuous support for GIS labs once the initial investment has been made, often by external donors or grants. The challenge of maintaining the labs with up-to-date equipment and software will therefore be quite formidable.

13.4.4 The Challenge of Trained Teachers

Availability of high school teachers trained in GIS is another challenge. As already stated, the country's teacher training institutions that are responsible for training secondary school teachers do not have GIS as part of their curriculum. Even if this were possible, the dearth of computer laboratory facilities makes it impossible for the would-be teachers to get the training in GIS that will make them competent teachers of the technology. This means that in order to get GIS established in secondary schools, GIS education should be part of the training of secondary education teachers. This in turn requires recognition and acceptance of GIS as part of the high school teacher education and secondary school curriculum.

13.4.5 The Challenge of Spatial Data Infrastructure

A GIS depends critically on the quality of the underlying dataset. This poses an almost insurmountable challenge. First, there is a dearth of spatial data on Ghana. Part of this is not because the data do not exist but rather because there is no centrally organized national data depository that provides quality spatial data. Not only that, no national metadata system exists that documents all the available spatial data. Second, much of the existing spatial data were created in connection with specific projects. Thus, when the project ended, the data collection and its subsequent upgrading also ended. Third, while there have been recent efforts at standardizing geologic mapping, there are no national data standards that govern the collection, processing, storage, and sharing of spatial data. For example, there currently is no mechanism for standardizing place names. The absence of this infrastructure will pose major problems for GIS education. Data obtained from different sources may be in different coordinate systems, while the same location may have multiple name versions and different locations may have identical names.

13.4.6 The Challenge of an Education Tradition that Emphasizes the Value of Postgraduate Education over Undergraduate Education

In an earlier section, we pointed out that due to lack of resources, GIS education with hands-on experience is limited to postgraduate students. However, a part of the explanation of this is due to an implicit belief that postgraduate education is more relevant to the job market than undergraduate education. This implicit belief has characterized GIS education even at the tertiary level. Thus, throughout most of Africa, GIS education is limited to the graduate level in contrast to places like the USA where GIS education is well established at the undergraduate level as well. The challenge that this reality will pose to getting GIS education into Ghana's secondary school is to convince the powers that be of the benefits of introducing GIS into the secondary education curriculum in a country where universities are struggling to get resources to train undergraduate students. Given all of these challenges, what are the prospects of GIS education in secondary schools in Ghana?

13.5 Prospects for GIS Education in Secondary Schools in Ghana

Notwithstanding the challenges, the prospects of introducing GIS education in secondary schools in Ghana are good for several reasons. First, the subject of geography is already represented in the high school curriculum. So, revising the geography curriculum, especially the map work section, to include introductory GIS concepts can form a starting point. Second, while there is a dearth of computer facilities in the schools especially in the less-endowed rural areas, those schools that do have computer facilities can be targeted for pilot programs. These schools will need GIS software and teachers able to teach GIS skills. Perhaps they can get help with demo or special software deals from GIS vendors such as Esri. For teachers, it is possible to follow a two-tier approach – one is to organize short-term training for teachers who are in the schools and the other is a long-term approach of incorporating GIS into secondary school social studies teacher education.

The current interest in ICT provides another opportunity for introducing GIS in Ghana's secondary schools. This will require a push beyond the use of the Internet, to focus on such technologies as GIS, and a champion to lead such efforts. Such leadership should come from GIS professionals that exist mainly in the universities in collaboration with the government, particularly the Ministry of Education.

13.6 Conclusion

The importance of GIS as a research, management, and decision-making tool is now universally recognized. In developed countries, this has led to demand for people with GIS skills in both public and private sectors, which in turn has led to a huge growth in GIS education programs in universities and colleges, and now in secondary schools. In developing countries such as Ghana, the story is entirely different. While there is a strong awareness of the importance of GIS, the application of GIS in both the private and public sectors remains nascent. Most GIS applications occur within the work of expatriate consultants and client-oriented GIS, both of which tend to be single-targeted and single-layered GIS projects. The limited nationwide demand for GIS skills has directly or indirectly led to a weak GIS education program in the country. Thus, GIS education is only available at the tertiary level – particularly the three premier universities where only graduate students are exposed to computer-based GIS due to limited laboratory facilities.

Introducing GIS education into secondary schools in Ghana, however, faces a number of challenges, including recognition of the importance of GIS, the weak national ICT infrastructure, the cost of computer hardware and software, and an education tradition that places more value on postgraduate education. There are good prospects, however, for introducing GIS education in secondary schools by building on the fact that there exists a base to foster curriculum change and also for introducing GIS as part of the drive toward the use of ICT in schools.

In the final analysis, however, it is good to introduce GIS technology into high schools, but caution must be exercised not to think of this at a very grandiose level. We cannot expect high school graduates to become GIS experts. At best, what can be done is to introduce computer applications to the map work section of the existing geography curriculum. However, rural schools need even more basic things – dependable electricity and computers that work.

References

Asiedu-Akrofi, A. (1982). Education in Ghana. In B. Fafunwa & J. U Aisiku (Eds.), *Education in Africa: A comparative survey*. London: George Allen & Unwin.

Dankwa, W. A. (1997). *SchoolNet: A catalyst for transforming education in Ghana*. Accessed October 20, 2003, http://www.isoc.org/isoc/whatis/conferences/inet/96/proceedings/c6/c6_1. htm

Graham, C. K. (1971). *The history of education in Ghana: From the earliest times to the declaration of independence*. London: F. Cass.

Hawkins, R. J. (2002). *Ten lessons from ICT and education in the developing world*. World link for development program, The World Bank Institute. Accessed March 23, 2010, http://www.cid. harvard.edu/archive/cr/pdf/gitrr2002_ch04.pdf

Ismail, M. (2002). *Readiness for the networked world: Ghana assessment*. Information Technologies Group, Center for International Development. Harvard and Digital Nations, The Media Lab, MIT.

Karikari, I. B., Stillwell, J. C. H., & Carver, S. (2003). Land administration and GIS: The case of Ghana. *Progress in Development Studies, 3*(3), 223–242.

Kerski, J. J. (2003). The implementation and effectiveness of geographic information systems technology and methods in secondary education. *Journal of Geography, 102*(3), 128–137.

Meyer, J. W., Butterick, J., Olkin, M., & Zack, G. (1999). GIS in the K-12 curriculum: A cautionary note. *Professional Geographer, 54*(1), 571–578.

Mfum-Mensah, O. (2003). Computers in Ghanaian Secondary Schools: Where does equality come in? *Current Issues in Comparative Education, 6*(1), 40–49.

Ministry of Education. (1974). *The new structure and context of education for Ghana.* Accra: Republic of Ghana.

Ministry of Education. (1999a). *Review of education sector analysis in Ghana 1987–1998.* Accessed October 20, 2003, http://www.adeanet.org/wgesa/en/doc/Ghana/chapter_2.htm

Ministry of Education. (1999b). *Comprehensive framework on education.* Accra: Republic of Ghana.

Oppong, J. R. (2010). Transport, communication and information technologies in Sub-Saharan Africa: Digital bridges over spatial divides. In S. A. Attoh (Ed.), *Geography of Sub-Saharan Africa* (pp. 243–264). New York: Prentice Hall.

Parthemore, J. (2003). *A secondary school computer lab in rural Brong Ahafo: A case study reflection on the future of secondary school computer literacy and computer based distance education in Ghana.* Accessed March 23, 2010, http://www.wess.edu.gh/lab/reports/paper.pdf

Share, P. (1997). *Telecenters, IT and rural development: Possibilities in the information age.* Accessed March 23, 2010, http://www.csu.edu.au/research/crsr/sai/saipaper.htm#top

Taylor, D. R. F. (1991). GIS and Developing Nations. In D. B. Maguire, M. F. Goodchild, & D. W. Rhind (Eds.), *Geographical Information Systems: Principles and Applications* (Vols. 1–2, pp. 71–84). New York: Longman Scientific and Technical.

US Department of Labor. (2007). *Geospatial technology.* Accessed March 23, 2010, http://www.doleta.gov/Brg/JobTrainInitiative/.

US Department of Labor. (2010). *The President's high growth job training initiative.* Accessed March 23, 2010, http://www.doleta.gov/Brg/JobTrainInitiative/

Chapter 14
Hungary: GIS in Natural Science Teacher Training

György Borián

14.1 Introduction

In Hungary, the national curriculum is called the Hungarian National Core Curriculum (NCC). The principal function of the NCC is to define the principles and the approaches that govern the content of public education. Allowing room for the autonomy of individual schools, it defines the general objectives of public education that should be pursued nationwide, the main domains that education must cover, the phases of public education in terms of content, and the development tasks that must be fulfilled in the various phases. The NCC lays down the foundations for the body of knowledge to be acquired in school, and thus creates unity in public education. The principles, goals, and tasks formulated in the National Core Curriculum are transferred into Local Curricula prepared by the teaching staff of the schools.

The NCC defines the content of public education in terms of cultural domains. The subject system of individual schools is defined in the Local Curricula, taking the various cultural domains into account. The 12 grades of compulsory education are a uniform process of development, broken down into four educational phases. The development tasks specified in the NCC are rendered to these phases. From a pedagogical point of view, the first six grades are uniform.

The Local Curriculum of schools also defines the conditions that must be fulfilled for a student to progress to a higher grade in school. The Public Education Act lays down the most relevant principle: Students may progress to a higher grade or a vocational grade in school if they have fulfilled all academic requirements. Whether such requirements have been successfully fulfilled can be determined on the basis of the evaluation criteria and concrete requirements defined in the school's Local Curriculum, which cannot prohibit the retake of a failed examination or repetition of the specific grade in the case of a failed retake examination.

In Hungary, secondary education is provided in four types of schools: (1) academic secondary schools (gymnasium) (2) vocational secondary schools

G. Borián (✉)
Danube-Drava National Park, Hungary
e-mail: gy.borian@freemail.hu

A.J. Milson et al. (eds.), *International Perspectives on Teaching and Learning with GIS in Secondary Schools*, DOI 10.1007/978-94-007-2120-3_14,
© Springer Science+Business Media B.V. 2012

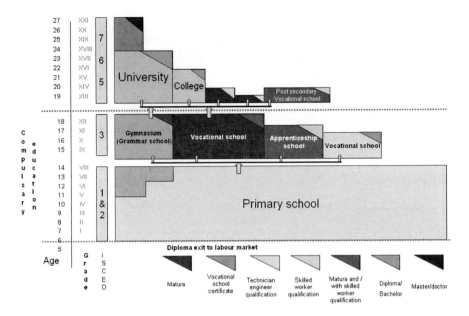

Fig. 14.1 The education system in Hungary

(3) apprenticeship schools, and (4) vocational schools. Schools often offer more than one program (e.g., academic and vocational courses) (Fig. 14.1).

The gymnasiums offer general education and a Secondary School Leaving Certificate and may be attended for four, five (e.g., bilingual secondary schools), six, or eight years. Secondary vocational schools provide general education in the first four years and award the Secondary School Leaving Certificates. Vocational education generally begins after the fourth year, although some introductory vocational subjects may also be taught during the first four years. The length of vocational courses may vary from one to three years. Vocational and apprenticeship schools do not award secondary school leaving certificates and the level of the vocational qualification is lower than that in the secondary schools.

14.2 Teacher Education

Teacher training has changed fundamentally in the past 15 years in Hungary because the whole school system has changed along with the volume and structural composition of the curriculum subjects. Before the 1990s, the school system was very rigid: compulsory primary school followed by four years of gymnasium/vocational secondary school or three years vocational school. Each school had to use one central curriculum. In addition, the method of training has been influenced by the revolutionary development of information technology. According to the 1996 Amendment of the Public Education Act, a university qualification is needed to teach academic subjects or a special field of the National Core Curriculum in secondary education. Secondary-school teachers are trained for five years in universities and specialize in one or two subjects. The recent reform of Hungarian higher education (including reforms in teacher training as well) began in September 2006 with the introduction of the linear higher education system, known as the Bologna system. According to the Bologna system, the master-level teacher training, which results in a master's degree in education, was started in 2009. In recent years, the lack of science (physics, chemistry, biology) teachers means a growing problem in the field of secondary education. In 2009 only one student started his study to be a teacher of physics and there are only three students at the master training in biology at ELTE (Eötvös Loránd Tudományegyetem – the biggest university of Hungary).

14.3 GIS in Hungarian Schools

GIS is not included as a requirement in the National Core Curriculum. However, the NCC encourages the use of new technologies – like GIS – in the chapters of Competences in Natural Sciences and Digital Competences (Ministry of Education, 2007).

> *Competences in Natural Sciences*: Having acquired the competences in science, the individual is able to activate his or her scientific and technological knowledge to solve problems at work and in everyday situations. He or she should be able to apply knowledge in a practical manner to get acquainted with and to operate new technologies and equipment, to utilize scientific achievements, solve problems, achieve individual and community goals, and to make decisions that demand technological literacy.

> *Digital Competence:* Digital competence refers to the understanding and extensive knowledge of the nature, role, and opportunities of Information Society Technology (IST) in personal and social life and work. Necessary skills comprise the ability to search for, collect and process information, use it in a critical way, and distinguish between real and virtual relationships. It includes the use of tools that promote the creation, presentation and interpretation of complex information, access to Internet-based services, conducting research with these tools, and the use of IST in critical thinking, creativity, and innovation.

In Hungary, the main obstacle to using GIS in secondary education is the lack of teachers who know this technology. Therefore, GIS is used only in the eight secondary vocational schools specialized in surveying. Recognizing the importance of GIS, the Ministry for Water and Environment Protection has been organizing courses for teachers of secondary vocational schools specialized in environmental and water sciences since 2005.

14.4 Case Study

In harmony with the EU Water Framework Directives, the Ministry for Water and Environment Protection together with GREEN Pannónia Foundation established a national environmental action program for secondary schools in Hungary in 2001.

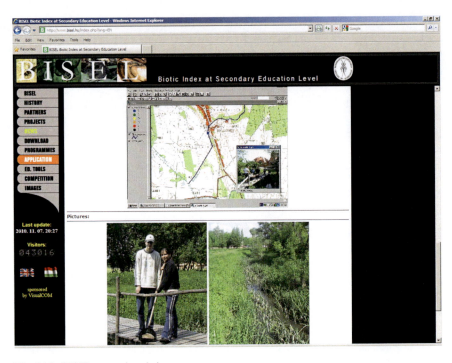

Fig. 14.2 BISEL network website

This is known as BISEL – Biotic Index at Secondary Education Level. In a short period, more than 70 schools joined and started to carry out water quality measurements on their local watercourses and uploaded their data on the www.bisel.hu website (Fig. 14.2).

In 2004, the members of the BISEL network (more than 200 teachers and national park educators) suggested that the huge database of the network should be connected to GIS. They realized that the integration of the biotic index and GIS could be a great step forward in the field of environmental education. In 2005, the website of the BISEL network was reconstructed so that it became possible for members to upload maps and photos and to insert the coordinates of the sampling points. The website also became available in English due to the increasing number of foreign partners.

Meeting the demand, the GREEN Pannónia Foundation and the Nieuwe Media School organized a training course (entitled BioGIS 2005) in Belgium with the support of the EU Leonardo program for the most active 40 members of the network. The Hungarian beneficiaries represented the teachers of secondary schools, the education program planners and managers, and those professionals who were interested in teacher education. (www.biogis2005.uw.hu)

With the help of three preparatory weekend courses and assignments, the beneficiaries became acquainted with GIS and learned the use of digital maps, such as editing the water quality results on digital map layers. Consequently, in Belgium, the participants could easily understand the use of GIS in many different areas: education, agriculture, nuclear industry, nature protection, and communal landfills.

After the course, the participants disseminated the results of the project broadly in Hungary. They held a detailed presentation about the project for 40 teachers in Jósvafő (May 25–26, 2006) and for 70 teachers and students in the VITUKI (May 29, 2006). Many of the participants used the results of the BioGIS 2005 in international school projects in Holland (Bobok Endréné), Germany (Durayné Vértessy Mária and Hublik Ildikó), Bosnia (Takáts Margit), Italy (Hebrang Ildikó and Fögler Dóra), Portugal, and Romania (Borián György and Borsos Sándor).

In 2006 Dr Kriska György, from the Eötvös Loránd University (Budapest), contacted the BISEL network in order to cooperate in the teacher training. Since the number of students of the biology faculty gradually decreased, he offered to organize "Mastercourses" for the best BISEL schools to motivate their students to become biology teachers. The BISEL program absolutely convinced him that GIS is a very attractive and useful tool for those students who like biology and scientific research work.

After a long preparation, in 2008 the GREEN Pannónia Foundation together with the Eötvös Loránd University organized the first Mastercourse (Integration of natural and technical sciences in the research work of secondary schools: use of GIS in biological water quality assessment) for the most successful BISEL schools (Fig. 14.3). Five schools (four students and one mentor teacher each) were selected to participate in this high-level course. The most important aim of this course was

Fig. 14.3 Students working in their project in the field

to motivate the students to continue their further studies as a science teacher or a scientific researcher. The Mastercourse was so successful that it had to be repeated for other five schools in 2009 and the third course is already being organized at the moment.

Reference

Hungarian National Core Curriculum. (2009). *Ministry of education and culture*. Accessed January 7, 2009, www.okm.gov.hu/english/hungarian-national-core

Chapter 15
India: Localized Introduction of GIS in Elite Urban Private Schools and Prospects for Diffusion

Chetan Tiwari and Vinod Tewari

15.1 Status of Secondary Education in India

Education policy in India changed substantially in 1976 via a constitutional amendment that required the central (federal) government and state governments to share responsibility with regard to the implementation of education policies in India. While elementary and secondary education has largely remained the responsibility of state governments, the central government via the Ministry of Human Resource Development (HRD) plays a larger role in terms of dictating broad guidelines regarding the content and process of education in India.

Following a constitutional mandate (86th Amendment) to universalize elementary education via programs like the *Sarva Shiksha Abhiyan* (education for all), the number of children in the 6–14-year age group who are not in school attendance has dropped to less than 5% of the total population (Government of India, 2009). This program was initiated by the Government of India in 2002 with targets of achieving universal elementary education in the country by 2010. Consequently, a surge in enrollment demand at the secondary school level is expected. The annual percent increase in secondary school enrollment in India grew from 2.83% during the 1990s to 7.4% between 2000 and 2003 (Government of India, 2009). Although the rates of secondary school enrollments have increased, there is still much effort needed in terms of reducing the percentage of "out-of-school" secondary school students (Fig. 15.1).

The success of programs like the *Sarva Shiksha Abhiyan* and the disparity in the percentage of "out-of-school" students at the elementary and secondary levels has paved the way toward the universalization of secondary education. A report of the Central Advisory Board of Education (CABE), the highest advisory body to advise the central and state governments in the field of education, recommends that the central and state governments jointly implement the agenda of universal and

C. Tiwari (✉)
University of North Texas, Denton, TX, USA
e-mail: Chetan.Tiwari@unt.edu

A.J. Milson et al. (eds.), *International Perspectives on Teaching and Learning with GIS in Secondary Schools*, DOI 10.1007/978-94-007-2120-3_15,
© Springer Science+Business Media B.V. 2012

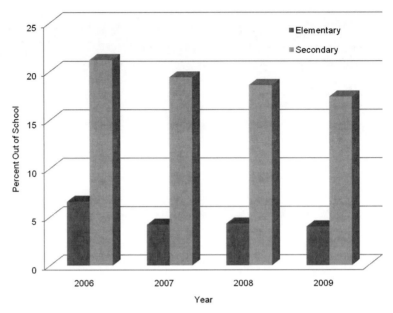

Fig. 15.1 Percentage "Out-of-School" students at elementary and secondary school levels in India

free secondary education by the year 2015 and then extend it to senior secondary education by 2020 (Central Advisory Board of Education, 2005).

15.1.1 Structure of Secondary Education in India

The general structure of education adopted by most states across the country follows a 10+2 system, which includes 8 years of elementary education (5 years of primary school; 3 years of upper primary/middle school) and 4 years of secondary education (2 years of general secondary; 2 years of senior secondary). The curriculum for school education in India is generally dictated by the National Policy on Education (NPE). There are three comprehensive versions of this policy document. The first was developed in 1968 and focused on the reconstruction of the educational system in postindependence India (Government of India, 1968). Keeping in line with economic and other developments in the country, a new national policy was formulated in 1986 (Government of India, 1986). Subsequently, certain provisions of the 1986 policy document were modified in 1992 (National Council of Education Research and Training, 2005). The National System of Education, as described in the 1992 version of the NPE, is based on a national curricular framework that contains a common core along with other flexible and region specific components. Using these guidelines as a basis, the National Council of Educational Research and Training (NCERT) develops the syllabi and instructional materials used in schools that are affiliated with either the Central or the State Boards of Education. Most schools in India are affiliated with one of these recognized boards of education. Presently, there are 23 state boards of education and 3 central (all-India) boards of education (Government of India, 2009). Note that NCERT's curriculum framework is a suggested framework that is not enforceable by law in individual states (National Council of Education Research and Training, 2005). State governments, therefore, can and do modify the syllabi based on region-specific objectives.

Due to these combinations of centralization and decentralization of education in India, in this chapter, we will examine the status of Geographic Information Systems education at the secondary and senior secondary level at those schools that are affiliated with the central (all-India) boards of education. The three central (all-India) boards of education are as follows:

- Central Board of Secondary Education
- Council for Indian School Certificate Examinations
- National Open School

Requirements for teacher qualifications are primarily influenced by the education level at which they are authorized to teach (Vadivelu, 2007). Teachers with an undergraduate degree in their area of specialization are qualified to teach at the secondary school level and those with a graduate degree are qualified to teach at the higher secondary level. Additional teacher training is provided by one-year programs that culminate in a Bachelor of Education (B.Ed.) degree. This degree program is often tailored to specific subject areas. The GIS components in secondary and higher secondary school curricula fall under the social sciences and are generally taught by teachers with a background in geography.

15.2 Status of GIS in the Secondary School Curriculum

In recent years, although GIS has been recognized as an important component of study in geography at the senior secondary school and higher education levels, its scope remains limited to specific topics as is evident from the following descriptions of the related course contents.

15.2.1 Central Board of Secondary Education

The Central Board of Secondary Education (CBSE) was initially established as a regional board in India in 1929. Following the growth and expansion of educational institutions in India, the Central Board of Secondary Education was reconstituted in 1952 and then again in 1962 with the main objectives of serving educational institutions that are within its jurisdiction. As of April 2007, 8,979 schools are affiliated with this board, including 141 schools outside India.

The secondary and senior secondary school curricula of schools affiliated with the CBSE are public documents that are categorized into two volumes. Volume I provides the curriculum for the main subject areas including mathematics, science, and social science. Volume II provides the curriculum for national and regional languages. These documents provide detailed guidelines on admission procedures, subject objectives and content, project/assignment requirements, and assessment procedures.

Geography at the secondary school level is part of the social science curriculum. This includes aspects of human, physical, political, and economic geography. However, there is no reference to either GIS as a subject area or their utility as pedagogical tools. After ten years of general education, geography is available for the first time as an elective subject area at the senior secondary school level. The syllabus for class XI focuses on the physical environment. Forty hours of practical lab work, which includes cartography and remote sensing, is required at this stage. Specific topic areas include the following:

- Map types, scale types, construction of simple linear scales, measuring distance, finding direction, and use of symbols
- Latitude, longitude, and time
- Map projection, typology, construction, and properties of projections: conical with one standard parallel and Mercator's projection
- Aerial photographs: types and geometry, vertical aerial photographs, difference between maps and aerial photographs, photo scale determination
- Satellite imageries, stages in remote sensing data acquisition, platform and sensors, data products (photographic and digital)
- Identification of physical and cultural features from aerial photographs and satellite imageries

The syllabus for class XII focuses on human and economic geography, data processing, and spatial information technology. Thirty hours of practical lab work, which includes the use of GIS, is required at this stage. Specific topics include the following:

- Sources of spatial data
- Use of computers in data processing and mapping
- Tabulating and processing of data including the calculation of averages, measures of central tendency, deviation, and rank correlation
- Representation of data including the construction of diagrams: bars, circles, and flowcharts, and thematic maps including the construction of dot, choropleth, and isopleth maps
- Introduction to GIS, hardware requirements, and software modules, data formats – raster and vector data, data input, editing and topology building, data analysis, and overlay and buffer

It should be noted that the emphasis at the senior secondary level goes well beyond the simple use of GIS software. The emphasis on the theoretical and practical aspects of cartography, remote sensing, and spatial data interpretation sets a sound basis for future studies in this area. While the CBSE syllabus places a fairly strong emphasis on GIS as a subject area, it attaches little importance to the applications of GIS and geographic data in subject areas other than geography. There is no reference made to the utility of GIS-based solutions like Google Earth and its use as a pedagogical tool for better describing and explaining the spatial dimensions of physical, human, and economic processes.

15.2.2 Council for Indian School Certificate Examinations

The Council for Indian School Certificate Examinations (CISCE) was initially established by the University of Cambridge Local Examinations Syndicate to standardize examinations to meet the educational needs of the country. In 1952, efforts toward replacing the overseas Cambridge School Certificate Examination by an all-India examination were initiated under the chairmanship of the Minister of Education. Following these efforts, the Council for Indian School Certificate Examinations was established in 1958 and registered as a society in 1967.

Like the CBSE, the secondary and senior secondary syllabi are available as public documents on the CISCE Website (http://www.cisce.org/). The syllabi for individual subjects provide a detailed list of topics, practical and project work requirements, and assessment and evaluation guidelines. Geography at the secondary school level focuses primarily on introductory principles, physical geography, and the geography of India. Although the interpretation of maps is a component of the practical lab work requirements, it does not specifically require the use of GIS software. Map interpretation in many schools, therefore, ends up being a series of exercises that primarily utilize printed copies of physical and thematic maps. At the senior secondary school level, the emphasis is on the principles of physical geography, human–environment interactions, and the physical, political, and economic geography of India. Practical lab work in class XI requires training in surveying, map properties including projections, and the interpretation of aerial photographs. Specific topics include the following:

- Elementary principles of surveying
- Map projections – uses, construction, and properties of the following: cylindrical equal area, simple conical with one standard parallel, and zenithal equidistant
- Aerial photographs – introduction, differences between maps and aerial photographs, uses, types, and advantages

Practical lab work in class XII covers an introduction to Geographic Information Systems and remote sensing. Specific topics include the following:

- Introduction to Geographic Information Systems
- Elements of visual interpretation of remote sensing maps/images
- Elementary principles of surveying using Global Positioning Systems (GPS)

It should be noted, once again, that greater emphasis is placed on the theoretical aspects of cartography, surveying, map interpretation, and remote sensing. The use of GIS technology as a pedagogical tool for describing the spatial dimensions of a variety of topical areas is neither required nor recommended. The emphasis on teaching about GIS rather than teaching with GIS can mainly be attributed to inadequate information technology infrastructure in most schools in India and the unavailability of spatial data layers.

15.2.3 National Institute of Open Schooling

The National Institute of Open Schooling (NIOS) was established in 1989 as an alternative board of education that allows students to choose subject combinations based on their needs and goals. This system is primarily designed to cater to the needs of children or adults who do not have access to traditional schools such as those affiliated with the CBSE, CISCE, or state government–affiliated boards of education. At the secondary school level, aspects of geography are included as part of the social science curriculum. Geography is available as an elective at the senior secondary school level. The main content focuses on physical geography, human–environment interactions, and the geography of India. Fieldwork in geography is an optional module that covers introductory topics in cartography, spatial sampling, and map interpretation with no specific reference to GIS.

15.3 Challenges in Adopting GIS Technology for Secondary Education in India

As discussed in the following section, the demand for GIS and related spatial data services is expected to grow at a rapid pace in India. Although GIS is now a specialized field of study in several institutions of higher education, its adoption in the secondary school curriculum is still at a nascent stage. Several challenges impede the adoption of such technologies both as a subject area and as a pedagogical tool. In this chapter, we outline some of the key challenges that need to be overcome before GIS and other spatial data technologies become commonplace in the Indian school curriculum.

15.3.1 Inadequate Access to Computer Hardware and Software

Although many schools, especially those that are managed by private entities, do have computer labs, their limited availability forces schools to restrict their use to subject areas that are directly related to the computer science curriculum. The social sciences in general are taught in a traditional classroom lecture format with little use of technology as a means for enhancing instruction. This problem is further exacerbated by expensive software and specialized hardware requirements that are generally required for the implementation of GIS, a requirement that remains unaffordable to a majority of schools in India.

15.3.2 Restricted Access to Spatial Data

Inadequate availability of spatial data, at a scale required to develop useful GIS, is another major hurdle for the widespread adoption of GIS and related spatial data technologies. Although spatial data for the country are routinely collected by a variety of government agencies including the Survey of India (SOI), National Atlas and Thematic Mapping Organization (NATMO), Department of Space (DoS), National Remote Sensing Agency (NRSA), and the Geological Survey of India (GSI), their availability for public use has been strictly controlled for reasons of national security. However, a revised national map policy established in 2005 calls for the liberalization of access policies with the goal of eventually making such data available for public use. These data will become part of a series of base maps, called Open Series Maps (OSMs), which will be made available via the Survey of India. Although a National Spatial Data Infrastructure for the country has been initiated, the amount of data presently available for public use is still quite limited.

The use of publicly available raster data collected by the National Remote Sensing Center of the Indian Space Research Organization is limited due to the costs involved in processing and using such data. In addition to the need for expensive hardware and software, procurement costs, which range from Rs. 30 to Rs. 70 per sq. km (http://www.nrsc.gov.in/), make the use of such data prohibitively expensive for many schools in India.

15.3.3 Poor Quality of Attribute Data

The quality of attribute data is usually quite poor too. For example, the attribute data on properties, household characteristics, road network, water supply network, and so forth in a city required to develop a GIS for urban management are quite poor, and therefore, even if good spatial data are obtained through air photographs or remote sensing, the resulting spatial data infrastructure is often of very limited use.

15.4 Scope of GIS in India

The development and use of geographically referenced databases in India, though initially limited to areas such as development planning and decision making, has been in vogue since the late 1960s. In the Growth Centre Project carried out in the Fourth Five Year Plan (1969–1994) of the Government of India, extensive use of such databases was made in identification of rural growth centers in the country. Subsequently, in the mid-1980s, with improved access to computational technology and GIS software, several research projects were funded by the Department of Science and Technology of the Government of India to encourage the use of GIS in areas such as natural resources management, school mapping and education microplanning, and location of rural services. However, until recently, due to government restrictions on access to spatial data, such applications had remained limited to the research and development activities of only government and semigovernment organizations that were permitted to access the spatial data inputs like remote sensing imageries, aerial photographs, and restricted topographical maps of the Survey of India. For instance, most GIS applications in research and development activities that required remote sensing data were invariably made by organizations that could establish collaboration with the government's National Remote Sensing Agency (NRSA). The Town and Country Planning Organization (TCPO) has assigned a number of projects for development of GIS databases using remote sensing imageries for selected cities in the country to NRSA since mid-1990s.

In the past ten years, GIS has increasingly been recognized in the country as an essential information system for all such planning, management, and governance activities that have a spatial connotation, be it infrastructure development, the establishment of e-governance systems in government institutions and agencies, or the marketing of products or services delivered by the private sector. The policy thrust of the Government of India on reforming governance using information and communication technology (ICT) requires applications of GIS. In an ongoing program of the Government of India, the Jawaharlal Nehru National Renewal Mission (JNNURM) that provides funding to state governments for urban infrastructure development in selected cities has made it mandatory for the local governments of all cities involved in the program to develop a GIS identifying all properties, utilities networks, transport networks, and land uses.

The increasing use of GIS by both public and private sector has created an enormous demand for GIS services that has been rising exponentially (Mishra, 2009). Consequently, a large number of GIS service providers have emerged in the country. These services include digitizing, software development, remote sensing, and complete end-to-end GIS solutions to help organizations scope, design, build, integrate, and manage their GIS infrastructure. This sudden growth in GIS service providers has in turn created a significant demand for a trained workforce, the availability of which is lagging far behind the demand.

15.5 Case Study: Mapping the Neighborhood – A Participatory GIS Project Involving School Children

Mapping the Neighborhood is a participatory mapping project involving school children from a number of rural and urban schools in Almora district in Uttaranchal, India. The project, which was implemented by The Centre for Science Development and Media Studies (CSDMS) and funded by the Government of India Department of Science and Technology, is an excellent case study that integrates information and communication technologies (ICTs) and GIS to map, analyze, and subsequently create a data repository of the socioeconomic, environmental, and ecologic conditions of local communities (Mallick, 2005). The project covered multiple educational and social aspects (CSDMS, n.d.). From an educational standpoint, students were trained in the proper use of mapping technologies including GPS, GIS, and remote sensing. A student's ability to understand and represent the spatial dimensions of geographic phenomena and objects was assessed via drawing competitions involving sketch maps of their local neighborhoods. Subsequent discussions on feature identification, map preparation, and interpretation reinforced concepts in thematic mapping. Students then identified problems in their local communities and used GIS and other technologies to create maps that represented the spatial dimensions of those problems. These maps covered a variety of societal issues including socioeconomic conditions, gender equality, health care, and environmental degradation. The spatial data infrastructure created as part of this project was made available to decision makers.

15.6 Conclusions

The education system at the secondary and senior secondary school levels has failed to keep up with the recent demand for GIS services. While there is recognition that GIS should become an important component of the secondary school syllabus, the educational system has not been able to respond adequately to this growing demand due to several challenges including data constraints, inadequate computer hardware and software availability, and education policy. However, the liberalization of spatial data access in India and the development of a national spatial data infrastructure will fuel the availability of high-quality spatial data in the country. A possible solution to the problem of limited availability of computing hardware and expensive GIS software is the use of open-source and Web-based GIS software, which has witnessed rapid growth in availability and usability in the recent past (Caldeweyher, Zhang, & Pham, 2006; Steiniger & Bocher, 2009). The availability of such software and support organizations like the India Chapter of the Open Source Geospatial Foundation (OSGeo-India) provide an alternate to proprietary GIS software that is often too expensive for a majority of schools to afford. The education system and policies in the country need to respond to the demand of a well-trained workforce by providing incentives and attractive facilities for imparting training in GIS. In this context, it is imperative that adequate financial allocations are made for introducing quality teaching using state-of-the art teaching material and pedagogy at the school level itself. At the same time, there is an urgent need to collect quality data inputs required for GIS development and make them easily accessible to the users. The current efforts made in this direction in the country under various programs of central and state governments and international development agencies must be monitored and evaluated regularly to ensure quality and sustainability of the efforts.

References

Caldeweyher, D., Zhang, J., & Pham, B. L. (2006). OpenCIS – Open source GIS-based web community information system. *International Journal of Geographical Information Science, 20*(8), 885–898.

Central Advisory Board of Education. (2005). *Universalisation of secondary education.* New Delhi: Ministry of Human Resource Development. Report nr F.2-15/2004-PN-1.

CSDMS. (n.d.). *Methodology for neighbourhood mapping exercise.* Accessed December 10, 2010, http://www.csdms.in/NM/project/methodology1.htm

Government of India. (1968). *National policy on education.* New Delhi: Ministry of Human Resource Development.

Government of India. (1986). *National policy on education.* New Delhi: Ministry of Human Resource Development.

Government of India. (2009). *Annual report of the department of school education and literacy.* New Delhi: Ministry of Human Resource Development.

Mallick, R. (2005). *Infusing map culture through participatory mapping.* Accessed December 10, 2010, http://www.gisdevelopment.net/magazine/years/2005/feb/infusing.htm

Mishra, S. (2009). GIS in Indian retail industry – A strategic tool. *International Journal of Marketing Studies, 1*(1), 50–57.

National Council of Education Research and Training. (2005). Education policies and curriculum at the upper primary and secondary education levels. In *Globalization and living together* (150p.). New Delhi: Discovery Publishing House.

Steiniger, S., & Bocher, E. (2009). An overview on current free and open source desktop GIS developments. *International Journal of Geographical Information Science, 23*(10), 1345–1370.

Vadivelu, V. M. (2007). Education system and teacher training in India. *Ethiopian Journal of Education and Sciences, 3*(1), 97–102.

Chapter 16
Japan: GIS-Enabled Field Research and a Cellular Phone GIS Application in Secondary Schools

Yoshiyasu Ida and Minori Yuda

16.1 Secondary School Education and GIS Trends in Japan

Japanese secondary schools are divided into lower secondary schools (Grades 7–9) and upper secondary schools (Grades 10–12). Although education is compulsory through the lower secondary level, the advancement rate in upper secondary has been greater than 95% since the early 1990s. The content of education in Japan is defined in the "National Curriculum Standards" (*Gakushu Shido Yoryo*) set by the government. For primary and secondary education, three national curriculum standards have been issued for elementary, lower secondary, and upper secondary levels. Only textbooks screened and approved by the Ministry of Education, Culture, Sports, Science, and Technology can be used.

A teacher's license for a specified subject is needed to be able to teach at the upper secondary level. To qualify for the license, 70 credits from a teacher-training course and specialized subjects in a university and 2–3 weeks teaching practice in a secondary school are required. Only a person who passes the prefectural teacher employment examination can be a public school teacher. But some who do not pass the examination for public schools can teach in private schools, if they pass an interview test by these schools.

While Japan's educational system has started to take notice of GIS since the late 1990s, its introduction into secondary school education is still in progress. This has been attributed to two reasons. The first is the lack of teachers who can understand GIS and associate lesson contents therewith. Although computers are now widely used by teachers in schools, their main usage has been limited to word processing, using spreadsheets for grading, and communicating through the Internet. Only a few teachers actually use GIS.

Geography is one subject area where GIS can be effectively utilized. Geography is included in Social Studies, a compulsory subject in lower-secondary school education. Unfortunately, many social studies teachers in lower secondary schools do

Y. Ida (✉)
University of Tsukuba, Tsukuba, Japan
e-mail: ida@human.tsukuba.ac.jp

A.J. Milson et al. (eds.), *International Perspectives on Teaching and Learning with GIS in Secondary Schools*, DOI 10.1007/978-94-007-2120-3_16,
© Springer Science+Business Media B.V. 2012

not specialize in geography; hence, they are hardly able to understand the significance and advantages of GIS. At the upper secondary level, "geography and history" is given as an elective course that only about 50% of the total students choose. Geography in upper secondary school includes geography A with two credits and geography B with four credits. Both subjects are elective courses. Furthermore, only a few motivated teachers try to use GIS at this level.

The second reason for the lack of GIS use in classes is related to the availability of GIS software, which is generally expensive. Since public school budgets are limited, it is hard for any school to acquire the software. Although GIS freeware and inexpensive software have been developed recently for educational purposes, their usage has remained limited. MANDARA (http://ktgis.net/mandara/) is a famous representative example of free GIS software. This is a freeware developed by Keiji Tani, Associate Professor at Saitama University, which can be used to visualize and analyze many kinds of statistic and geographic data on the map.

In this context, the new "National Curriculum Standards" issued a notification through official gazettes in 2009 including utilization of GIS in upper secondary education. This chapter provides some practices in using GIS at the secondary school level and discusses prospects for utilization in Japan.

16.2 Cases of GIS Utilization in Secondary Schools

16.2.1 Neighborhood Survey Using GIS in Lower Secondary Schools

16.2.1.1 GIS as a Tool for Decision-Making

The first uses of GIS in classes have focused on producing maps from data for presentation purposes (Akimoto, 1996; Kobayashi, 1999). Recently, however, some experiments using GIS for fieldwork and decision-making have been conducted. Ota (2006) conducted classes wherein students get to learn their own community through GIS-enabled field research. As one example of using GIS for decision-making, GIS was utilized in social studies classes at Azuma Junior High School in Tsukuba City, Ibaraki Prefecture, in September 2007. To prepare the classes, a project team was organized by university professors and graduate students majoring in Social Studies Education. The project was spurred by many sightings of suspicious persons around the school. In cooperation with the police, the Parents' Association made a map showing areas where suspicious persons have been spotted and distributed it to each household. The initiative later provided the basis for these associations to request the municipal government to provide additional street lights in dark places. The appearance of suspicious persons in the area is a serious social issue to the community, yet students were not as aware of the danger as their parents. Therefore, the aim of the GIS-supported classes was to make the students realize the dangers in their area and learn to behave as community members.

16.2.1.2 Introduction of Fieldwork and Steps Teaching Plan for the Classes

Fieldwork is one of the interesting aspects of geography. Field research was there-fore introduced into the practice because the study area was the students' living area. The project team aimed to combine fieldwork and GIS, which are attractive elements for a geography class. Fieldwork has been recommended in the curric-ula of lower and upper secondary education, but many teachers have not actually conducted it. Ida (2008) clarifies the issues of fieldwork in geography classes and explains the method of field research from the point of view of teachers. In this context, the introduction of fieldwork is needed to introduce GIS into geography education.

The content of each class was as follows. First, the teacher showed students a map on appearances of strangers as immediate threat and then asked them to guess where and what kind of locational distribution was observed on the map. The students were instructed to anticipate the characteristics of places where strangers would likely appear. They then went to the field and conducted research to test their hypotheses. Students were divided into ten groups of three or five people depending on survey content or study area. Each group plotted positions of street lights and bushes as their findings in the survey on the computer map. All information collected by the groups was put together in the computer.

Using the overlay function in GIS, the students were later instructed to observe the relationship between places where suspicious persons appeared and the data they collected. Students could easily overlay different data layers on the map using GIS; however, it was difficult for beginners to read such overlaid maps and find relations among different elements. To give them advice on how to read the maps, the grad-uate students majoring in Social Studies Education who participated in the project came to a class and joined the groups to support them. This became possible because the project involved not only students and teachers in lower secondary school, but also professors and graduate students at the local university. Through the project they could utilize local human resources and contribute to each other; the lower sec-ondary school could use GIS with support from university staff and students, the university could provide knowledge and skills on GIS to a local school and do an experiment using GIS, and university students as future teachers gained experience working with junior high school students. Students in each group then discussed the results observed in the overlaid maps and made a presentation using these maps.

Students obtained the following results:

(a) Places where suspicious persons have been sighted are related to the distribution of street lamps and bushes. However, even if there is a street light, strangers appeared in places where bushes contribute to the darkness.
(b) Suspicious persons appeared in darker areas lacking street lights.

Since students found these tendencies of appearances of strangers, they real-ized that it is better not to pass the darker points to protect themselves against such dangers. Lastly, they submitted a petition for improvement of the environ-ment in their community to the local government, attaching their research results pointing out places where repair of street lights is needed and more street lights are warranted.

In this exercise, one of the immediate problems in students' living area was chosen, so students were motivated in the classes. This practice dealing with local social issues covered collecting information and data, plotting them on maps, analyzing and understanding phenomena observed on overlaid maps, judging value and making decisions by the results, and participating in society. This series of lessons using GIS is a method for developing the problem-solving skills of students. Through it, they learn to understand a social phenomenon, collect, analyze and interpret related data, and make a decision, allowing them to participate in society.

16.2.1.3 Prefecture-Wide Expansion of Utilization of GIS

To share the data collected in each school, an incorporated nonprofit organization (NPO) to promote GIS was established in Ibaraki prefecture in 2004. Smap Promotion Council is running the website, "Smap Safety map." The concepts of the council are sharing maps as tools to analyze problems by working with schools and square (community area) with Smart IT. The council has been organized with the aim of supporting the safety of areas to share information on dangerous points in the neighborhood with people in the entire community. One of the activities of the NPO is to provide a children's safety map. The system developed by the NPO shows maps with data from students in the prefecture to support children's safety. The following elements are shown on the map: locations where accidents involving cars and bicycles potentially happen, points where suspicious persons appeared, streets where many cars park, and areas where outbreaks of hornets are observed (Fig. 16.1). On the basis of the information sent by students from registered schools, updated data are provided on maps uploaded in the NPO's website.

This project enables students not only to conduct research in their neighborhood, but also to do a comparative study of other areas. When such classes are conducted in many schools, it is possible to collaborate with each other. The environment supporting utilization of GIS for fieldwork is improving not only in some specific schools but also in general secondary schools.

16.2.2 Cellular Phone GIS to Improve Geography Classes in Upper Secondary Schools

16.2.2.1 Integration of Fieldwork, GIS, and Cellular Phones

Fieldwork is also recommended in geography education in upper secondary education. In conventional fieldwork, paper maps and notebooks are the media for recording land use and the number of stories of buildings in the research area. If we use GIS to analyze many kinds of data collected in fieldwork, we only have to input such data into a computer after the fieldwork.

Fig. 16.1 Neighborhood crime map. **a** Screen shot of Smap safety map showing dangerous points. **b** Explanation of a symbol by animation and information on the point
Source: From Smap safety map website with permission of Smap promotion council

Teachers pointed out the limitation of lesson hours, as well as lack of time to learn to use GIS and to prepare for lessons, as reasons not to use GIS in classes (Yuda, Itoh, & Johansson, 2009). If GIS could be brought into use everywhere and connected to a network to allow inputting, editing and processing of data in the field, then there would not be a need to make a new database or move data into desktop GIS after fieldwork. The efficiency of the fieldwork would thus be greatly improved.

One idea to realize these possibilities is a GIS application for cellular phones. Mobile phones are widely used in our daily lives as communication tools and hand-held terminal devices. Cellular phone technology itself has made spectacular progress. Currently, almost all (96%) of upper secondary school students have cell phones (Cabinet Office, 2007). Furthermore, almost all mobile phone users connect to the Internet via their mobile phone. These conditions indicate that cellular phones with GIS have the potential to be a tool that will assist in education. To investigate this potential, the authors implemented a project to develop a system called "Cellular Phone GIS." This section explains Cellular Phone GIS and its applications in upper-secondary school education.

16.2.2.2 Outline of Cellular Phone GIS

"Cellular Phone GIS" refers to a system that includes the GIS application for mobile phones (hereinafter called the Cell Phone GIS Application), an Internet browser-based viewer for PCs (hereinafter called a PC viewer), and the data server that connects them. With the Cell Phone GIS Application, data can be input and edited outdoors. These data can be viewed on both the cellular phones and the personal computers via the Internet (Fig. 16.2).

The Cell Phone GIS Application is a Java-based application for a Japanese 3G cellular phone. This application accesses the database server where the map data are stored and retrieves the necessary map data. Users can plot data on the map displayed on the mobile phone. The plotted data are displayed as icons on the map. In addition to collecting attribute data by choosing options from a drop-down menu, the user can take notes by inputting text with the application. Also, the user can add image data taken with the built-in camera on the phone. The PC viewer on which the user can see the data input from the Cell Phone GIS Application on the PC is a Web-based application using Google Maps. The PC viewer retrieves the data from the database server shared with the Cell Phone GIS Application and displays the data on Google Maps. From the PC viewer, the data can be viewed and edited. It is notable that the PC viewer only needs an environment with an Internet connection and Web browser and the functions do not depend on the type of OS or the PC's performance.

16.2.2.3 Application of Cellular Phone GIS in an Upper Secondary School

Classes using Cellular Phone GIS in upper secondary schools were implemented in the unit, "Research in Neighborhood," in a "geography B" elective course at Takasaki High School, Gunma Prefecture, in September 2007. Gunma prefecture used to be famous for sericulture and mulberry fields, however, sericulture has declined. In this research study, students compared topographic maps showing the area near the school between 1975 and 2000 and found places where land use had changed from mulberry fields within 25 years. Then they surveyed the land use of former mulberry fields at the time of the survey in September 2007. The content of each class was as follows. In the first period, the students compared old and new topographic maps from around 1975 and 2000 and found changes in land use. In the second period, the students used the Cellular Phone GIS Application and PC viewer in the classroom and learned how to operate them and confirm the route of

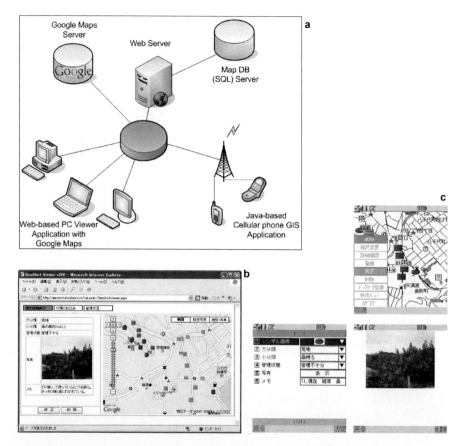

Fig. 16.2 Structure of the cellular phone GIS system and images of PC viewer and cell phone GIS applications, **a** Structure of the cellular phone GIS System, **b** Screenshot of the PC viewer, **c** Display images of the cell phone GIS application

the survey. In the third period, the students conducted fieldwork with the Cellular Phone GIS Application. In the fourth period, the students reported the collected data in each surveyed area using the PC viewer and shared the information with the other students. Then in the fifth period, each student considered the actual condition of the mulberry fields and the tendencies in distribution in the surveyed area by choosing the layers to display on the PC viewer.

The study area of the fieldwork, which is within 5 km from school, was divided into six subareas. Each group collected data from seven to ten points in each subarea. For the survey, the students went to their assigned research areas as indicated in the Cellular Phone GIS Application. They then input the data on land use and mulberry fields while in the field, and took images with the mobile phone camera. Also, they interviewed local people about the area and learned more about the changing face of the mulberry fields. After the fieldwork, the next lesson was held in the classroom. Displaying the maps and collected data with the PC viewer, each group reported what they had found and all the students shared the information on the conditions at the research points and the mulberry fields there. Then, in the next class, the students considered the distribution and tendencies in land use in the study area from many points of view such as geological, historical, and economic.

16.2.2.4 Benefits of Introducing Cellular Phone GIS and Possibilities of Its Utilization in Upper Secondary School

According to the results of the students' evaluation of Cellular Phone GIS, overall students' satisfaction with the classes in fieldwork was high. Students also actively participated in the survey. From the point of view of the teachers who designed the course, it was reported that a lot of time can be saved by using Cellular Phone GIS. The operation of the mobile phone itself hardly needs to be explained to the students. It is possible to complete the data input during the fieldwork, therefore, teachers do not need to arrange extra lesson hours or ask students to stay after school. Furthermore, they have time to think about how to develop ideas for the next step of the class. Mobile phone use is widely accepted, so it is expected that the Cell Phone GIS Application will be easily accepted by upper secondary school students. Regarding the PC viewer, only an Internet connection and Web browser are required, so it should be possible to introduce it into schools. The combination of these technologies will surely be helpful in upper secondary school education.

16.3 Prospects and Issues: GIS and Secondary School Education in Japan

Recently, the environment for utilization of GIS in secondary school education is drastically changing. Under the new National Curriculum Standards issued in 2008 and 2009, the proactive use of Information Technology (IT) for collecting and processing information and data on the neighborhood for geography education in lower secondary schools was mentioned. Although the term GIS itself was not explicitly written, the instruction manual of the national curriculum suggests utilization of GIS. Meanwhile, utilization of GIS is clearly stated in the national curriculum for upper secondary school. The national curriculum standards have a binding force and teachers have to teach the contents covered in the national curriculum standards. This means that GIS has to be introduced into upper secondary school education by 2013.

Since the Government of Japan promoted the reinforcement of IT-driven education in 2001, all schools have installed computers on the initiative of the government. Today, every school in Japan has computers with access to the Internet. But not all teachers of geography can understand and use GIS in classes to date. The biggest challenge for the future is that not only a few motivated teachers but also many ordinary teachers acquire the skills to utilize GIS. The introduction of GIS into the new national curriculum standards will trigger a change in teachers' minds.

Some education groups promoting GIS such as the Commission of Geographical Education of the Association of Japanese Geographers have held workshops and seminars on use of maps and GIS. Furthermore, the Ministry of Land, Infrastructure and Transport has started to support teachers to use GIS and develop education materials of GIS for teachers. Thus, GIS-related associations and government are trying to expand the use of GIS in education.

In Japan, there are many teachers who have to teach geography at the secondary school level, even if geography is not their specialty. To encourage them to use GIS in classes, a GIS freeware that is simple to operate and requires less time to use is needed. If such software would be introduced in textbooks, teachers would try to use it. Geography Education in Japan has traditionally emphasized the acquisition of knowledge. As a next step, it is necessary to develop a user-friendly GIS that includes traditional geography education so that teachers can realize its effectiveness. Also, it is necessary to provide opportunities to enjoy geography and experience the usefulness of GIS in classes at the higher education level. Through these experiences, they will be able to convey the importance of this subject and good points of GIS to the younger generation when they return to secondary schools as teachers. We have to design curriculum in higher education to make such a good circulation in education. This chapter shows how teachers in secondary school could introduce GIS into their classes. It is hoped that these practices can inspire them to develop more advanced lessons using GIS.

References

Akimoto, H. (1996). GIS (chiri joho shisutemu) to koko chiri kyoiku. *Shin chiri, 44-3*, 24–32.
Cabinet Office. (2007). Dai 5-kai johoka to seishonen ni kansuru ishiki chosa ni tsuite (sokuho). Accessed December 1, 2010, http://www8.cao.go.jp/
Ida, Y. (2008). Mijika na chiiki shirabe – chiiki no tokucho wo miidasu hoho. In K. Asakura, J. Ito, & Y. Hashimoto (Eds.), *Chugakko shakai wo yori yoku rikai suru* (pp. 131–136). Nihon bunkyo shyppan.
Kobayashi, T. (1999). Hyo keisan sofuto niyoru chiri joho shisutemu. *Tsukuba shakaika kenkyu, 18*, 61–70.
Ota, H. (2006). Chizu to GIS de sekai to kyodo wo shiro. *Chizu, 44-4*, 36–39.
Yuda, M., Itoh, S., & Johansson, T. (2009). Geographic Information Systems in upper secondary school education in Japan and Finland: A comparative study. *The Shin-Chiri (The New Geography), 57*, 156–165.

Chapter 17
Lebanon: A Personal Journey from Professional Development to GIS Implementation in an English Language Classroom

Rawan Yaghi

17.1 The Structure of the Lebanese Curriculum

The system of public education in Lebanon has been a concern of reformers since the conclusion of the Lebanese war (circa 1989). Since 1998, the public schools in Lebanon have followed a national curriculum designed by the National Center of Education, Research and Development (NCERD) under the leadership of the Minister of Education, Youth and Sports. The Lebanese system of education is divided into cycles: 3 years of kindergarten (KG) for children above three years; Cycle 1 that includes Grades 1, 2, and 3; Cycle 2 that includes 4, 5, and 6; and Cycle 3 that includes Grades 7, 8, and 9. In Grade 9 (Brevet), students have an official exam in Arabic, English, mathematics, physics, chemistry, biology, geography, history, and civics in which they are supposed to get total marks of 140/280, or 50%, in order to get into the secondary level. If the student does not pass the Brevet, he or she can take another test in September, which follows the same testing system and rubrics, but contains different questions. The student can also opt to fail the year and take the test again by the end of the academic year. He or she also has the choice to leave academic education and change to a technical school. At the secondary level of education (Grades 10, 11, and 12), students have the choice to be in the scientific or literary section. Grade 12 has 4 sections: life sciences, general sciences, socioeconomics, and humanities, in which the courses or their level of difficulty might completely differ. At the end of Grade 12, students in both public and private schools must pass an official exam in all the required subjects of their section, which enables them to go to the university. The major they choose should be related to their final year at school (literary or scientific). They then have an entrance exam to the major they choose, and it is then up to the university to select them or not. High marks on the official exams, SAT scores, TOEFL scores, and applications are all taken into

R. Yaghi (✉)
Nabil Adeeb Sleiman Secondary Public School, Bednayel-North Bekaa, Lebanon
e-mail: rawan.yaghi@gmail.com

A.J. Milson et al. (eds.), *International Perspectives on Teaching and Learning with GIS in Secondary Schools*, DOI 10.1007/978-94-007-2120-3_17,
© Springer Science+Business Media B.V. 2012

consideration. The Lebanese University (Public), however, doesn't require the Test of English as a Foreign Language (TOEFL) or Standardized Administrative Test (SAT).

The Lebanese curriculum, designed by NCERD, assigns the courses, identifies the objectives and competencies, publishes national books, sets assessment means, and is responsible for training teachers. KGs curriculum includes Arabic, English or French, some math, physical education, and arts. In addition to these, Cycle 1 has science and introductions to history and geography. Cycle 2 adds computer basics and civics. In Cycle 3, science is categorized as biology, physics, and chemistry. The secondary level includes courses about civilizations, philosophy, psychology, economics, and sociology according to chosen sections in Grades 11 and 12.

Most public schools are traditional with classes organized in rows because, in most of the cases, they are small rented buildings that do not provide enough room for group-focused classes or interactive patterns. In the typical public schools, students might have the chance to go to the class where the subject is taught in modern ways or in the laboratory. Secondary public schools have separate science laboratories and libraries. In addition, students are encouraged to participate in sports, extracurricular activities, and community service. Almost all secondary schools have access to the Internet and LCD projectors, which set the groundwork for using GIS in education.

Public school teachers in the third cycle and in the secondary level must have a BA or BS in the course they teach. Unfortunately, this is not a necessity for the first and second cycles despite the fact that there is an excess of bachelor degree holders. However, in order for teachers to have a permanent job in the Ministry of Education, they have an exam in the subject they intend to teach in addition to a general knowledge test that requires teachers to write an essay in Arabic about the given topic that is usually a universal issue (e.g., globalization, the environment, or immigration). Upon success, they receive a 1- or 2-year training of 10 courses; with each course taught 14 hours per semester (four months). The courses are mandatory and include child psychology, methodology, curriculum design, assessment and evaluation, technology in the classroom (using overhead projectors, LCD projectors, Microsoft Office, some Internet), and other courses relevant to the subject taught. This happens in various intervals of time, which can vary between 2 and 15 Years! In certain cases, the schools might need more teachers, which opens the door to contractors who are usually of lower levels of expertise because they did not undergo the training of permanent teachers.

The new curriculum mentioned technology, but it did not require it. Hence, there are computer labs that work on IT, but its integration with other subjects ranges from limited to active. Creative teachers are usually ready to share their new information with others. Technology became crucial also in management such as in official records, report marks, and some announcements from and to the Ministry of Education. So, it is up to teachers to include technology in their classes knowing that there is an encouragement from school principals to use computer labs. It is rare to find a secondary school that does not have LCD projectors, overhead projectors, or access to the Internet. However, teaching based on new trainings that integrate subjects with technology is related to intrinsic motivation in which teachers apply to training workshops by themselves without a recommendation from school principals. On the other hand, there are some workshops that are imposed on schools by the Ministry of Education (www.higher-edu.gov.lb) and the NCERD (www.crdp.org) in what they call "sustainable development." However, these do not satisfy ambitious, capable teachers. NCERD sends a fax or mail to announce some trainings abroad like The University of Middle East Project's TEI (Teachers of English Institute) that takes place in Boston, USA, or some training in Malaysia, but the number of teachers who apply in order to get benefit of such programs is still limited.

Most universities in Lebanon teach through GIS in different domains, especially architecture, and GIS in universities has a noticeable history since 1990. Some universities that implement GIS include American University of Beirut, Beirut Arab University, Lebanese American University, Haigazian University, Al Balamand University, Notre Dame University, Saint Joseph University, and others. Some secondary private schools are implementing GIS, but the statistics are not available. On the other hand, engineering companies like Alami and Khatib as well as Dar Al Handasa and Oger International implement GIS.

17.2 A Personal Case Study of GIS in the Classroom

I was one of those teachers who took every single opportunity to improve myself. I had an intensive training in Boston through TEI. I took two online courses: English for Law taught by The University of Tennessee and Shaping the Way We Teach English through Oregon University, both as a scholarship from the American embassy. The embassy also announced a training institute in Tunis entitled "My Community Our Earth" (MyCOE) through the Association of American Geographers, to which I applied and was selected (AAG, 2008). This training, a part of the International Geographic Union's annual conference, brought together 24 teachers from the Middle East and North Africa. It included three teachers from Lebanon (two geography teachers and me). I think I was selected on the basis of the history I had of implementing whatever I was trained on, and I did not let them down. GIS, which was the basic part of training, was so fascinating for me. It can be widely implemented in the English classroom because our English curriculum follows the thematic approach.

When I came back from the MyCOE workshop in Tunis in 2008, I contacted one of the geography teachers in my school, Nabil Adeeb Sleiman Secondary Public School, to train her. Unfortunately, her limited IT knowledge proved to be a major challenge. She has sufficient geographic knowledge and is able to motivate her students to do certain projects in geography. She was motivated to try GIS, but getting her to pursue it was a very difficult task. She had a course in computer basics, but still she was unable to handle the higher level of computer skills needed for GIS. This year, we have a new teacher who has better computer skills and is willing to implement GIS in his teaching.

However, I successfully taught using GIS in different classes. I used it on one computer in a classroom rather than in a computer lab. The reasons were that only 10 computers existed in the lab for 25 students, and my emphasis as a teacher of English was not on computer skills. I also successfully used GIS in discovering different natural disasters, which is a common theme in the English course I teach in Grades 9 and 12. I demonstrated the layers that show Africa, fires in Africa, and the strongest fires on my laptop and the LCD projector in the classroom. I also used the same technique to show earthquakes in the Middle East, including the strongest ones. This is part of a six-lesson unit on natural disasters that lasts for two weeks. It focuses on different English skills, grammar, and writing, but to help students better understand reading and broaden their horizons, I found GIS to be a perfect teaching tool. I also used it to investigate demographics in the Middle East in a relevant unit for Grade 12 too. The number of students in the class varies between 18 and 25, but I can say that with such a technology there is no lazy student or a troublesome one. They are all attentive and breathless. The school provided a room for all lessons that require technology with access to Internet and an LCD projector. Now, the biology teacher, the geography teacher, and I are asking for an active board that will be provided in a month.

In Grade 9, we teach a unit on natural disasters and another on environmental problems where GIS, ArcMap, and other parts of training proved to be of a great benefit and could be implemented successfully to an extent that one of my students made a project on GIS where she talked about its history and development

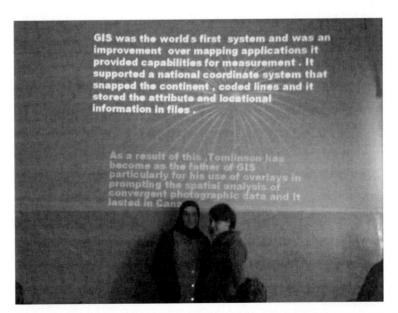

Fig. 17.1 Zahraa Dirani and Alissar Samaan from Bednayel Secondary Public School, Grade 9, presenting their project about GIS

(Fig. 17.1). In Grade 10, it can be used in units regarding environment and population growth, child labor, travel and tourism, and natural wonders. If we want to use GIS in subjects like sociology, economics, and even Arabic, we will have a genius generation that knows about life skills and technology and one that knows about implementing theories and understands the subjects through concrete examples.

The lessons are quite good, although I believe that we are still unable to navigate thoroughly for themes we need because we need further training ourselves. The other hindrance is the inability of our public schools to provide a room with a computer for each student to discover by himself or herself. GIS teaching is thus limited to teachers' presentations, and students are still in the phase of astonishment without real hands on experimenting.

With such a humble experience that does not satisfy a teacher, it proved to be an original and very new method to teach students. They were amazed by the information displayed. They were able to understand lessons presented using GIS in an unprecedented manner. They started asking more intellectual questions and even the weakest students were so much more interested and were able to understand. Students presented their projects to their peers with great enthusiasm (Fig. 17.2). I discovered hidden potential in most of them. Does it prove that most learners are visual? In a technology era, all our educational processes and curricula need to be revised and edited in a way that tackles technology. Learning will only be able to take place using Powerpoint presentations, assimilations, active boards, online self tutoring, hands-on projects, and researching using new technologies. Our role as educators should be limited to facilitating the process of learning, tackling

Fig. 17.2 Grade 9 students amazed by their own project about GPS

knowledge through ways that prove to create such a difference in our students' mentality in order to discover their potential and prepare them for higher education and the marketplace.

In brief, GIS can be promising in Lebanon since the themes in our curriculum (such as civilizations, environment, child labor, and economics) can provide a large space for implementation. However, people who have access to it are mostly engineers in different domains who do not teach, which hinders the spread of GIS. Teachers' GIS skills are based on intrinsic motivation, and they are still not proficient. If we want to see GIS implemented prosperously in education, we should keep in our minds the successful industrial examples of Oger International, Dar Al-Handasa, and Alami and Khatib. A meeting with NCERD came to the conclusion that teachers of geography, sociology, sciences, statistics, and English should be trained on using GIS efficiently and that this training is going to be followed up by the Ministry of Education inspectors in building the cornerstone in the GIS situation we would like to see in Lebanon.

Finally, I hope that all teachers would benefit from the trainings provided worldwide. I would also like to thank the American embassy for its generosity and concern in developing teachers' abilities in rural areas. I would like to thank Esri and National Geographic for their dedicated manner in training us efficiently and introducing us to a new world of GIS.

Acknowledgments Introduction to the National Curriculum by Miss Samya Abou Hamad. MyCOE training CD.

Reference

Association of American Geographers. (2008). *Supporting basic math and science education in the Muslim world*. Accessed December 8, 2010, http://www.aag.org/cs/mycoe/middle-east-north-africa

Chapter 18
Malta: GIS and Geography Teaching in the Context of Educational Reform

Maria Attard and John A. Schembri

18.1 Introduction

Malta is made up of three inhabited islands with a total area of 316 km^2 and a population of just over 400,000. The population density is over 1,200 persons per square kilometer, making it the country with the highest population density in Europe and one of the highest in the world (National Statistics Office, 2009). This demographic situation is partly attributed to the strategic position of the islands in the central Mediterranean. Figure 18.1 shows the position of Malta as the southernmost boundary of Europe and in the middle of the Mediterranean Sea. The chronological timeline later displayed in Table 18.1 shows the continuous interests for the islands to be governed by a series of occupiers as an indication of the archipelago's strategic value. However, it was during the tenure by the Knights of St John and the British that contributed to the population growth and the development of the educational system along southern European lines in the sixteenth to eighteenth centuries and along British lines from the nineteenth century. As a result, most Maltese can now speak fluently and write in the native Semitic tongue, in Romance Italian, and in Anglo-Saxon English.

18.2 Developments in the Education System

Education policy has been one of the most contentious issues in Maltese society. Table 18.1 provides a timeline of the main events that shaped the spread of education in Malta. Although this was done in spurts followed by periods of stagnation, in general the public education of the Maltese started in earnest in the mid-nineteenth century and by 1870 all villages had a local primary school. Although secondary schooling had its origins in the sixteenth century, it was not until 1970 that it was introduced for all, with compulsory schooling increased to age 16 in 1974.

M. Attard (✉)
University of Malta, Msida, Malta
e-mail: maria.attard@um.edu.mt

A.J. Milson et al. (eds.), *International Perspectives on Teaching and Learning with GIS in Secondary Schools*, DOI 10.1007/978-94-007-2120-3_18,
© Springer Science+Business Media B.V. 2012

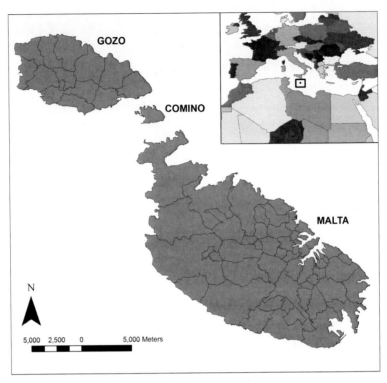

Fig. 18.1 The three islands of Malta, Gozo, and Comino commonly referred to as Malta and the respective geographic position in the Mediterranean Sea

In the current system, education is compulsory for all children between the ages of five and sixteen. This is subdivided into a six-year primary cycle, from the age of five to ten years, and five years of secondary education from eleven to sixteen. An overall policy of inclusive education throughout the whole educational cycle ensures that children with special needs are also integrated into the mainstream (Ministry for Education, Youth and Sport, 2009).

These developments were running in conjunction with the 'language debate' that saw the exclusion of the local vernacular by the Italian and French languages by the local nobility and intelligensia during the centuries of the Knights' presence in Malta; Italian being replaced by Maltese in court proceedings in the 1930s, and English becoming a second language during the nineteenth and twentieth centuries. Ironically, since independence from Britain in 1964, a number of families opted for English as their first and only language of communication, isolating themselves into an ethnic social enclave. This is also reflected in some church schools that insisted on children using only English as a means of communication. Today this trend has transferred to privately owned, independent schools.

Education was the medium through which socioeconomic development in Malta gained rapid momentum. Even though the island received limited financial aid, it

Table 18.1 A chronological timeline of salient aspects of education policy in Malta

Year(s)	Occupier in the Islands	Overall policy (language of instruction)	Educational value	Further information
870–1090	Arabs	(Semitic)	–	–
1090–1530	Normans, Swabians, Angevins, Aragonese, and Castillians	Educating for the leadership of the islands (Italian)	European values and alphabet Monastic orders religious education	Universita' financed by Cathedral chapter First Public school
1530–1798	Knights of St John	(Italian and Latin)	Religious and public education	1592 Founding of Collegium Melitense by the Jesuits University for General Studies Collegio d'Educazione, a Grammar School Sc[u]ola Infima – a preparatory school Schools of Navigation, Naval Architecture and Cartography 1769 Public University
1798–1800	French	The State had practically full control of education (French)	–	Introduction of elementary education All private schools closed down University reduced to 'ecole centrale'
1800–1964	British	To promote the affection of the people to be drawn more closely to the British Crown (English replaced Italian as language of instruction at secondary and post-secondary levels. Maltese was encouraged)	The start of the Anglicization of Malta	1870s Primary education institutionalized Schools opened in every town and village Education was free of charge
1964–	Post-colonial	(Maltese and English)	Education Acts	1970 secondary Education for all was introduced 1974 Education Act 1987 Education Act

Monastic Orders: Franciscan Conventuals (c 1350), Carmelites (1418), Dominicans (1450), Augustinians (1460), Friars Minor (1492), Benedictine Nuns (1480/1495)

fashioned its own educational opportunities in the arts, medical, social and engineering sciences, law, and business (Caruana, 1992). A number of Parliamentary Acts and Legal Notices were then introduced to provide a strong legal basis in education and to have a measure of control over the methods of instruction and conduct of examinations throughout the publicly owned institutions, the independently run schools and others run by the church authorities. Future policies aimed at consolidating past gains and the development of long-term financial forecasts linked with growing student numbers moving into higher education (National Commission for Higher Education, 2009).

18.3 The Current Situation in Secondary Schools

At the secondary school level, there are three different kinds of state schools in Malta. These include the Junior Lyceums, the Area Secondary Schools, and schools for very low achievers. Alongside the state schools the church manages a number of schools, mostly administered by specific religious orders (nuns and priests). Since the early 1990s, however, there has also been the development of privately owned, independent schools. These schools are either foundation schools where parents have the main shareholding or are privately run businesses. In Malta, these are referred to as private or independent schools and require parents to pay relatively high registration and attendance fees.

Currently the local educational system is undergoing radical changes, proposed by the latest administration elected in 2008 and published in the same year by the Ministry (Grima et al., 2008). The most important change is the transition procedure from primary to secondary education with the abolition of the 11+ entry examinations in 2011. The policy also extends to the provision of more support for teachers and educators in state schools, and, more recently, the expansion of a number of church schools to include the primary sector or the secondary sector where they did not exist. This expansion project in church schools, which is expected to cost €20 million, will see a 40% increase in the number of pupils in church schools (Ministry for Education, Youth and Sport, 2009).

Until 2010, children in primary schools had to undergo an examination to place themselves in a different state or church secondary school. Apart from assigning students to particular schools these examinations also streamed students in between attending Junior Lyceums or Area Secondary schools in the state run system or, in particular, church schools. According to the policy document For All Children To Succeed (FACTS) (Ministry of Education, 2005), the transition is described as staccato and abrupt. Following on the lines of an all inclusive education, which also included the removal of streaming, children are placed on an educational path as from primary school age and continue their studies in what are termed 'Colleges,' thus removing the need to 'compete' for a place in the secondary school pool.

Starting in 2011, all children will be placed in colleges as determined by geographic distribution (Fig. 18.2) and will be assigned a primary and secondary school.

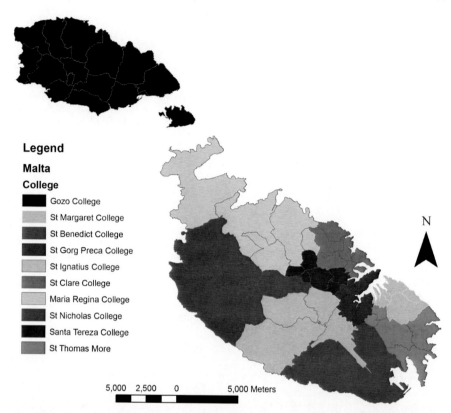

Fig. 18.2 The geographic distribution of the newly set up public colleges in Malta (http://www. education.gov.mt/edu/schools/state_colleges.htm)

Unfortunately in this reform, changes to the allocation of time dedicated to particular subject and pedagogy are also being proposed. One of the proposals is for geography to be amalgamated with history and social studies, therefore reducing the actual time allocated for these core subjects in the first two years of secondary school. Geography will be replaced with environmental studies in the core subjects list for the remaining three years of secondary school, unless geography is chosen as an option for further study (Grima et al., 2008).

18.3.1 The National Minimum Curriculum

The National Minimum Curriculum was formalized through three legal notices for the primary level in 1989, secondary level in 1990, and post-secondary level in 1991 (Cassar, 2003, 2009). Debates regarding its mission statement however ensued among teachers, policy makers, and academics. Borg, Camilleri, Mayo, & Xerri (1995) argue that the contents of the local National Minimum Curriculum can

Table 18.2 The principles and objectives of the National Minimum Curriculum (Ministry of Education, 1999)

Curricular principles	Educational objectives
1. Quality education for all	Each objective has subheadings of
2. Respect for diversity	knowledge/information, skills, and attitudes
3. Stimulation of analytical, critical, and creative thinking skills	1. Self awareness and the development of a system of ethical and moral values
4. Education relevant for life	2. The development of citizens and a
5. Stable learning environment	democratic environment
6. Nurturing commitment	3. Developing a sense of identity through
7. Holistic education	creative expression
8. An inclusive education	4. Religious education
9. A more formative assessment	5. Strengthening of gender equality
10. The strengthening of bilingualism in schools	6. Education on human sexuality
11. Gender equality	7. Preparing educated consumers
12. Vocation and competence	8. Media education
13. The importance of learning environments	9. Effective and productive participation in the world of work
14. Increasing participation in curriculum development	10. Education for leisure
15. Decentralization and identity	11. Wise choices in the field of health
	12. Greater awareness of the role of science and technology in everyday life
	13. Competence in communication
	14. Preparation for change

be related to national curricula in western countries, including UK and the USA in that it is 'underpinned by a conservative ideology' that discriminates against women, gives insufficient value to non-European cultures, and promotes a hierarchical form of education that is geared to a capitalist concept of work. They argue that the curriculum, in its present form, cannot serve to promote a genuinely democratic education. Nonetheless, the Legal Notices related to the National Minimum Curriculum were published in a complete document in 1999 (Ministry of Education, 1999). Table 18.2 lists the educational objectives and the curricular principles within the National Minimum Curriculum.

18.4 Geography in Secondary Schools

The geography syllabus in the three types of secondary schools found on the islands varies slightly. State schools follow the syllabus issued by the Department of Curriculum Management and eLearning within the Directorate for Quality and Standards in Education of the Ministry of Education, Youth and Sport (Gilson, De Battista, & Quintano, 2009) whilst church and independent schools are free to formulate their own syllabus. These syllabi ultimately reflect the Matriculation examinations which are equivalent to Ordinary Level GCSE's in the UK. These

Matriculation examinations are necessary for any student to move to tertiary education and eventually to a higher education institution.

The aim of the Matriculation examination syllabus is to provide teachers with a choice of materials that should suit a variety of approaches. It also provides teachers and students with an opportunity to look at environmental problems (both physical and human) that are closer to home as is implied in the examples taken from the Mediterranean. The Geography of Europe continues to occupy a key place in the examination due to Malta's connection (past, present, and future) with this continent, whilst giving importance to developing countries due to their geographic interactions between them and the developed world (University of Malta, 2009).

The geography syllabus also lists the abilities that candidates will be examined on as well as the knowledge, skills, values, and attitudes that candidates are expected to demonstrate. In the area of knowledge and understanding, candidates are expected to:

- Recall specific facts in connection with the syllabus content;
- Show understanding of geographical concepts, ideas, and principles contained in the syllabus; also their application in the context of the physical and human environments;
- Show understanding of the spatial patterns and interactions within these environments; and
- Demonstrate locational knowledge applied to the Maltese Islands, the Mediterranean region, the rest of Europe, and the developing world.

In the area of skills, candidates are expected to:

- Observe, record, classify, and interpret data collected in the field or from secondary sources, to form conclusions and communicate ideas;
- Read, interpret and use maps, photos, and statistical data; and
- Represent geographical information in simple map form (sketch-maps), graphs or diagrams, and to write in a coherent manner.

And in the areas of values and attitudes, candidates are expected to:

- Demonstrate awareness of environmental issues in terms of the conservation and the protection of both the physical and the human environment; and
- Form reasonable judgments in relation to environmental issues of a geographical nature (University of Malta, 2009).

This syllabus does not include any reference to Geographic Information Systems (GIS) but rather a reference to the representation of geographic information. On the other hand, the state school syllabus contains a specific reference to GIS as a specific ICT tool for which students should develop skills:

develop the student skills in the following ICT toolkit namely word processor, spreadsheet, presentation software e.g. PowerPoint; desktop publishing (DTP) software; internet browser/e-mail; electronic encyclopedia; geographic information system (GIS); automatic data logging weather station; digital camera. (Gilson et al., 2009)

This reference to GIS is unique in the syllabus since nowhere else is there any reference to how or what should be taught about it. The focus is on the development of skills about using GIS rather than on why and how GIS is applied. More importantly there is no reference to any of the aspects of data handling in GIS, data quality issues and application, and only a single reference to how teachers can apply ICT in geography by using 'geographic data and themes that lend itself easily to work in ICT' (Gilson et al., 2009).

18.4.1 What Do Geography Teachers Teach?

In late 2009, the Geography Division of the Mediterranean Institute at the University of Malta carried out a survey of secondary school teachers to assess the use of GIS in classrooms. The results of this survey are being published in this chapter for the first time. The questionnaire was developed focusing on three aspects including the details about the teacher, a description of the school, and questions about their teaching of geography (and GIS). There are 76 secondary schools in Malta including state, church, and independent schools. The response rate was of 30% with 17% of the responses coming from church schools and the remainder from state schools. Most schools have a student population ranging from 500 to 1000 students; however, only 13% of the schools had relatively large numbers of students (40–50) enrolled in geography, both as a general subject and as an option. Only 30% of schools reported having a geography room where studies in geography are conducted and 22% of the total number of schools reported using computer facilities during geography. It was very encouraging to see that many of the respondents had knowledge and some exposure to GIS in their university education; however, this might have also affected the type of teachers answering the questionnaire in the first place.

The teachers were asked the number of lessons dedicated to local geographic information, to the visualization of geographic information (cartography) and to GIS. Results show that overall the teachers give importance to visualization and cartography, and local geographic information with sources of such data focusing on the National Statistics Office (www.nso.gov.mt) and the Malta Environment and Planning Authority (www.mepa.org.mt), which provides an online GIS Server with some data; however, very few schools dedicated lessons specifically to GIS (Fig. 18.3).

Over 80% of the respondents use Web tools such as Google Earth in their lessons establishing at the very least the foundations of what is digital cartography. In many cases teachers reported using virtual globes to support lesson plans as well as visualization of places and aerial imagery. These tools are also used to locate particular places and events that are explained in the lesson. This is probably the reason for

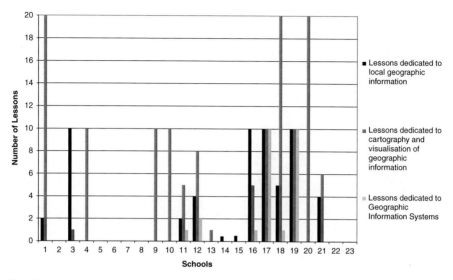

Fig. 18.3 Lessons dedicated to geographic information and GIS. (Survey conducted by the authors 2009)

reporting a 50% share of students being aware of digital mapping tools and software. Unfortunately, none of the teachers surveyed reported using GIS in class, despite some having the software installed on their personal computers.

When asked about the use of computers with geographic tools, only 35% of the schools replied that students get the opportunity to use computers with geographic tools, most of which refer to virtual globes. At the same time, no school requires students to try GIS at school or at home as part of their learning. This means that no student in the secondary school is required to learn how to use GIS. In one instance a teacher reported that GIS 'was too difficult anyway.'

Overall, there is agreement between teachers that GIS is an important tool for geography students to learn (96% of respondents) and 91% agree that it should be an essential part of the curriculum at the secondary school level. Finally, teachers were asked to identify the barriers and opportunities for the introduction of GIS in secondary schools. Some of the comments pointed toward the lack of teacher training, on how GIS should be taught, and the lack of materials and data for class sessions; others complained about the lack of computer/computer time for geography, with some pointing toward the fact that the geography syllabus is already 'packed' and if GIS had to be introduced, other subjects should be removed; however, not at the expense of fieldwork or map reading. In one case a teacher reported that GIS is for professionals so it should not be taught at the secondary school level, with the syllabus focusing on more generic geographic concepts and facts.

The current lack of teaching about GIS in secondary schools should not however act as a barrier. An opportunity lies in the extensive experience of teaching GIS at the University of Malta. The GIS Laboratory at the University was set up in 1996 and since then GIS has been taught to various faculties and departments within the same University. Attard (2009) developed a framework for undergraduate study units with problem-based learning approaches that allow for students to interact with the technology but also apply it to real world scenarios. This approach allows for students to learn how to handle the technology (software) without compromising the learning of the principles of geographic information and GIS. The opportunities for the development of GIS in secondary schools are based on four important developments. These are as follows:

– the growing interest and use of virtual globes;
– the availability of open source software;
– the availability of free data (e.g., OpenStreetMap); and
– the role of the University of Malta and the Education Department to develop materials using local case studies and data.

18.5 Conclusions

In 2008, Malta embarked on a strategy that will see the island develop into an ICT hub with very well developed ICT infrastructures and communications (Ministry for Investment, Industry and Information Technology, 2008). This strategy contains a strong emphasis on the deployment of ICT in education. Alongside this, there is the use of GIS in the provision of data, mapping services and an overall Government framework for GIS (Streams 1.16, 4.8 and 5.29).

The importance of GIS is growing in Malta and there is recognition of its importance as a tool to support improved quality of life, infrastructure, and enhanced governance. This priority will eventually filter through the education system and whilst today GIS is formally taught at tertiary level only, there is great potential and willingness for Malta's secondary schools to include such topics in secondary school syllabi.

References

Attard, M. (2009). Developing undergraduate GIS study-units. In K. Donert (Ed.), *Using geoinformation in European geography education* (pp. 102–111). Rome: Società Geografica Italiana, International Geographical Union.

Borg, C., Camilleri, J., Mayo, P., & Xerri, T. (1995). Malta's National curriculum: A critical analysis. *International Review of Education, 41*(5), 337–356.

Caruana, C. M., (1992). *Education's role in the socio-economic development of Malta.* New York: Praeger Publishers.

Cassar, G. (2003). Politics, religion and education in nineteenth century Malta. *Journal of Maltese Education Research, 1*(1), 96–118.

Cassar, G. (2009). Education and schooling: From early childhood to old age. In J. Cutajar & G. Cassar (Eds.), *Social transitions in Maltese society* (pp. 51–74). Malta: Agenda Publishers.

Gilson, E., De Battista, R., & Quintano, A. (2009). *Geography syllabus. Department for Curriculum Management and eLearning, Directorate for Quality and Standards in Education.* Ministry of Education, Youth and Sport, Malta. Accessed January 5, 2010, http://www.curriculum.gov.mt/docs/Geo_Gen_JL_1_5.pdf

Grima, G., Grech, L., Mallia, C., Mizzi, B., Vassallo, P., & Ventura, F. (2008). *Transition from primary to secondary schools in Malta: A review.* Floriana, Malta: Ministry for Education, Youth and Sport.

Ministry of Education. (1999). *National minimum curriculum.* Valletta: Ministry of Education.

Ministry of Education. (2005). *For all children to succeed.* Malta: MEYE.

Ministry for Education, Youth and Sport. (2009). *Malta: A guide to education and vocational training.* Accessed January 2, 2010, http://www.education.gov.mt/edu/edu_03.htm

Ministry for Investment, Industry and Information Technology. (2008). *The smart island the national ICT strategy for Malta 2008–2010*, Malta. Accessed January 21, 2009, http://www.thesmartisland.gov.mt

National Commission for Higher Education Malta. (2009). *Annual report 2008.* Valletta: National Commission for Higher Education.

National Statistics Office. (2009). *Malta in figures 2009.* Valletta: National Statistics Office.

University of Malta. (2009). *SEC syllabus (2010): Geography.* Matriculation and Secondary Education Certificate Examinations Board, Malta. Accessed January 20, 2010, http://www.um.edu.mt/matsec/docs/syllabireports/syllabi2010/sec/SEC15.pdf.

Chapter 19
The Netherlands: Introduction and Diffusion of GIS for Geography Education, 1980s to the Present

Tim Favier, Joop van der Schee, and Henk J. Scholten

19.1 Introduction

The digital revolution is one of the most important trends in education in The Netherlands, and digital maps and GIS are seen by geographers in The Netherlands as essential for enriching geography education. Modern secondary education in The Netherlands emphasizes the development of knowledge as well as skills and attitudes. The attention to learning through inquiry-oriented problem-solving activities around authentic issues using real-world data is growing. GIS fits very well into this development.

Secondary education in The Netherlands is structured in three levels: VBMO (pre-vocational education), HAVO (senior general education), and VWO (pre-university education). A central test in the students' primary school and the advice of the primary school teachers determines which type of secondary school a student will attend. Schooling is compulsory until the age of 16. The Dutch Ministry of Education has a great impact on the main lines of what happens in a school through the national curriculum. Within this framework, teachers have quite a lot of freedom to organize their curriculum. Nevertheless, almost all teachers use the schoolbooks that are based on the national curriculum. The Royal Dutch Geography Society has substantial input into developing new geography curricula.

In most schools for secondary education, geography is a subject in itself. Geography is obligatory from 1st Grade (12–13 Years) to 3rd Grade (14–15 Years) and optional for the higher grades. The national curriculum directs what happens in the schools. The Dutch system has school exams as well as central exams at the end of the secondary education. Digital maps and GIS are not yet an official part of these exams, but there is discussion about expanding computer-based testing in this direction in the near future. Some geography teachers already use assignments with digital maps for school exams.

T. Favier (✉)
VU University, Amsterdam, The Netherlands
e-mail: t.favier@ond.vu.nl

A.J. Milson et al. (eds.), *International Perspectives on Teaching and Learning with GIS in Secondary Schools*, DOI 10.1007/978-94-007-2120-3_19,
© Springer Science+Business Media B.V. 2012

Teacher training for secondary education is organized by teacher training institutes at universities and colleges. A master's degree is required to become a teacher in the upper levels of secondary education, but at the lower levels a college-level degree is sufficient. In-service training organized by universities and colleges is available for teachers to update their knowledge and skills. Most of the younger generation of teachers are familiar with GIS, and a part of the big group of middle-aged teachers try to keep up with new developments via in-service training.

19.2 The History of GIS in Schools in The Netherlands

The introduction of GIS in secondary education in The Netherlands started within geography education, and up to today, GIS is still mainly used within the geography domain.

19.2.1 The First Step (The 1980s and 1990s)

A careful first step in the introduction of GIS in secondary education was made in the 1980s, when a researcher at the VU University Amsterdam developed software that allowed students to construct thematic maps (choropleth and chorochromatic maps) and charts (histograms and pie charts) on the basis of blank maps of The Netherlands, Europe, and the World and tables with thematic data. At the end of the research project, the software was sold to the publishers of the Bosatlas, which is the national school atlas in The Netherlands. The Bosatlas put the software on a CD, together with some lessons, and sent it to all schools. However, only a small fraction of the teachers actually used the software.

19.2.2 Limping with One Leg (The Late 1990s and Early 2000s)

By the end of the 1990s, a few geography teachers started experimenting with professional Desktop GIS packages in student research projects. They organized projects in which students combined data collection in the field with data visualization in GIS and projects in which students used Web-based data download portals (such as the National Census data download portal) to construct maps. Currently, these teachers can still be seen as the champions of educational innovation, running far ahead of their colleagues. One of them, Willem Korevaar, became the primary advocate for GIS in secondary education in The Netherlands by spreading the word about his successes in magazines for geography teachers (e.g. Korevaar, 2003, 2004) and in workshops at the annual Dutch Conference for Geography Teachers. Although many teachers acknowledged that such GIS-based student research projects were interesting ways to enhance students' geographic literacy, they also found out that these projects were very difficult to organize. This phase in the diffusion of GIS can therefore be seen as the limping-with-one-leg phase.

19.2.3 Walking on Two Legs (The Mid 2000s)

In order to stimulate the spread of GIS beyond a couple of innovative teachers, more work needed to be done on the software and data infrastructure. In 2003, the publishers of *Wereldwijs*, one of the main schoolbooks for secondary geography education in The Netherlands, included a chapter on GIS in their book for 3rd Grade (14–15 Years) HAVO and VWO. The publishers also supplied teachers with a CD with a Dutch version of Esri's ArcExplorer Java Edition for Education (AEJEE), some lessons, and data sets. Around the same time, the Bosatlas developed a WebGIS, called the Bosatlas online (www.bosatlasonline.nl). This is a sort of online version of the Bosatlas CD, but much more accessible and user-friendly. The Bosatlas online allowed students to create maps and charts by themselves in a simple online environment. The most important stimulus to get things walking, however, was the development of an Internet portal for secondary education with GIS called EduGIS (www.edugis.nl). This Internet portal was developed with the support of the Dutch government. The unique selling point of EduGIS is the cooperation of data suppliers, governments, research institutes, teachers, teacher trainer institutes and publishers of geography schoolbooks, and the embedding of the activities in a national and international network. Central in the EduGIS portal is a free WebGIS based on open standards. This WebGIS contains hundreds of digital map layers, including topographic maps, elevation maps, historic maps, and census maps (provided by the Topographic Survey, the National Census Office, and a number of other government agencies) and tools for analyzing these map layers. EduGIS also provides lessons on topics like spatial planning, water management, and environmental pollution, so that teachers could use the WebGIS right away. Publishers of schoolbooks also developed lessons for the WebGIS. Later, EduGIS began supporting other GIS means too, such as Google Earth and ArcGIS, by providing data and lessons. In order to get teachers familiar with the EduGIS, WebGIS, and other GIS means, the EduGIS team organized numerous workshops and courses for teachers throughout the country. These projects helped GIS to begin walking in schools in The Netherlands, but more was needed for GIS to begin running.

19.2.4 Running (The Late 2000s)

Because of the presumed benefits of GIS for teaching and learning geography and because of the rising importance of GIS in professional geography and in society in general, the Ministry of Education following the advice of the Dutch National Geography Society decided to include GIS in the National Geography Standards for secondary geography education at each of the three levels. For HAVO and VWO (KNAG, 2003), students should be able to use GIS to select, process, and present geographic information. However, the standards are not further specified. The new standards for VMBO also mention GIS (KNAG, 2008). Students should be able to work with Google Earth, EduGIS, and simple Desktop GIS. Although GIS is in the National Geography Standards, it is not tested yet in the national exams.

Due to the fast introduction of computer hardware in schools, the increased availability and performance of the educational WebGIS and the rise of Google Earth, more and more teachers started using GIS in their classes. At the end of the decade, accessible GIS like EduGIS, the Bosatlas online and Google Earth were frequently used by teachers and students. However, not all teachers were able to use these GIS in their classes because they had limited ICT skills. Thus, while about half of the teachers are running with GIS in their classrooms, other teachers have not even made a first step.

19.3 A Case of Geography Education with GIS

This case describes an innovative student research project that combines fieldwork with data visualization in GIS. The student research project focused on the topic of 'services & customers,' and was conducted with a VWO 5 class (16- to 17-Year-old students) at the Merewade High School in Gorinchem, a city in the center of The Netherlands. The project covered seven hours of supervised lessons and about seven hours of students' own time.

The student research project started with an introduction in which the teacher explained economic-geographic concepts like 'range' and 'market area.' Every service has its own market area from which the service draws its customers. Its size and shape depend on a large number of spatial factors, such as the population density around the service, the accessibility of the service, and the situation of the service in respect to competitive services and complementary competitive services. It also depends on a number of nonspatial factors, such as the price level of the service and the quality level of the service.

After the introduction, students engaged in a short GIS training session in which they learned how to map the market area of their school on the basis of Microsoft Excel files with data of the home addresses (postal codes) of the students of their school. In the student research project, the market area was seen as a simple circle around the school, in which the radius is the median distance between the home addresses of the students and the school. Fifty percent of the students live inside this circle while the other 50% live outside this circle. The teacher then told the students that they were going to do a research project on the market areas of services in their town. Students had to choose four services of the same category and formulate hypotheses about the size of the market areas of these services and the causes of differences in the sizes. Nadhie (16) and Bart (17) chose to investigate four gyms in their town: the Procare, Breedveld, Fit-4-Less, and Living Well. They expected the Procare gym to have the largest market area. The students reasoned that the gym had a high service level, and that people are willing to travel further to go to a gym with a high service level and therefore bypass gyms with a low service level. They expected the Breedveld gym in the city center to have the smallest market area, as its accessibility is low, especially by car. In order to explain the differences in the size of the market areas, the students had to formulate additional survey questions. Nadhie and Bart formulated the following questions: "What is your age?" "How often do you visit this gym per month?" "Which kind of transport means do you use to go to this gym (car/bicycle/walking)?" and "What is the main reason to visit *this* gym (good service/good price/good atmosphere/nearby/other)?".

Next, students went to the service locations and interviewed 20 customers at each location. They asked the customers for their postal codes and conducted the survey. Back at school, the students entered the data into Micosoft Excel. They converted the postal codes into X and Y coordinates with the help of an online converter and entered the survey data into the same sheet. Next, they calculated the median distance customers live from each service and summarized the survey data. In the following phase, students visualized their data in GIS. They constructed dot maps of the distribution of the customers of the different stores and used the buffer tool to construct a map layer of the market areas of the different services. They also constructed quantities maps and pie chart maps on the basis of the survey data. Figure 19.1 shows the maps created by Nadhie and Bart.

In the next lesson, the students presented their research to the teacher and their classmates. The students explained their research approach, showed their maps, and presented their conclusions. Nadhie and Bart explained that they found out that the size of the market areas of the four gyms depended on the accessibility and the price. "The Living Well has the largest market area, probably because it's situated near the highway. Most people get here by car. . . . The Fit-4-Less also has a large market area. This is the cheapest gym. Many of the visitors come from the western part of the town, which has a low per capita income." The teacher then encouraged the students to reflect on their research approach. The central question was, "If you would do the same research again, would you choose the same services? Would you ask the same survey questions?" Nadhie and Bart replied,

> I think the survey question about the transport means was very useful to get insight in the factors which determine differences in the size of market areas. The question about the main reason to visit this gym was useful too. Maybe we could have skipped the question about age. We could have asked whether they leave from home, work, school or somewhere else, just like Sophie and Sanne did. That might have been useful too.

After the presentations, the teacher started a discussion in which the students had to summarize what they learned. The teacher then drew a symbolic representation of the theory of the factors that influence the size of market areas of services on the blackboard and asked the students for input. Through this process, the teacher tried to raise students' thinking to a higher abstraction level. This student research project shows how GIS could be used not only to stimulate the development of students' geographic knowledge and thinking but also to stimulate the development of students' skills in conducting simple geographic research. In such a way, this student research project enhances all components of students' geographic literacy.

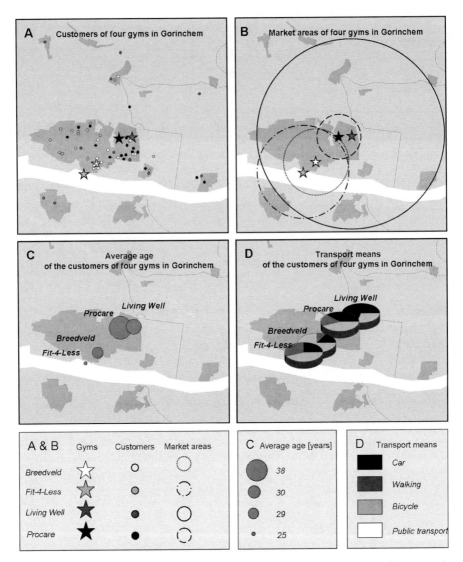

Fig. 19.1 Maps constructed by Nadhie and Bart: **a** distribution of the customers of four gyms in Gorinchem, **b** market areas of the four gyms, **c** average age of the customers, and **d** transport means of the customers

19.4 The Future of GIS in Schools in The Netherlands: Taking a Leap (The Next Decade?)

One of the limiting factors to the diffusion of GIS in secondary geography education is the limited knowledge of teachers about how to teach geography with GIS in a viable and effective way. The question is, What knowledge do geography teachers need in order to be able to teach successfully with GIS? In the past two decades, the pedagogical content knowledge (PCK) concept has become an accepted academic concept to describe the knowledge that teachers use and need to teach specific declarative, procedural, and conditional domain knowledge (Berry, Loughran, & Van Driel, 2008). In order to be able to teach successfully, teachers need to have domain knowledge, also called disciplinary content knowledge (CK), and general pedagogical knowledge (PK). However, just adding CK and PK together does not make an effective teacher. Teachers need to know how to teach the specific subject knowledge and methods of a domain in a legitimate, viable, and effective way. Shulman (1986) introduced the term "Pedagogical Content Knowledge" to refer to the knowledge that exists at the intersection of PK and CK. According to Shulman (1986), teachers' PCK consists of three things: (1) domain knowledge, which is transformed in such a way that it is accessible for students; (2) knowledge about common preconceptions of students; and (3) knowledge about common learning difficulties of students. Successful teaching with technology requires the teacher to have general ICT skills (technological knowledge) and knowledge about how to teach with technology in general (technological pedagogical knowledge). In order to teach with technology, teachers should have knowledge about how to use the technology to gain knowledge in that domain (technological content knowledge). Also, the teacher should have knowledge about how to use the technology in instruction in such a way that students develop knowledge and skills in that domain. This is called 'Technological Pedagogical Content Knowledge (TPCK)' (Mishra & Koehler, 2006). The TPCK model (Fig. 19.2a) can be specified for different domains, including for geography education with GIS (Fig. 19.2b).

As teachers can be seen as the "gate keepers of educational innovations" (Wallace, 2004), the successful diffusion of GIS in secondary education depends on, among other factors, teachers' motivation to do geography with GIS and whether teachers know how to teach geography with GIS in such a way that it raises students geographic thinking to a higher level. At present, many teachers in The Netherlands lack a sufficient basis of GIS TPCK. This is not surprising, as GIS still does not have a structural place in teacher education programs (van der Schee & Scholten, 2009) and as innovations in education take time. Some teacher educators in The Netherlands have included short Google Earth and EduGIS courses in their programs, but most are still trying to find out how to do this. One of the problems is that teacher educators as well as researchers in the field of education, geography, and GIS are looking themselves for the best way to shape geography education with GIS. Research on how to teach geography with GIS is very limited. After reviewing the relevant literature, Bednarz (2004) concluded that "scant attention has been paid to issues related to pedagogy and GIS." In order to take a real leap in the diffusion of GIS in secondary geography education, more research on how to teach geography with GIS is a must. Recent research in The Netherlands, which is closely connected to the EduGIS project, aims to develop design principles for legitimate, viable, and effective GIS-based instruction methods that aim to enhance students' geographic literacy (Favier & Schee, 2009).

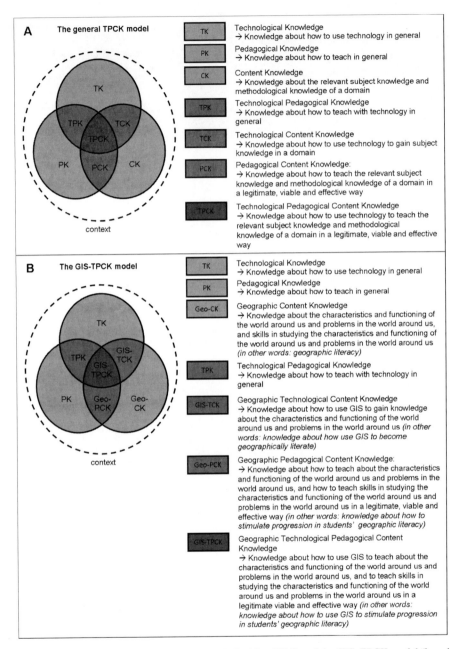

Fig. 19.2 The general TPCK model (Mishra & Koehler, 2006) and the GIS–TPCK model (based on Doering & Veletsianos, 2007)

We see a bright future for geography with GIS in The Netherlands. GIS opens new opportunities to show how geography can help people understand what is going on with planet Earth. GIS is an exciting means to teach and learn geography and will become increasingly important in preparing young people for the world of today and tomorrow. The EduGIS project combined the activities that are needed to stimulate the diffusion of GIS in secondary education, and will continue to do so in the future.

References

Bednarz, S. W. (2004). Geographic information systems: A tool to support geography and environmental education? *Geojournal, 60*, 191–199.
Berry, A., Loughran, J., & Van Driel, J. H. (2008). Revisiting the roots of pedagogical content knowledge. *International Journal of Science Education, 10*, 1271–1279.
Doering, A., & Veletsianos, G. (2007). An investigation of the use of real-time, authentic geospatial data in the K12 classroom. *Journal of Geography, 106*, 217–225.
Favier, T. T., & Schee, J. A. v. d. (2009). Learning geography by combining fieldwork with GIS. *International research in geographical and environmental education, 18*, 261–274.
KNAG. (2008). *Kijk op een veranderende wereld: Voorstel voor een nieuw examenprogramma aardrijkskunde VMBO*. Utrecht: Koninklijk Nederlands Aardrijkskundig Genootschap.
KNAG. (2003). *Gebieden in perspectief: natuur en samenleving, nabij en veraf. Rapport Commissie Aardrijkskunde Tweede Fase*. Utrecht: Koninklijk Nederlands Aardrijkskundig Genootschap.
Korevaar, W. (2004). Modern aardrijkskundeonderwijs met GIS op de kaart gezet. *Geografie, 13*.
Korevaar, W. (2003). GIS in het voortgezet onderwijs. *Vi Matrix, 11*.
Mishra, P., & Koehler, M. J. (2006). Technological pedagogical content knowledge: A new framework for teacher knowledge. *Teachers College Record, 108*, 1017–1054.
Shulman, L. S. (1986). Those who understand: Knowledge growth in teaching. *Educational Researcher, 15*, 4–14.
van der Schee, J. A., & Scholten, H. J. (2009). Geographic Information Systems and geography teaching. In H. J. Scholten, R. Van der Velde, & N. Van Manen (Eds.), *Geospatial technology and the role of location in science* (pp. 287–301). Dordrecht: Springer.
Wallace, R. M. (2004). A framework for understanding teaching with the internet. *American Educational Research Journal, 41*, 447–488.

Chapter 20
New Zealand: Pioneer Teachers and the Implementation of GIS in Schools

Stephanie A. Eddy and Anne F. Olsen

20.1 Introduction

Over the last decade, New Zealand's education system has undergone significant change. This has resulted in increased teacher workloads. Debates and discussions have also become commonplace amongst educators throughout New Zealand. These are also exciting times with the advancement of technology, bringing with it renewed teaching practices and pedagogy. Spatial technologies such as Geographic Information Systems (GIS) allow students to explore and manipulate information directly (Fig. 20.1), bringing further possibilities and innovations in education.

In New Zealand, school attendance is compulsory for ages 6–16. However, it is customary for children to start school at age 5. Parents in New Zealand can choose the type of school they wish their children to attend or they can home-school. While the school-leaving age is 16, parents of students aged 15 can apply to the Ministry of Education (MoE) for an early leaving exemption. There are three types of schools in New Zealand: (1) state schools, which are self governing and fully funded by the government; (2) state integrated schools, which have a special character, including all Roman Catholic and some other religious schools, where the government pays all costs associated with tuition. Such schools must meet the same regulations as state schools; and (3) independent schools, which are funded by parent fees with a small per capita grant from the government.

The most recent national curriculum for secondary students was launched in 2007 and was progressively implemented until February 2010 when it became mandatory and replaced all previous curriculum documents. The New Zealand Curriculum is a statement of official policy relating to teaching and learning in English medium New Zealand schools. A parallel document, Te Marautanga of Aotearoa, serves the same function for Maori medium schools. The New Zealand Curriculum specifies eight learning areas: English, the arts, health and physical education, learning languages,

S.A. Eddy (✉)
GISMAPED, Auckland, New Zealand
e-mail: stephanie@gismaped.co.nz

A.J. Milson et al. (eds.), *International Perspectives on Teaching and Learning
with GIS in Secondary Schools*, DOI 10.1007/978-94-007-2120-3_20,
© Springer Science+Business Media B.V. 2012

Fig. 20.1 A student engages directly with data using GIS

mathematics and statistics, science, social sciences, and technology (Ministry of Education, 2007).

The National Certificate of Educational Achievement (NCEA) is New Zealand's national qualification for senior secondary students. The New Zealand Qualifications Authority (NZQA) is the governing body that coordinates qualifications in secondary schools. NCEA is a standards based qualification that replaced School Certificate, Sixth Form Certificate, and Bursary in New Zealand schools. Students are assessed against preassigned standards. NCEA involves both internal and external assessments. The first year that students sat level three NCEA, which is the highest level of the qualification, was in 2004. NCEA differs considerably from the system it replaced; School Certificate, Sixth Form Certificate, and Bursary, where a student's performance was judged relative to other students through norm referencing (Mahoney, 2005).

This change in qualification system prompted much discussion and debate in New Zealand secondary schools. The introduction of NCEA also caused a significant increase in classroom teacher workloads. A number of schools have chosen to offer alternative qualifications, alongside NCEA, namely Cambridge International Examinations and International Baccalaureate. The ongoing changes to NCEA, as it is refined and polished, have meant that ever-increasing demands have been placed upon the classroom teacher, and debates and discussions continue.

In New Zealand, all secondary teachers must complete a specialist subject degree followed by a one-year Graduate Diploma of Teaching or complete a combined specialist subject degree and secondary teaching qualification. A teacher becomes fully registered when he or she has successfully completed two years of teaching at a New Zealand secondary school and has met the Registered Teacher Criteria (RTC). Teachers during their first two years of teaching are closely mentored by a professional leader, are provided with a high level of support, and are actively engaged in a professional development program. The New Zealand Teachers Council provides an important control on the quality of teachers. This includes a requirement for appropriate professional development.

Overall, there exists a high level of support for professional development in New Zealand secondary schools. The MoE is the principal adviser to the Government on the education system and influences the arrangements for professional development for teachers. The Education Review Office (ERO) has estimated the ministry's spending on professional development as being more than $200 million NZD a year (Ministry of Education, 2008). The MoE provides operational funding to schools, some of which schools can use for the professional development of their teachers. It is the school leaders who make decisions about how much operational funding they will use for teacher professional development.

Consequently, the level of support that teachers are given for professional development varies between secondary schools.

In 1999, advancement in ICT professional development began with the Information and Communication Professional Development (ICT PD) cluster program with 23 groups or clusters of schools. Each cluster worked together to explore and foster innovative use of ICT in the classroom to support teaching and learning. The initiative was developed as a part of the Ministry of Education's ICT strategy Interactive Education released in October 1998. Over 60% of New Zealand schools have participated and shared their experiences in integrating e-learning activities into their professional practice (Ministry of Education, 2006).

The development of the ICT PD cluster program has been beneficial to teachers gaining professional development in GIS. A number of teachers have attended GIS workshops and GIS institutes facilitated by GISMAPED (a company that provides professional development in the use of GIS for teachers) through funding from the ICT PD cluster program. A few one-day school-specific GIS workshops have also been funded through the ICT PD cluster program.

20.2 GIS – The Beginning

There has been an interest in GIS in schools in New Zealand since 1994. In 1994, Murray Brown, then a computer advisor with the Palmerston North College of Education offered to help Phil Parent at Eagle Technology who was keen to see GIS used in schools. A project called the "Eagle Geography" project (Brown, 1994) was set up with Murray Brown as the national coordinator. A number of schools received GIS software, but it is unclear how many actually used it successfully. With Murray Brown's departure in 1995, others, particularly AURISA, Critchlow Associates, and Eagle Technology, continued to support the idea of GIS being used in schools. From 1997 the major format for promoting the use of the software was through a competition for senior students. Critchlow Associates (MapInfo agents) provided a trophy and a number of companies including Eagle Technology supported the competition with prizes. The competition was co-coordinated by Pip Forer of Auckland University until Lex Chalmers of Waikato University took over the role in 1999. At the same time, in Auckland, a teachers' support group set up to encourage the use of GIS and to develop teaching material was established by Tony Batistich of the Auckland Regional Council.

Since 2000, the focus of GIS in schools has moved toward integrating GIS as a teaching tool to enhance the teaching of geography. In 2004, Anne Olsen and Stephanie Eddy, both geography teachers, formed a company called GISMAPED, which provides quality professional development for teachers in the use of spatial technologies. Since 2000, they have facilitated over 30 workshops throughout New Zealand and 45 secondary schools have purchased the software on completing a workshop (Olsen & Eddy, 2010). This approach gave teachers the vision of how GIS can be used and the confidence to begin with simple lessons in the classroom. In 2006, Anne Olsen, Stephanie Eddy, Nick Page, and Peter Arthur, all geography teachers, published a CD called IMAGIS – Interactive Mapping with GIS (Olsen, Eddy, Page, & Arthur, 2006), which comprised 22 New Zealand curriculum–based GIS lessons.

20.3 Case Study: Botany Downs Secondary College

Botany Downs Secondary College (BDSC) is located in Manukau City, Auckland. Auckland is the most populous city in New Zealand with a population of 1.3 million. Botany Downs Secondary College is a relatively new school opening in 2004 with 300 Year 9 students. It is a co-educational state school with a current enrollment in 2010 of 1,820 students from Years 9–13 (ages 13–18). An important facet of BDSC's philosophy is to have an ICT-rich school with twenty-first century technology incorporated into the learning process.

GIS was introduced to all Year 9 students in the school's foundation year. Stephanie Eddy, the Head of Social Sciences, integrated GIS into Social Studies, a compulsory subject for all Year 9 students. This was successful as all Social Studies teachers were aware of Stephanie's passion in GIS and her aim to integrate GIS into the Social Studies program. They were willing to learn the necessary GIS skills and BDSC had a plentiful supply of computers for student use and a robust computer network.

Within a unit titled "Asian Tsunami," students learned what processes produce a tsunami, historical tsunami patterns, and details about the Asian Tsunami of 2004. The integration of GIS within this unit was predominately through the teacher displaying the GIS layers and maps using a data projector. All students in Year 9 displayed a very good understanding of both the global pattern of where earthquakes and tsunamis are located and the processes that produce them.

Year 9 students gained hands-on experience with GIS in a unit titled "Access to Resources." Students worked individually through the lesson "ArcView: The Basics" in Mapping Our World (Malone et al., 2002). This was completed during class time. Some students chose to work in pairs and this worked particularly well when a student did not have English as their first language. The students then learned how to create and save their own basic GIS projects. Using a data projector, individual classroom teachers demonstrated how to execute the necessary GIS skills.

In an assessment, students chose one Less Economically Developed Country and one More Economically Developed Country. They presented information about each country's development level and made some basic comparisons. The students used GIS to create three maps; one world map locating each case study, and a map for each case study locating both cultural and natural features. This approach to mapping resulted in nearly 100% student completion in the mapping task of this assessment. It was also interesting that some of the least able students completed only the mapping task.

In the final term of the year, students worked together in small groups to create their own Pacific Island. They used ArcView 3.2 to create maps of their island and used text, skills, and visuals to describe their island. Their presentation comprised of a minimum of six maps. The compulsory map was to show the location of their island in the Pacific. The other five maps needed to show important geographic features of their island, both cultural and physical. For example, the relief of the island (most groups also created a 3-D model of the relief), vegetation types, population, transportation networks, and any natural hazards their island may experience.

The Pacific Island projects produced by the majority of the students were outstanding and demonstrated a high level of understanding about Pacific Island geography. High order thinking skills were also evident. For example, students demonstrated an understanding of how contour lines show the geomorphology of the land and how human settlement will be greater where the land is flat and the soils fertile.

Spatial concepts were introduced into the Gifted and Talented Program that ran at BDSC during 2004–2005. This program allowed teachers to work for one day with gifted and talented children, completing activities that would extend the students' learning. Social Studies students worked in teams on their choice of a Community Atlas Project. The Community Atlas is a project in which students research an aspect of their community. The characteristics and patterns of the community are displayed primarily through maps, supported by written descriptions and other visuals. The maps are produced by the students using GIS software.

This aligned with a Community Atlas Competition operating at the time, sponsored by Eagle Technology (New Zealand's Esri Distributor). One day was only enough time to plan the project, so students worked on their projects after school. In 2006, the Gifted and Talented Program ceased and was replaced with an overall extension class. In its first year, this class in Social Studies worked for two terms in teams on a Community Atlas Project. This provided opportunity for the students to learn many new skills, for example how to use a global positioning systems (GPS), aerial photograph interpretation, census data interpretation, team work, and a range of GIS skills.

The Community Atlas Projects produced were outstanding. The projects were on a range of local issues; for example Howick (the local community) Past and

Present, Graffiti at Lloyd Elsmore Park, Rubbish Location in BDSC, Dannemora—A Suburban Region of Change, BDSC Demographics, and Shopping Patterns at Botany Town Centre. All projects were presented as a Web page. In 2004 and 2006, students from BDSC won the national Community Atlas Project Competition. The winning team and their teacher were rewarded with a flight to Wellington to present their project at the New Zealand Annual Esri Users Group Conference.

In Year 11, geography students are introduced to the theory of plate tectonics by working through "The Earth Moves: A Global Perspective" from Mapping Our World (Malone et al., 2002). Students then study in more detail the earthquake hazard in New Zealand. From IMAGIS, the lesson titled "New Zealand Earthquakes" is completed during class time. Students demonstrated a good understanding of the theory of plate tectonics and where and why earthquakes occur in New Zealand.

From 2008 onward, teaching and learning with GIS in Years 9 and 10 became optional. This decision was made by Alison Derbyshire, the current Head of Social Sciences out of necessity. The large number of teachers (14) teaching junior Social Studies, many of whom were not specialist social scientists, made teacher training more challenging, and the growth of the schools (to 1,800) meant that there was no guaranteed access to computers. Having made this decision, the numbers of students choosing to use GIS in their Access to Resources assignment is approximately 75%. This is a healthy number considering the real barriers they had to overcome. In the Pacific Islands project the number is about 33%, but of those 33% choosing to use GIS, the work they are producing is exceptional.

In 2007, BDSC piloted a project titled "What's Up in Our Streams?", which looked at how we manage our water resources for social and environmental sustainability in the Auckland region. The project was funded by the Auckland Regional Council. This project used GIS maps and aerial photography to explore and understand the complexity of the drainage catchment. The school was fortunate to have the authors of this project (Tony Batistich and Sarah Wakeford) provide support and guidance in the first year. In 2008, the unit continued to be taught to a similar number of students of a range of abilities. In 2009, only one class worked on this project due to Auckland Regional Council funding being no longer available. This funding was needed to produce the students' workbooks.

In 2009, Sally Brodie, head of geography, taught an optional course called "Investigative Geography." The class comprised of a total of 25 students. Students worked in teams learning a range of practical geography skills with ArcGIS being a predominant component. In the first term, the students learned various GPS and GIS skills including Google Earth. In teams they created a project displaying an inventory of the trees on their school campus. The students really enjoyed this process. The inspiration for this project came from the "1, 2, Tree" in Community Geography: GIS in Action (English & Feaster, 2002). In another exercise, students created a global-scale map showing the location of deserts using ArcMap. The students learned important cartography skills. Sally stated that a word which was used frequently by the students in her course was "fun." For 2010, this course has 41 students enrolled, which is evidence of the course's success.

In senior geography there has been less use of GIS due to time and resource commitments to NCEA. In Year 12, all students worked through three lessons from Mapping Our World: (Malone et al., 2005). "The Basics of ArcMap," "The Earth Moves: A Global Perspective," and "The Wealth of Nations: A Global Perspective." Year 13 students could choose to use GIS to create their maps in assessments.

20.4 Case Study: Chilton Saint James School

Chilton Saint James School (CSJ) is an independent school for girls. It was established in 1918 and is located in Lower Hutt. The school has a roll of 700 students from preschool to Year 13. The students at the school achieve excellent results in the national examinations with the school results consistently in the top 10% of New Zealand schools. CSJ was one of a number of schools that received a free copy of ArcView GIS software as part of the Eagle Technology GIS in Schools program in 1997. The geography teacher (Penny Hunt) was keen to use the software but getting the software installed and finding the time to develop her skills were problems.

In 1998, the school timetable allowed a semester-based GIS program to be trialed with Year 10 students. The program was based on an inquiry model with students working on projects that used GIS as a problem-solving tool. The teacher of this class (Anne Olsen) and the students learned the required GIS skills together. The following year a small group of students in an after-school activity entered the AURISA GIS in Schools Competition and won the New Zealand and Australian competitions. Their project looked at fire risk in the vegetation around Lower Hutt City.

From these beginnings, the Social Sciences Department has developed a program so that all Year 8 and Year 10 students have some exposure to GIS. GIS is used in a unit of work planning a new recreation facility for the city. Students spend about 10 hours using GIS to map suitable locations for their proposed facility (Fig. 20.2). In Year 10, students complete a spatial analysis of the vineyards in a nearby region calculating the proportion of land used for winegrowing.

Fig. 20.2 Students work collaboratively planning a new facility for their city

Those students who elect to study geography in their senior classes all have the opportunity to complete credits from GIS standards toward their National Certificate of Educational Achievement. Currently at NCEA level one, students analyze tropical cyclones using material that was developed by the SAGUARO project (Hall-Wallace, 2002). This has been modified to meet the requirements of the New Zealand achievement standard. At NCEA level two, students survey other students to find their mode of transport to school. This data is mapped and conclusions are drawn as to the factors that influence the pattern of student travel to school.

Over the years, the issues with using GIS at CSJ's have changed. Initially the main issue was teacher skill and confidence. As a department-wide program was implemented the issues were ensuring all teachers felt confident, had sufficient skill and support, and access to computers. Resources and lessons have not been a problem as the teacher in charge has had the opportunity to gain writing skills in GIS. In New Zealand, we are fortunate that data sets are freely available for school use. The ongoing problem remains access to computers. This is largely due to competing demands for a limited resource.

While the use of GIS has become a part of the regular program, its potential to encourage inquiry-based learning is not fully utilized. This is largely due to teacher skill, a need to focus on the requirements of the standards, and access to computers. In student course evaluations there are usually many positive comments about using GIS; although this is balanced by students who find the program complex and frustrating. A number of former students have commented favorably on the value of being exposed to GIS at secondary school.

BDSC's and CSJ's integration of GIS is not the norm in New Zealand. They are examples that illustrate both the potential and the issues of integrating GIS at secondary school. BDSC has the unique situation of a new school, highly motivated staff, and very good access to computers. The vision and expertise of Stephanie Eddy meant that GIS was successfully integrated into the school curriculum with relative ease. CSJ's success also depends on a committed teacher, staff who are prepared to train and use new approaches, and fairly good access to computers. Both schools face the problem of decreasing computer access as the demand for computers increases in all subject areas and the challenges of meeting the national qualification requirements and some have gone on to professional careers where they use GIS.

20.5 Prospects of Learning with GIS

In a survey completed in 2004 it was found that about 20% of New Zealand schools had GIS software (GISMAPED and Eagle Technology records). Since 2002, five or six schools on average have purchased the software each year with the peak year being 14 schools in 2005. Purchases have declined over the last two years. Current purchases include schools upgrading to ArcView 9 (Olsen & Eddy, 2010). The number of students gaining GIS credits has increased but remains a very small percentage of geography students. Few if any other subjects use GIS, although a few science and history teachers have attended GIS training workshops.

Currently, students in New Zealand can gain credit toward their National Certificate of Educational Achievement by completing GIS unit standards. These include Standard 11087: Use a Geographic Information System for a Specific Task, Standard 11088: Use a Geographic Information System to Derive a Solution to a Specific Task, and Standard 11089: Derive a Solution Using Geographical Information Systems. The number of students completing unit standard credits using GIS is small but has grown (Table 20.1).

Beginning 2011, students will be able to complete Achievement Standards that include the use of GIS. Achievement Standards are credit-based and allow students to have their credits endorsed as merit or excellence. There will be three standards available, one each at level one, level two, and level three. These achievement standards focus on using GIS for spatial analysis to solve a geographic problem. While there was some increased interest in GIS due to the impending implementation of the new standards, at this time, it is difficult to predict how many schools will actually include these achievement standards in their courses.

Technical issues noted in the case study schools will continue to limit the number of schools using GIS. These include limited computer access, insufficient network strength, teachers finding it difficult to manage GIS data files, and problems with school network managers mishandling the software. One solution to the problems of locally installed GIS software is the development of a Web-based GIS portal. This is an exciting development that could remove some of the school based technical difficulties but is not without issues. Claire Thurlow of Eagle Technology (personal communication, 2010) listed the following concerns:

a. Knowledge transfer to teachers and how a Web-based GIS can enhance lessons and how to teach/utilize this resource.
b. Availability of reliable broadband to all schools utilizing Web-based GIS systems.
c. Availability of computers to access the Web.

Table 20.1   The number of students who have sat GIS unit standards over the past six years and their success rates (New Zealand Qualification Authority)

		2009	2008	2007	2006	2005	2004
11087	Numbers attempting	176	128	111	83	76	178
	% achieved	69.9%	68.8%	80.2%	95.2%	97.4%	70.2%
11088	Numbers attempting	141	183	93	88	20	32
	% achieved	69.5%	38.8%	63.8%	52.3%	80%	93.8%
11089	Numbers attempting	36	Nil	Nil	Nil	Nil	Nil
	% achieved	100%	–	–	–	–	–

d. Availability of resources to develop a Web-based GIS in the first place.
e. Ownership of a Web-based GIS.
f. Development of lessons and incorporation/maintenance of these lessons in a Web-based GIS application (in a similar way that this is an issue with a locally installed GIS).
g. Limitations of Web-based GIS technology – currently Web-based GIS technology is not designed to replace all functionality of a locally installed GIS.

A more serious obstacle to the use of GIS in schools is the attitude that GIS is not particularly relevant; that it is not needed in tertiary study of geography or in the workplace. While it is true that a degree in geography can be gained without any GIS courses, it is not true that GIS is irrelevant in the workplace. Teacher attitudes and awareness, along with a pedagogically very broad curriculum, means that GIS is seen merely as a tool rather than a tool that revolutionizes the way spatial data are manipulated, analyzed, and presented.

Few teachers are aware of the career opportunities in GIS and it must be said that this is generally true within the workplace. GIS in New Zealand has been what a lot of people have "fallen into" because they have seen it somewhere and been fascinated enough to pursue it; usually after a person started paid employment. Employees have many times indicated to Eagle Technology that university graduates have limited knowledge of GIS and need to be trained from scratch (Thurlow, 2010, personal communication, *GIS Training Manager, Eagle Technology*).

The advent of Web-based mapping systems such as Google Earth has brought GIS to the forefront of many people's minds, and they are pursuing it earlier and earlier in schools and tertiary institutes. Unfortunately, especially in tertiary institutes, only now are people seeing the need to have a program dedicated to GIS – hence the cooperation of many New Zealand universities in the GIS masters program. Yet, until there is some structure to GIS education in New Zealand, students are coming out of schools and universities with varying degrees of knowledge and scattered understandings of a variety of functions within a GIS (Thurlow, 2010, personal communication. *GIS Training Manager, Eagle Technology*). The future of GIS in New Zealand secondary schools looks promising with the advent of Achievement Standards in 2011. However, the development of software specific to the needs of schools and easy access to data and lessons would assist the growth of GIS technologies and make it easier for more teachers to integrate GIS into their teaching program.

References

Brown, M. (1994). Presentation material NZFSS conference, Hamilton. Unpublished.
English, K. Z., & Feaster, L. S. (2002). *Community geography: GIS in action*. Redlands, CA: Esri.
Hall-Wallace, M. K. (2002). *Exploring tropical cyclones*. Brooke Coles-Thomason Learning.
Mahoney, P. (2005). *National certificate of educational achievement executive summary*. New Zealand Parliamentary library.

Malone, L., Palmer, A., & Voigt, C. (2002). *Mapping our world: GIS lessons for educators, ArcView edition*. Redlands, CA: Esri.

Malone, L., Palmer, A., & Voigt, C. (2005). *Mapping our world: GIS lessons for educators, ArcGIS desktop edition*. Redlands, CA: Esri.

Ministry of Education. (2006). ICT Professional Development Clusters – schools. Accessed June 2010, www.tki.org.nz.

Ministry of Education. (2007). *The New Zealand curriculum*. Wellington, New Zealand: Learning Media Limited.

Ministry of Education. (2008). *Supporting professional development for teachers*. Performance Audit Report.

Olsen, A., & Eddy, S. (2010). *GISMAPED company records*.

Olsen, A., Eddy, S., Page, N., & Arthur, P. (2006). IMAGIS: Interactive mapping with GIS. (CD), Wellington.

Chapter 21
Norway: National Curriculum Mandates and the Promise of Web-Based GIS Applications

Jan Ketil Rød, Svein Andersland, and Arne Frank Knudsen

21.1 Introduction

Norway is a pioneer in establishing GIS in lower education because it made GIS a compulsory part of geography and geoscience. Norway's schools have met challenges related to technology, time, learning, and skills. As Norwegian schools in general are well equipped with computers and wireless Internet, and the teachers have free access to GIS-based Web applications and data, the major technical obstacles for the first generation of integrated use of digital maps in schools have been overcome. This chapter includes a discussion of some of the implications of the 2006 revision of the Norwegian national curriculum. Prosperous case examples of a beta version of a new Web atlas for Norway as well as the more established Google Earth are presented. As very few Norwegian school teachers may possess the necessary competence to create GIS-based teaching material, we believe that the use of Web-based atlases constitute an important first step toward a successful implementation of GIS in upper secondary education. This first step would not need major changing in existing teaching practice but may challenge some teachers to look at their subject through new spatial lenses.

Pupils start school when they are six years old in Norway, and they have ten years of compulsory schooling. Upper secondary education (pupils aged 16–19 years) is not compulsory, but pupils are entitled to three years upper secondary education, which can be extended to five years, if necessary, before the age of 25. Pupils can freely choose schools, but must compete for the best or most popular schools. During the 2009–2010 school year, Norway had 442 upper secondary schools with 189,352 pupils, including both vocational and university-preparatory schools. Almost 62% of the upper secondary schools are university-preparatory schools (272 in total) where about 58% of the pupils study (http://www.udir.no/). Secondary education is almost universal for Norwegian young people; at least this is what the number tells us if we count the number of pupils entering upper secondary

J.K. Rød (✉)
Norwegian University of Science and Technology, Trondheim, Norway
e-mail: jan.rod@svt.ntnu.no

A.J. Milson et al. (eds.), *International Perspectives on Teaching and Learning with GIS in Secondary Schools*, DOI 10.1007/978-94-007-2120-3_21,
© Springer Science+Business Media B.V. 2012

school the same year they finish lower secondary school. From 1993 to 2003, the percentages varied between 95 and 97%, but a large number of pupils, especially from the vocational training schools, give up upper secondary education (Falch, Borge, Lujala, Nyhus, & Strøm, 2010).

Curriculum reforms are the main instrument that the Norwegian Government can use in order to influence teaching in schools. The latest reform in Norway is the Knowledge Promotion Reform 2006 (hereafter referred to as K06). K06 is a national curriculum, and it applies to the curriculum for any subject taught in every upper secondary school in Norway. For a geography course and a geoscience course, it explicitly states that pupils should know how to use digital maps and Geographical Information Systems (GIS). The geography course is taught two hours per week and is compulsory for all students in first Grade. The geoscience is elective for both second and third Grades and is taught either three or five hours per week. All university-preparatory upper secondary schools in Norway offer geography subjects and about one third offer geoscience subjects. One of the key documents in the curricula for these subjects is called "the toolbox," where GIS is mentioned as an essential part. The aim is to qualify pupils to use digital tools for collection, measuring, manipulation, and visualization of geoinformation.

Although the curriculum reforms since the 1990s are designed to be governing instruments, it is not easy to understand their content. What is said about aims relating to competence and skills, as well as which tools or methods to use, is so general that the content may be interpreted in several ways. For example, K06 pronounces that pupils should know how to use GIS, but does not provide any guidelines on how this is to be achieved. This is wise in relation to educational freedom: Let the teachers integrate GIS in their own way in their own practice. It is for the geography or geoscience teachers to decide what kind of GIS application they may use, but many of them may feel frustrated because the reform documents lack the information by which teachers can make a decision.

The reason why the curriculum is so general and minimalistic in the way it mentions GIS, we believe, is to avoid claims being made toward the Ministry of Education about economic support in order for the schools to fulfill the reform. However, the integration of GIS into secondary schools need not and should not be very expensive. Fitzpatrick and Maguire (2001) stated that schools can have GIS applications appropriate for classroom needs with only a modest investment. It is unnecessary to purchase a powerful and expensive GIS. The vast majority of tasks that pupils need to perform to accomplish their given assignments, such as investigating natural hazards or mapping personal activity areas, can be handled with a reasonable number of basic operations. Baker (2005), who summarizes the main barriers that school researchers have identified in the United States, mentions computer systems as a barrier since these constitute a variety of hardware and software often incapable of meeting the robust demands of desktop GIS.

Baker (2005) also recognizes that the required commitment in terms of time is beyond the reach of schools to widely adopt GIS into their teaching. However, by using Web-based GIS applications, Baker (2005) claims that teachers may avoid technology and time barriers. Many Norwegian geography teachers have already managed to integrate GIS in their teaching. They have done so without spending much money on computer resources or much time to master the technology because they have adopted GIS-based Web applications. Most, if not all, Norwegian secondary schools have high-speed Internet access, and, often, pupils have their own laptop. There may be a diversity of computer systems used, but they have nevertheless general access to GIS-based Web applications that display digital maps. The integration of GIS into Norwegian classrooms is less impeded by a lack of time than by a lack of support on how to teach geography and geoscience with GIS to fulfill curricular goals. We will return to this particular issue after having discussed what we consider the prospects of using GIS in schools.

21.2 Using GIS in Teaching

To be a geography teacher in an upper secondary school in Norway, one needs to have a bachelor degree with an additional one year of practical pedagogic training. The bachelor degree must include a minimum of one year of geography studies. In Norway, student teachers can take a one-year study course in geography at the three universities with geography departments (Bergen, Oslo, and Trondheim) and at one regional college (Steinkjer). At all of these places, students may study GIS, but it is only at the Norwegian University of Science and Technology (NTNU) in Trondheim that GIS is a compulsory part of the one-year course unit in geography. A majority of Norwegian teachers are therefore unlikely to be prepared for teaching GIS in the secondary schools in Norway. Nevertheless, the majority of teachers in Norway were positive regarding the K06's statement that pupils should be introduced to GIS (Rød, Larsen, & Nilsen, 2010). However, a significant number did not like the inclusion of GIS. Disagreement on whether or not pupils should learn GIS was also found among the reform committee members (Mikkelsen, 2009). Whereas some of the members regarded GIS as a highly advanced tool not fitting naturally into school curricula, others viewed it as facilitating a better spatial understanding of data, which could fit very well within the school curricula. Although Norwegian school reforms generally do not provide guidance on methods or tools to be used in order to achieve competence aims, an exception is made for essential subject-specific methods or tools. Since GIS has been recognized as a specific tool for geography and geoscience, the statement that pupils should learn to use GIS was included in K06.

An advantage of using GIS in teaching recognized by many Norwegian geographers is how GIS emphasizes the synthesizing nature of geography. Geography, as it is taught in Norwegian schools, was called a synthesis subject in the curriculum reform as far back as 1994 (L94, 1994) because it combines natural sciences and social sciences and because it includes elements from several disciplines: 'The view that geography is a synthesizing subject has always been basic to the philosophy of the discipline' (Holt-Jensen, 1981, p. 110). It is presented as such also in school textbooks (see, for instance, Dokken, Johansen, & Øverjordet, 2004).

Likewise, geoscience is a synthesis subject consisting of elements from physical geography, geology, geophysics, and geochemistry. In addition, GIS can be a powerful ally in efforts to pull together elements of other subjects in the school curriculum, such as nature studies, mathematics, and statistics, which are all relevant to geography and geoscience. GIS is an integrative technology that allows bringing together not only data from a variety of sources but also various perspectives in which GIS becomes the common denominator (Sinton, 2009). Just as we have seen GIS facilitate interdisciplinary research, should we not expect GIS to facilitate interdisciplinary teaching? The ongoing work on the Web application *Kart i Skolen* or Map in the School (http://www.kartiskolen.no/) is very promising in this respect. Map in the School is initiated by the Norwegian Mapping Authority and the Norwegian Centre for Science Education and aims to facilitate the use of mapping and GIS applications in schools by making digital data more available and by producing teaching resources.

21.3 Case 1: Web Atlas

Map in the School is a Web atlas for Norway. Digital maps and other data that are displayed using Map in the School are taken directly from the national geographical infrastructure where various providers of data make their data available. As a consequence, the atlas is continuously updated. The providers of data include the Norwegian Mapping Survey, the Norwegian Geological Survey, Statistics Norway, and many others. Schools have free access to these data as a result of the Norwegian Ministry of Education and Research's decision in 2007 to take part in the national geographical infrastructure called Digital Norway (Persson, 2007). A teacher in geoscience may use Map in the School to visualize where various sorts of sill deposition are and superimpose on top historic landslide events. The light, medium, and dark blue parts in Fig. 21.1 show the marine deposits and the small red circles shows historical quick clay slides. Geoscience pupils would see that the quick clay slide events tend to colocate with the areas having marine deposits. A history teacher may want to use the Web atlas to browse for historical maps, protected sites for cultural heritage, or to look at the landscape paintings that are made available through hyperlinks. A biology teacher may use the Web atlas to see where the habitats for various birds are located along the coast, or show which areas hold the wild reindeer population, or where in Norway we have high or low biodiversity. A teacher of Norwegian may let his or her pupils speak a certain sentence in their Norwegian dialect, record it, and insert it as an mpg file at the location where the student lives, or they may listen to other dialects already registered. None of these teachers have to alter much of the content of a lesson by using Map in the School. The teachers and pupils who want to explore this Web atlas may do so without rewriting any syllabus or sacrificing several hours of course preparation time. Along with the Web atlas, prepared lessons are available from the Web and teachers are encouraged to share their own lessons based on this Web atlas. Teachers who want to use the Web atlas will, hopefully, find a lesson suitable or, with minor preparation, adapt existing

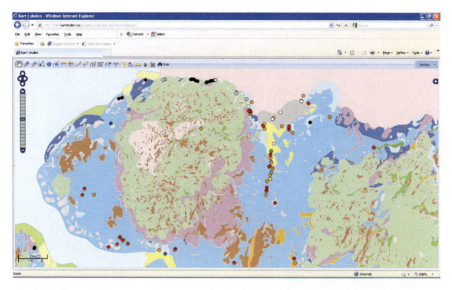

Fig. 21.1 From the Map in the School atlas map server: map showing sill deposition superimposed with historical landslide events

lessons to this Web atlas. Therefore, teachers do not need to change the content of what they usually teach. As a result, pupils will have a visual appreciation on how a simple GIS-based Web application may tell the story through spatial lenses.

21.4 Case 2: Google Earth

Google Earth may affect atlas and globe businesses in a way similar to how Wikipedia has affected encyclopedia salesmen (Lisle, 2006). Google Earth has not been welcomed by the military, some authoritarian governments, traditional atlas and globe businesses. Pupils and teachers, however, salute its arrival. Figure 21.2 shows the use of Google Earth in the classroom in which pupils work on the task of presenting their home. Google Earth is free, and through visualization, it portrays and describes places and allows you to add features to such description if provided in a format known as Keyhole Markup Language (KML). Further, the added information can be shared with others. By making use of updated overlays for mapping disaster zones, this ability has turned out to be very useful for relief operations (Nourbakhsh, 2006). Most GI scientists would not consider Google Earth as a GIS but rather consider it as a Web-based virtual globe with much GIS functionality included (see, for instance, Goodchild, 2008). Educators in Norwegian upper secondary schools, however, would not make this distinction and therefore consider Google Earth as their platform by which they implement the use of GIS in school education. Several geography and geoscience teachers in Norway use Google Earth generally as an electronic atlas or more specifically as a visual platform to present interesting areas. Figure 21.3 shows one pupil's resulting work from the task of

Fig. 21.2 The use of Google Earth in the classroom

Fig. 21.3 Perspective view of Bergen with information such as home, school, nearest bus stop, and so on superimposed

presenting her home. The pupil has drawn a polygon enclosing the place she inhabits daily as well as point symbols representing home, school, bus stop, home for best friend, where she works, and so on. Also linear features can be added, such as the path from home to the nearest bus stop. Using basic operations in Google Earth, pupils can also measure the path distance from home to school and thereby determine whether they have rights to a free bus pass from the local government (Andersland, 2005). Thus, with this exercise, pupils are learning how GIS represents reality by abstracting real phenomena into points, lines, or polygons. Furthermore, properties can be added to the map and an oblique photograph of their home can be added.

21.5 Educational Constraints

GIS is believed to motivate interest in geography and geoscience. GIS makes the phenomena easier to understand, but technical problems often arise when working with GIS. Technical problems are unfortunate as pupils thereby become distracted. From surveys on how Norwegian geography teachers could be best helped in implementing GIS in their teaching (Rød et al., 2010), three elements are emphasized in prioritized order: (1) available ready-to-use teaching material, (2) training in GIS, and (3) a textbook.

What geography teachers should be equipped with, therefore, are pedagogic solutions that lead to a greater appreciation for the themes or subjects that the pupils should learn about without any distraction from the computer technology itself. If students are to be educated about certain issues, it is important that they have the issues in focus and are not confronted with some technical puzzle as to which command to use next. It is therefore important to find ways of emphasizing instruction while making the GIS technique as transparent as possible. Ideally, GIS technology should be so transparent that it could be integrated into the curriculum without an isolated, dedicated place in the schedule. How should we design ready-to-use GIS-based learning packages in which the subject issue is in focus and the technology is a non-distractive element? More research is needed to answer this question properly, but we are convinced that educational packages should be designed with guidance on what should be taught, why, and how.

Unfortunately for Norwegian geography teachers, while the K06 states that pupils should be introduced to GIS (what) and most teachers may understand why, many are left alone with the issue of how. However, the situation may not be too difficult to resolve. At Wilfrid Laurier University (Ontario, Canada), a model curriculum has been developed and evaluated for the integration of GIS into the geography classroom. The model curriculum is based on teachers' recommendations, as revealed from surveys and focus group discussions. Among the recommendations are generic exercises that are not site specific and thus allow teachers to adapt lessons to their locale, that contain detailed step-by-step instructions, and that are sufficiently flexible to fit into a variety of time schedules. Ideally, the exercises should be accomplished within the standard time frame of a geography class (Sharpe & Best, 2001). In Norway, we do not have a similar model curriculum. If teachers had available relevant instructional materials for fulfilling the learning goals of the curriculum and were being supported by GIS, then the integration would be easier. In Kerski's words, 'the approach to GIS should not be, How can we get GIS into the curriculum? but, how can GIS help meet curricular goals?' (Kerski, 2003, p. 135). In our opinion, the status of implementing GIS into the Norwegian K-12 curricula is less impeded by the lack of time to teach GIS than by poor guidance on how to use GIS to fulfill curricular goals (see Meyer, Buttterick, Olkin, & Zack, 1999 for support of this claim regarding the situation in the USA).

Although K06 does not tell teachers how to use GIS in their teaching, K06 presents GIS as a tool for collecting, measuring, manipulation, and visualization of geoinformation. Andersland (2006) stresses the importance of input functionalities. He recognizes that pupils are more motivated when they are able to work with data they have collected themselves. Although a Web-based GIS application seldom has input facilities, Google Earth is a welcoming exception.

21.6 Conclusion

We believe that a successful integration of GIS into upper secondary school curricula will happen through evolution rather than revolution. Norwegian geography teachers seem, to a large extent, to have taken the first step of integrating GIS to their curricula. For upper secondary school curricula in Norway, it is only for geography and geoscience that GIS is mentioned as a basic tool that pupils should learn to use. However, since GIS is used widely within several disciplines at the university level, we may hope that curricula reforms for standard science courses such as biology also see the benefit of introducing the use of GIS for biology already at schools.

The general availability of GIS-based Web applications and their extensive use mean that the time, resource, and technology constraints are far from insurmountable. The main hindrance is related to educational constraints: How can GIS help meet curricular goals? Teachers urgently need teaching resources such as textbooks and ready-to-use teaching materials, but they should also be empowered with GIS skills enabling them to make such textbooks and teaching materials themselves. This step may also require Norwegian teachers to reach a next level regarding GIS skills, which for many will be to use free GIS data viewers or even fully featured GIS.

References

Andersland, S. (2005). GIS i geografiundervisning. In R. Mikkelsen, & P. J. Sætre (Eds). *Geografididaktikk for Klasserommet. En Innføringsbok i Geografiundervisning for Studenter og Lærere* (pp. 213–229). Kristiansand: Høyskolerådet.

Andersland, S. (2006). Eg syns ArcView var eit skamtøft program. Om GIS og geodata i skule og undervisning. *Kart og Plan, 99*(3), 195–200.

Baker, T. R. (2005). Internet-based GIS mapping in support of K-12 education. *Professional Geographer, 57*(1), 44–50.

Dokken, Ø., Johansen, O. I., & Øverjordet, A. H. (2004). *Geografi: Landskap, ressurser, mennesker og miljø*. Oslo: Cappelen.

Falch, T., Borge, L. E., Lujala, P., Nyhus, O. H., & Strøm, B. (2010). *Årsaker til og konsekvenser av manglende fullføring av videregående opplæring. SØF-rapport nr. 03/10*. Accessed September 10, 2010, http://www.sof.ntnu.no/SOF-R%2003_10.pdf

Fitzpatrick, C., & Maguire, D. J. (2001). GIS in schools. Infrastructure, methodology and role. In D. R. Green (Ed.), *GIS: A sourcebook for schools* (pp. 62–72). London: Taylor & Francis.

Goodchild, M. F. (2008). What does Google Earth mean for the social sciences? In M. Dodge, M. McDerby, & M. Turner (Eds.), *Geographic visualization. Concepts, tools and applications* (pp. 11–23). Chichester: Wiley.

Holt-Jensen, A. (1981). *Geography: Its history and concepts: A student's guide*. London: Harper & Row.

Kerski, J. J. (2003). The implementation and effectiveness of geographic information systems technology and methods in secondary education. *Journal of Geography, 102*(3), 128–137.

L94. (1994). *Læreplan for videregående opplæring. Geografi. Felles, allment fag i studieretning for allmenne, økonomiske og administrative fag*. Oslo: Kirke, utdannings- og forskningsdepartementet.

Lisle, R. J. (2006). Google Earth: A new geological resource. *Geology Today, 22*(1), 29–32.

Meyer, J. W., Buttterick, J., Olkin, M., & Zack, G. (1999). GIS in the K-12 curriculum: A cautionary note. *Professional Geographer, 51*(4), 571–578.

Mikkelsen, R. (2009). Geografi i K06 – læreplanprosess, utfordringer og endringer. In O. Fjær & E. Eikli (Eds.), *Geografi og Kunnskapsløftet. Rapport fra Norsk Geografisk Selskaps konferanse i Trondheim; Sted, levemåter og sårbarhet 27–28. mars 2008* (pp. 9–25). Acta Geographica – Trondheim. Serie B, No 16. Trondheim: Department of Geography.

Nourbakhsh, I. (2006). Mapping disaster zones. *Nature, 439*(16), 787–788.

Persson, A. (2007). *Digital Norway. GIM international*. Accessed September 14, 2010, http://www.gim-international.com/issues/articles/id816-Digital_Norway.html

Rød, J. K., Larsen, W., & Nilsen, E. (2010). Learning geography with GIS: Integrating GIS into undergraduate geography curricula. *Norwegian Journal of Geography, 64*(1), 21–34.

Sharpe, B., & Best, A. C. (2001). Teaching with GIS in Ontario's secondary schools. In D. R. Green (Ed.), *GIS: A sourcebook for schools* (pp. 73–86). London: Taylor & Francis.

Sinton, D. S. (2009). Roles for GIS within higher education. *Journal of Geography in Higher Education, 33*(Supplement 1), S7–S16.

Chapter 22
Portugal: Experimental Science Learning, WebGIS, and the ConTIG Project

Madalena Mota

22.1 Introduction

In Portugal, schooling is mandatory and universal until Grade 9 (or 15 years old), although it is soon expected to be mandatory until Grade 12 (or 18 years old). The Portuguese National Education Ministry (www.min-edu.pt) defines the national curricula. Geography is a mandatory subject from Grades 7–9 and optional for humanities and economics course areas in Grades 10–12. Portugal has different courses during middle school (Grades 7–9) and high school (Grades 10–12): the regular courses with the national curricula and the professional and career-oriented courses (e.g., informatics technician, arts-oriented courses, and cartography). These last courses were developed for students with high drop-out potential and low school skills. The curricula for these courses are defined by the National Agency for Qualification (under the Education Ministry and also the Job and Social Solidarity Ministry). To teach at a public school, teachers must have a university degree with a specialization in education, which can be integrated in the university course or completed after a regular course in arts or science. This is changing with the Bologna process (http://www.ond.vlaanderen.be/hogeronderwijs/bologna/), which rearranged university courses in Europe in order to establish the European area of higher education and to promote the European system of higher education world-wide. New teachers in Portugal are now expected to have a master's degree to enroll in a teaching job.

Throughout Europe, educational reforms aim at experimental learning supported by communication and information technologies. In spatial analysis and geography, this should require some use of GIS in the classroom. One can learn geography by locating places on the map, understanding why things are where they are, and identifying patterns and relationships between different phenomena. One can also get kids to think about how planning problems can be solved.

M. Mota (✉)
Pinhal Novo Secondary School, Pinhal Novo, Portugal
e-mail: madalenamota@gmail.com

A.J. Milson et al. (eds.), *International Perspectives on Teaching and Learning with GIS in Secondary Schools*, DOI 10.1007/978-94-007-2120-3_22,
© Springer Science+Business Media B.V. 2012

In the European Union (EU), there is a recognized need for developing and improving young people's skills in spatial analysis and in general vocational training. It is also a goal of the European Commission on Education and Training to follow a cooperation policy with EU members, which means that they develop lifelong learning policies with the aim of enabling countries to work together and to learn from each other. There is a Lifelong Learning Program that aims to enable individuals at all stages of their lives to pursue learning opportunities across Europe. "Erasmus," "Comenius," "Leonardo," and "Grundtvig" programs are examples of this policy.

In Portugal, there is not a lot of experience in using GIS in middle and high schools. There are some courses that will provide a professional diploma in GIS along with the Grade 12 diploma, but very few schools have these programs due to GIS training not being extended to the majority of teachers (at least older ones). Also, few students are aware of these courses or even know about GIS and its potential.

In 1999, the GEOLAB project, funded by the *Ciência Viva* Program (from the Portuguese National Agency for Scientific and Technological Culture), was a first official attempt to get GIS into the classroom. The aim was to build a network of schools and laboratories to interest students in GIS. There was some teacher training but it was clearly not enough because it lacked continuity. In addition, it is possible that other programs may be in practice in other schools throughout the country, however, there is no record or information about them. The ConTIG team had some contacts with some teachers, sporadically, who were interested in trying to use GIS in their classrooms, but we never got feedback nor have we seen any publications or presentations about them except during the trainings for teachers.

22.2 Case Study

Aiming to promote experimental teaching methods in middle and high schools using Geographic Information Technology (GIT) tools, the Portuguese *Instituto Superior de Estatística e Gestão de Informação*, New University of Lisbon (ISEGI-UNL), developed a project that, in part, was originated from the idea presented in a Master's thesis (Mota, 2005) and was initially financed by the Portuguese National Agency for Scientific and Technological Culture – *Ciência Viva* VI program. Among the goals of the project, are the following:

- Develop spatial analysis skills among students,
- Develop research and group work skills,
- Develop critical sense about spatial planning,
- Alert students to the importance of geographic information in daily life,
- Create maps for various purposes,
- Share the learning experiences and materials, and present and discuss the project in seminars and in teachers meetings.

The ConTIG project also aims to demonstrate the potential of GIT to better organize and manage the curricula, in accordance with the guiding principles of Portuguese standards and national curricula. Among other things, it states that a diversity of methods, activities, and strategies must be used in education, especially using information and communication technologies, to promote and develop skills for lifelong learning. The project title, "ConTIG," has a double meaning in Portuguese. It can stand for "*Com Tecnologias de Informação Geográfica*," or "with GIS," but also "*contigo*," which means "with you."

To achieve the project goals, the ConTIG team produced materials and guiding tutorials (for teachers and students), created a WebGIS tool to support some of the learning activities, and got students to experiment with it. Results were disseminated on the ConTIG Web page, as well as at conferences and meetings, including the 2008 Esri Education Users Conference in San Diego, California, USA (http://gis. esri.com/library/userconf/proc08/papers/papers/pap_1963.pdf). Financing of the project was completed in June 2008 with the end of the *Ciência Viva* VI program. However, materials are available online (http://ubu.isegi.unl.pt/labnt-projects/ contig), in Portuguese, and are being used by other schools.

Learning experiences were designed to lead students to develop different skills through experimental learning using GIT. Because both natural and human phenomena occur in a common element (space), GIS is the tool that can emphasize the process of research leading students to a better understanding of the spatial dimension in different disciplines. Teachers prepared materials for their classes, using different scales and promoting the acquisition of different spatial analysis skills. To define a learning experience or lesson plan – which consists of a package containing written guide for teachers and students, often with geographical data – a set of criteria must be established, such as:

- Exploring issues that are significant to the curricula, but also to local reality.
- Working at different scales (local, regional, European, and worldwide).
- Emphasizing the importance of geographic information.
- Using different technologies to collect, process, view, and analyze spatial information.
- Including tasks that promote multidisciplinary activities.

The definition of criteria on a form in the ConTIG's website is the first step to create a learning experience:

- Subjects involved;
- Curricula contents;
- Scale;
- Necessary equipment and resources;
- Expected results (a report, a map, a poster, a flyer, etc).

After a learning experience is created, it can be changed, updated, and edited online. It is also possible to upload materials such as tutorials, guidelines for teachers and students, images, geographic data, and results. The WebGIS tool allows students to undertake data editing and visualization in order to analyze and process collected or produced information. With everything online, the work, data, ideas, maps, and results can be shared.

During the 2007–2008 school year, ConTIG carried out activities with students and teachers at four schools: (1) *Escola Secundária Maria Amália Vaz de Carvalho* in Lisbon, (2) *Escola Secundária de Pinhal Novo* in Pinhal Novo, (3) *Escola Profissional de Ciências Cartográficas* in Lisbon, and (4) *Escola Secundária de Palmela* in Palmela. The school year started mainly with meetings and workshops to outline the activities. The first activity held with students consisted of a GIS Day in November 2007 in the ISEGI campus with students from two secondary schools. The students experimented with ArcPad (some basic functions such as visualization and adding locations on maps) and GPS for the first time, and had a great time with the activity.

During the school year, teachers developed classroom activities, using ArcGIS technologies and published them (as well as the results) on the ConTIG Web page. There were also field trips and outdoor activities. One of the field trips (to Arrábida Natural Park) involved two schools, several subjects, and different levels of students. It covered analysis at different scales and focused on a variety of skills (such as data acquisition, visualization, cartographic representation, analysis tools, among others). The activity was organized by geography, biology, geology, philosophy, and computer teachers from different levels (Grade 11, Grade 9, and also professional courses).

Learning with GIS through the ConTIG project can mean solving a simple assignment by searching the Web and some WebGIS sites (for beginners) or it can be a field trip where students use ArcPad and GPS to locate points of interest (Fig. 22.1). Data that they create in the field will later be managed along with other data in the classroom using ArcGIS.

Fig. 22.1 Tenth Graders learning to use ArcPad during a field trip

22.3 Prospects

During 2008–2009, the ConTIG project continued with more schools, teachers and students involved. The school year ended with a student's meeting held at the university (ISEGI-UNL). Middle and high school students were able to present their work to each other and to an audience with university teachers and representatives of official entities that supported the ConTIG project and the meeting itself (like the Portuguese Geographical Institute – www.igeo.pt, the Portuguese Army Cartographic Services – www.igeoe.pt, the Portuguese Geographers Association – www.apgeo.pt, the Portuguese Youth Institute – www.juventude.gov.pt/portal/ipj, the Ciência Viva program – www.cienciaviva.pt, Esri-Portugal – www.esriportugal.pt, among others) (Fig. 22.2).

The meeting was a success and three prizes were given to the best presentations. The auditorium was full of teenagers, high school teachers, university professors, and representatives of official entities. The teenaged speakers did a great job of presenting their work and capturing the interest of the audience. The youngest students (Grade 8) presented a science project about bird watching. The audience was amazed to see them tell their story about how they built and put on the school trees wooden houses for the birds, how they georeferenced them with ArcPad, and how they created and analyzed the maps. The older students (Grade 12) presented different projects: there was one about analyzing demographic, economic, and social data from the UN database and others about public participation. They built maps according to the results of inquiries about the environment (e.g., the location of eco-spots in Portugal (ecoponto) where one separates garbage for recycling, the traveling habits to and from work, or the location of the new Lisbon Airport and the new bridge over the Tagus river). The presentations were very different in their themes, showing

Fig. 22.2 The ConTIG seminar in June 2009

how versatile working with GIS can be, but they all had in common the feeling that the students had fun learning and were passionate about showing others their work. One could understand by listening to the students that they learned about science, geography, and urban planning while having fun with GIS.

Given the positive results and the potential of ConTIG, ISEGI continues to support this project. The ConTIG website (http://ubu.isegi.unl.pt/labnt-projects/contig/index.php) is a database of ideas, activities, and geographical information. It is expected that the teachers involved in creating learning experiences will continue to use them and adopt these teaching methods in their classrooms. It is also expected that other teachers and schools throughout the country will adopt some of these methods.

The success of the project and the methods of learning are yet to be assessed, but it can be seen that students who were involved in it are eager to repeat the experience. Unfortunately, not many teachers and schools are open to change the traditional way of teaching and it has been a struggle to get them to try the technology in the classroom. The ConTIG project is also committed to promoting training for teachers, but sometimes there are not many opportunities to continue to use the technology, and quite often teachers give up.

The goal of the ConTIG project is to use GIS as a tool for regular middle school and high school courses ("teaching with GIS" instead of just "teaching GIS"). In 2002, I was one of the teachers who received training and then pursued additional training on my own. From my personal interest in this field, the master's course in science and GIS (at ISEGI-UNL) was available, and with it came the ConTIG program. So, in short, challenges are huge and advances are still very small. The ConTIG project is in its third year, with the support of ISEGI-UNL and several schools with a few very interested teachers willing to go the extra mile in getting kids interested in GIT. The biggest challenge is having enough time to prepare materials and improve teachers training so they can be at ease with the technology and use it in the classroom. Financial support for ConTIG in Portugal was always sporadic. The project has gone through and made progress thanks to the effort of a few teachers who still believe in the advantages that students can gain from these techniques and methodologies. However, greater efforts should be made by schools officials in order to integrate the project in their educational objectives and goals.

Reference

Mota, M. (2005). *Concepção de curricula em análise espacial para o terceiro ciclo do ensino básico*. M.Sc. thesis, 181 pp. Lisbon ISEGI-UNL, Accessed May 15, 2010, http://www.isegi.unl.pt/servicos/documentos/TSIG/TSIG0007.pdf. ISEGI-UNL

Chapter 23
Rwanda: Socioeconomic Transformation to a Knowledge-Based Economy Through the Integration of GIS in Secondary Schools

Martina Forster, Theodore Burikoko, and Albert Nsengiyumva

23.1 Introduction

Rwanda's high altitude and its proximity to the equator result in a moderate climate suitable for diverse cultivation. Traditionally, more than 80% of the population relies on subsistence agriculture and farming. However, Rwanda is striving for a socioeconomic transformation to a knowledge-based society that excels in science and technology with a particular focus on the use and adoption of Information and Communication Technology (ICT) at all levels of the society and specifically in the education sector. Theodore Burikoko, former teacher of Construction at ETO Gitarama secondary school, was trained in ICT and GIS through governmental programs and instructs pupils to explore the world through GIS. After we introduce the principles of Rwanda's education system in face of its recovery from the war in 1994, Theodore tells how he experiences the socioeconomic transformation to a knowledge-based economy through integration of novel technologies in secondary school. As he talks about his educational work with GIS, he also reveals how GIS started in schools in Rwanda. The chapter closes with an outlook for GIS in Rwanda's secondary schools in the future.

23.2 Education Principles Before and After the War

Before the genocide in 1994, the education system discriminated pupils at school entry according to ethnic and regional criteria (Obura, 2003). Good performance did not count much and the situation in schools was probably stimulating the destructive movements in society. As a major concern after the war, the government reopened schools and focused on the reestablishment of universities. Obura (2003) describes in her report for the International Institute for Educational Planning that

M. Forster (✉)
Esri Deutschland, Kranzberg, Germany
e-mail: m.forster@esri.de

A.J. Milson et al. (eds.), *International Perspectives on Teaching and Learning with GIS in Secondary Schools*, DOI 10.1007/978-94-007-2120-3_23,
© Springer Science+Business Media B.V. 2012

"the main goal of the post-war education policy has been to promote national unity and reconciliation, prioritizing equity and provision and access, and encouraging a humanitarian culture of inclusion and mutual respect." During rebuilding of the education system, the Ministry of Education had to deal with critical bottlenecks, such as a low number of trained teachers and missing teaching and learners' materials. Due to the families' low income, pupils interrupt school for work, drop out, or repeat classes. For a country of about 10 million people, of which 50% are aged below 18, solid and efficient basic education is crucial for the future of the society (estimates from: The World Factbook, 2009; UNICEF, 2007).

In the Education Sector Strategic Plan (ESSP) 2006–2010, the Ministry of Education outlined concepts of access for all and equity in education quality (Ministry of Education, 2006). The plan focuses on the expansion of free basic education from six to nine years and on the prioritization of science and technology as key areas in human resource development for Rwanda. The ESSP is "designed to assist in reducing poverty within Rwanda while creating a foundation for our vision of economic development based upon applications of development skills and technology," said former Education Minister Mujawamariya (Ministry of Education, 2006).

23.3 Characteristics of Rwanda's Education System

The schooling system of Rwanda consists of six years of primary school, three years lower secondary school, three years upper secondary school, two to six years tertiary education, and a postgraduate program. It follows a national curriculum with national exams through primary and secondary education. For secondary schools, the National Curriculum Development Center has the governmental mandate to publish curricula for scientific, technical, or vocational options and for both lower and upper secondary education (National Curriculum Development Centre, 2010). However, due to the complete revision of the education system since 1994, the publication of curricula takes time, and schools readily assist in course development and sometimes pioneer through associations such as the Technical and Vocational Schools Association or the Workforce Development Authority. Especially in the field of ICT, schools adopted programs available from other countries while waiting for national guidelines and anticipating teachers' guides and learners' books that can be used in computer classes.

In 2002, about 1.5 million children enrolled in primary school and a tenth of that, 157,210 pupils to be exact, enrolled in secondary education. The same year, 7,224 students started tertiary education (Ministry of Education, 2002a, 2002b). If the ratio between enrollment and examination candidates did not change much over time, it could be approximated that about 1/8 of initially enrolled secondary students complete school with an examination. In 1995/1996, the number of examination candidates in secondary schools was 8,906 of the 73,767 pupils enrolled (Obura, 2003). Unfortunately, no recent statistics about examination candidates is known to the authors.

According to the above numbers, gender equity in primary and secondary education seems to be achieved, but women are still underrepresented at institutions of public higher education (Mukama & Andersson, 2008). The class compositions at secondary level usually show a wide range of ages. Secondary students mix from across the country and stay in boarding schools. Good performance in primary or lower secondary respectively relates to open choices of school and options.

The war resulted in the loss of many teachers, and, after 1994, only about a third of all secondary teachers were qualified (Obura, 2003). Teacher development is driven by the Kigali Institute of Education that educates secondary teachers at degree level (A0), and coordinates teacher training centres, offering A1 diplomas, and colleges of education (A2). Professional training for in-service teachers is organized by the Teacher Service Commission and includes short and intensive courses in school management, English, science, ICT, and GIS. At the same time, teacher motivation is crucial for high-quality education and is thus considered in the revised ESSP 2008–2012 (Ministry of Education, 2008).

23.4 Catch a Glimpse of a Teacher's Mind

Theodore tells his story about the use of GIS at *Ecole Technique Officiel* (ETO) Gitarama:

ETO Gitarama is the foremost technical secondary school located in Rwanda's Southern Province, in the Butansinda Cell of Kigoma Sector, Nyanza District. The school has enough space with 49 rooms for students' accommodation, with 12 students per room. 14 classrooms, a library, a computer lab, 7 workshop rooms, an office and staff building, teachers' and staff hostels, a refectory and a play ground area. The school counts 450 students – 400 boys and 50 girls – and about 20 teachers. The school offers five options of upper secondary such as Construction, Public works, Electrical work, Electronics, and Automobile mechanics.

The students study 8 hours a day, of which 40% are theory and 60% practical work. This school gives the certificate of A2 level in all those options, providing a solid knowledge to meet the basic needs of Rwandan society. As a preliminary requirement, the students of the school pass the national examination of lower secondary level (9th school year) with good results. They come from different Rwandan backgrounds, but for most of them their families are farmers with a simple house of three bedrooms and one living room. However, the students normally live at school during the scholastic period. For better control and to keep them busy with their studies, nobody is allowed to live outside the school area. Only during holidays do many students spend time with their family, but some of them pass their holidays at school: they can do vocational jobs in different projects of the school in order to familiarize with practical work. A small number of students spend their holidays outside of Rwanda in neighboring countries – others never leave Rwanda because of their family's low income.

I studied Construction in my secondary school and finished in 2006. In the end of that year I was employed by ETO Gitarama to be the one in charge of all construction projects and to be assistant lecturer in some courses of construction and public work options. In response to my good results at school, in 2007 I was promoted by my Headmaster to participate in a GIS training for secondary school teachers. This was organized by Martina Forster, who initiated GIS use at secondary schools in her role as supportive staff of the GIS Center of the National University of Rwanda. From then on, I was in charge of the GIS courses at my school and I started teaching GIS to the students. At times I helped in GIS teacher trainings for other secondary schools as well. I was responsible for all GIS and construction projects at the school.

23.4.1 Collecting Data, Query and Create Maps with GIS

The students of construction and public works options at ETO Gitarama learn GIS as a distinct subject by using the Rwandan GIS textbook starting from an introduction to GIS concepts, lessons on basic GIS functions, on the geography of Rwanda and basic knowledge about the Global Positioning System (GPS). Then they do further practices like preparing a nice map and collecting geographic coordinates with basic GPS handhelds in the field. They import GPS data into the GIS to analyze, combine and present them on the map. During teaching we normally use the GIS textbooks and a projector to show the students how the things must be done before they do it themselves. We use GIS in other subjects as well, especially in surveying, where the students learn more about contours, distances, superficies, the area of our school, or other places where the school has a job to do.

In our school, we started mapping the school for the localization of the existing buildings, empty space reserved for proposed buildings, the school's green space like forest and grassland, and to map water supply through pipelines and septic tanks (underground structures). We also use GIS in other school's activities such as in preparing a map showing the locations where the school conducts a project (e.g., when constructing domestic biogas digesters). Or we use it to map the location where our students do their practical training in companies all over the country. We use it everywhere, where we feel responsible to have a map showing us information in "space" for planning or to make wise decisions (see example in Fig. 23.1).

The students are very interested to learn with and about GIS as they understand well the importance of this tool and that it will be used at work after their studies. One of my students, a graduate from ETO Gitarama who lives in Kigali, started earning some money by using GIS knowledge in collection of geographic data with GPS. He produces maps for people showing their land (size, shape, position and the boundary with their neighbor). Today he advises students, who are also learning GIS, to be serious with it because it can help them in the future. Personally, I note that today I reflect more about the development of Rwanda in the future and I believe it stems from my experiences from GIS. GIS is somehow stimulating Rwandan students to think more and it makes work easier and fantastic.

As their teacher, I can tell that the students improve their computer skills during GIS lessons and I believe it has to do with the many functions, buttons, and data. They learn on one hand to manage data – as files and folders on the computer, but also as tables of an application – and on the other hand to prepare maps showing

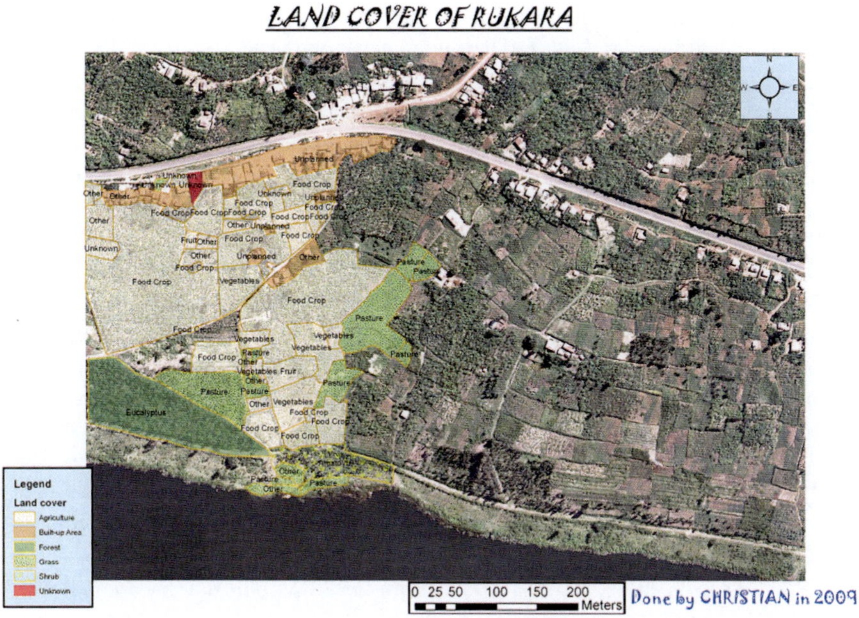

Fig. 23.1 Mapping of land use at Lake Muhazi in eastern province of Rwanda

the data they have. Through this multipurpose-driven work with data, they acquire a deeper understanding of computers. I have noted such improvement by comparing two students, where one learnt GIS and the other didn't. Those two students have quite a different capacity in regards to computer skills.

When I think about why my school decided to adopt GIS in class, it is because through GIS the students grow their knowledge in the field of ICT through a practical and authentic approach and they gain basic knowledge of managing a database. The students can do some work for the school using GIS for planning of future work or future development of the school. GIS also helps the school administration with decision-making. All of these reasons through demonstrations of real examples convinced our school administration to support GIS as an important tool for our students."

Because of the elevated value of a high level diploma for job security, Theodore continues his studies and started a higher learning course in civil engineering at Kicukiro College of Technology in Kigali since the end of 2008. There, he expands his knowledge about the use of GIS in order to see whether he can bring a substantial contribution to Rwanda's urban planning and infrastructure development.

23.5 Promising Setting for GIS in Schools

After reading Theodore's story, one might ask how GIS gained a foothold in secondary education in Rwanda at all. Obviously there was a promising setting after the GIS Centre of the National University of Rwanda won a Special Achievement award at the Esri User Conference 2006, which brought Esri President & CEO Jack Dangermond and his wife Laura to Rwanda later in the year.

During their visit, Rwanda's strong efforts to transform into an IT-literate society and the thousands of children living on those famous Thousands Hills gave Jack and Laura Dangermond the idea to grant GIS software to every secondary school. To ensure that GIS was started in secondary schools, Esri Germany sponsored a GIS specialist and teacher, Martina Forster, to the GIS Centre. Finally, the GIS Centre together with the Ministry of Education 2007 initiated the introduction of GIS at Rwandan secondary schools with the "ArcGIS in Secondary Schools" project.

Since its early days the project aims at capacity building for Rwanda's youth overall. The use of maps and graphics helps students to familiarize with computers; and, at the same time, it fosters spatial reasoning, and students gain a comprehensive understanding of their environment and its challenges in a fast-changing world (Forster & Mutsindashyaka, 2008).

Applying a snowball dispersion model, in 2007, only 10 secondary schools were installed and trained for the use of GIS as a teaching and learning instrument. ETO Gitarama was one of them. The subsequent teachers-train-teachers training approach and the integration of GIS into secondary curricula are key to the successful rollout of GIS to more schools, as seen with the 30 schools that had been trained in 2008.

In parallel to the curricular integration, a GIS textbook was developed on the basis of a Swiss GIS textbook for secondary schools. With support from Esri Germany, the Swiss book was translated into English and French and then localized with data and content relating to Rwanda and Eastern Africa. The textbook has been reviewed by the GIS Centre of the National University of Rwanda and the National Curriculum Development Centre, and has so far been distributed to the 40 trained schools.

23.6 Outlook for GIS in Rwanda's Secondary Schools

The case of ETO Gitarama is an encouraging example for GIS integration at a technical school. However, in scientific or vocational options of upper secondary schools, GIS is used differently, namely as a teaching and learning method to explore and understand curricular content from various disciplines such as geography, history, politics, or biology. Currently, the 40 trained secondary schools are using GIS based on the Rwandan GIS textbook, but in different courses (Fig. 23.2).

The government of Rwanda fosters ICT use at all levels of education and aims at equal chances for young people through access to information and key technologies. GIS is part of this objective and was recently integrated into the National Curriculum for lower secondary education as a chapter of ICT. The integration of GIS as a teaching and learning method in other subjects, such as geography, is foreseen to follow within the next years in the form of learning modules for selected topics to be treated with GIS methodology. At the same time, the support to schools and teachers is

Fig. 23.2 Permanent secretary of the Ministry of Education officially launches the Rwandan GIS textbook and hands it over to a Catholic girls' school near Kigali in September 2009 (courtesy of Claudio Pajarola)

critical for successful distribution to more schools and requires long-term strategies to answer existing bottlenecks. A detailed view of teachers' experiences and challenges for GIS in schools can be found in Forster and Mutsindashyaka (2008).

Theodore sees GIS as a critical component of secondary school in Rwanda, and we close this story with his vision for responsible and well-educated citizens: "Rwanda will have young people who can explore the world, and who can make good decisions according to our situation. So, with GIS in all secondary schools of Rwanda we will be on a good level to help the Rwandan society with proposals for decision makers, and students will develop a good understanding of geography and the situation of our country so that they actively participate and think about future development of our country."

References

Forster, M., & Mutsindashyaka, Th. (2008). *Experiences from Rwandan secondary schools using GIS*. Proceedings of the Esri Education User Conference 2008. Accessed February 1, 2010, http://proceedings.esri.com/library/userconf/educ08/educ/papers/pap_1119.pdf

Ministry of Education of the Republic of Rwanda. (2002a). *EFA plan of action*. Kigali: Ministry of Education.

Ministry of Education of the Republic of Rwanda. (2002b). *Annuaire statistique 2001/2002*. Kigali: Ministry of Education.

Ministry of Education of the Republic of Rwanda. (2006). *Education sector strategic plan 2006–2010 (Draft)*. Kigali: Ministry of Education.

Ministry of Education of the Republic of Rwanda. (2008). *Education sector strategic plan 2008–2012*. Kigali: Ministry of Education.

Ministry of Education of the Republic of Rwanda. (2009). *Official launch of GIS textbook for secondary schools in Rwanda*. Online article. Accessed February 1, 2010, http://www.mineduc.gov.rw/spip.php?article484

Mukama, E., & Andersson, S. B. (2008). Coping with change in ICT-based learning environments: Newly qualified Rwandan teachers' reflections. *Journal of Computer Assisted Learning, 24*, 156–166.

National Curriculum Development Centre. (2010). *Online curricular programs*. Accessed January 29, 2010, http://www.ncdc.gov.rw

Obura, A. (2003). *Never again: Educational reconstruction in Rwanda*. Paris: UNESCO International Institute for Educational Planning.

The World Factbook: Rwanda. (2009). *General information about Rwanda*. Online resource. CIA. Accessed January 28, 2010, https://www.cia.gov/library/publications/the-world-factbook/geos/rw.html

UNICEF Rwanda Statistics. (2007). *Resource document*. Accessed January 28, 2010, http://www.unicef.org/infobycountry/rwanda_statistics.html

Chapter 24
Singapore: The Information Technology Masterplan and the Expansion of GIS for Geography Education

Yan Liu, Geok Chin Ivy Tan, and Xi Xiang

24.1 Education in the Twenty-First Century in Singapore

Before the turn of the century, education in Singapore primarily focused on "quantity," with the target of providing education for the masses and reducing attrition rates in schools. At the turn of the new century, for Singapore to redefine herself to be competitive regionally and globally, education has been identified as one of the keys to the success of the new economic strategy (Goh, 2001). Hence, there is a need to transform education and learning from focusing on "quantity" to providing "quality" education.

In 1996, the Ministry of Education in Singapore commissioned an External Review Team to examine the prevailing school curriculum and make recommendations for educational reform (Ministry of Education, 1998). The report highlighted the urgent need to deal with the over-crowded curriculum. Such an overcrowded curriculum could not provide the time and space to generate a positive learning culture. There was also a heavy emphasis on drilling students to get the right answers for the examinations. Students spent a lot of time in completing workbooks and very little time on creative learning. One of the review team's recommendations was to provide all students with the experience of self-directed learning which can take the form of project work or open-ended assignments. There was also a need to reduce the didactic, whole-class, teacher talk and teaching so as to provide more time for learner-centered activities that involved more interaction between the students and the teacher and with other students in cooperative efforts.

In response, in 1997, the then Prime Minister of Singapore, Mr. Goh Chok Tong, spoke about Singapore's vision of meeting future challenges as encapsulated in the concept of "Thinking Schools, Learning Nation" (Goh, 1997). Following this vision, new initiatives emphasizing Thinking Skills, Information Technology, Project Work, National Education, Teach Less Learn More, and Innovation and Enterprise have been introduced into schools (Ng, 2008, Tan, 2002).

Y. Liu (✉)
University of Queensland, St. Lucia, Queensland, Australia
e-mail: yan.liu@uq.edu.au

A.J. Milson et al. (eds.), *International Perspectives on Teaching and Learning with GIS in Secondary Schools*, DOI 10.1007/978-94-007-2120-3_24, © Springer Science+Business Media B.V. 2012

The Information Technology Masterplan (IT Masterplan) was launched to integrate the use of Information Technology (IT) in the classrooms (Ministry of Education, 1997). Within five years (1997–2002), it was expected to achieve a teacher–notebook ratio of 2:1, a pupil–computer ratio of 2:1, and that IT be used up to 30% of curriculum time (Deng & Gopinathan, 2005). Schools were upgraded with facilities like computer laboratories, media resource libraries, and IT learning resource rooms. In 2002, IT Masterplan II was launched with the key focus on redesigning the curriculum and creating a more student-centered learning environment. Schools were given more autonomy in using funds for IT in the classrooms. The importance of and need for professional development of teachers to help them integrate the use of IT were also recognized in the IT Masterplan II. A more customized approach was adopted to develop teacher competency in using IT for teaching and learning. These approaches included getting teachers to partner with one another in co-planning and co-teaching lessons, guiding teachers in lesson design and classroom delivery, and introducing a platform for replicating the best practices in professional development. Numerous other workshops on using IT to support teaching and learning were organized for teachers and principals (Deng & Gopinathan, 2005).

24.2 History of GIS Use in Secondary Schools

24.2.1 EduGIS

In 1998, in response to the IT Masterplan, the Humanities and Social Studies Education (HSSE) Academic Group of the National Institute of Education (NIE), in collaboration with the Educational Technology Division (ETD) of the Ministry of Education, conducted training workshops on IT tools for the teaching and learning of geography. Geographical Information Systems (GIS) was first introduced to Singapore secondary schools and junior colleges as a teaching and learning tool for geography through those workshops.

The collaborative effort between NIE and ETD resulted in the production of a GIS-based resource package called EduGIS, which essentially contained ready-to-use GIS datasets and GIS-based geography lesson plans and activity worksheets stored on CD format. This GIS learning package was developed based on the use of the commercial GIS software, Esri's ArcView GIS and its free Web-based version ArcExplorer. Joint workshops by NIE and ETD were conducted initially to train teachers on the use of ArcView or ArcExplorer to encourage the use of GIS. After 2001, GIS training was mainly provided by ETD instructors for secondary school and junior college teachers on the use of EduGIS package and other GIS tools for project work. Some schools had also invited GIS software vendors to give talks and conduct GIS sharing sessions at individual schools or in school clusters. In 2002, the Ministry offered every school a subsidy of S$2000 for the purchase of GIS software.

24.2.2 Early Adoption of GIS in Schools

In December 2002, a group of researchers from NIE conducted a study to determine the status of GIS availability and use in schools (Yap, Tan, Zhu, & Wettasinghe, 2008). Two questionnaire surveys and three focus group discussions were used to gather data for the study. The first questionnaire was completed by the Heads of Department (HOD) to assess the availability of GIS software and GIS teaching packages in schools. The second questionnaire was completed by the geography teachers. Some teachers were invited to take part in the focus group discussion.

In the assessment from the first set of questionnaires, more than half (56.2%) of the 89 HODs stated that they did not have any GIS software in the school. A similar proportion of schools (58.4%) did not have any GIS-based resource package. Of those with GIS-based resources, only 3% of the HODs claimed that the schools designed their own GIS resources; the remainder were very dependent on the EduGIS resource package prepared by NIE and ETD. From the schools with GIS software, 33.7% had installed ArcExplorer and 25.8% had purchased ArcView GIS software. Those installing both ArcExplorer and ArcView software were 16.9%. Lack of funds was cited as one of the reasons for not acquiring any GIS software although the Ministry of Education had offered a subsidy of S$2000 to every school for the purchase of GIS software.

A total of 323 geography teachers responded to the second set of questionnaires. Less than half of the respondents had GIS training and most of the training occurred within the last five years of the IT Masterplan. Even with GIS training, only 38 of them (11.8%) had ever conducted GIS-based lessons using EduGIS. The three most important factors that discouraged the teachers from using GIS included insufficient curriculum time, need for extra preparation time, and the lack of suitable instructional packages. The teachers were concerned about the time-consuming nature of GIS-based lesson preparation and implementation. Many of them also felt constrained by the lack of suitable GIS resources. Some felt that the GIS software and resource packages available to them were also difficult to use. Other discouraging factors that surfaced at the focus group discussions included hardware problems, lack of GIS training, and incompetence in use of GIS software. Overall, the integration of GIS in the teaching and learning of geography subjects had been slow in the first decade since it was first introduced into Singapore secondary schools, despite the financial and training support given to schools and teachers by the Ministry of Education.

24.2.3 Spatial Challenge Competition

Most recently, there has been an increasing interest in teachers who wish to expose their students to new methods and skills, including GIS technologies. This increased interest in GIS use in schools can be attributed to a number of factors, including the increased number of trainee teachers who received GIS training from NIE and research on the effectiveness of GIS use in schools (Liu, Bui, Chang, & Lossman,

2010; Liu & Laxman, 2009). In addition, community-based GIS programs such as the SLA Spatial Challenge competitions also play an important role in promoting GIS in schools.

The SLA Spatial Challenge is an initiative of the Singapore Land Authority (SLA) with support from the Ministry of Education (MOE) and other industry partners such as Esri South Asia with the aim of exposing students to GIS technologies and engaging students to seek, process, and apply knowledge infused in their curriculum on real life issues through the use of GIS. First launched in 2008, the Spatial Challenge competitions have received overwhelming responses from the schools. A total of 27 schools and more than 200 students participated in the competitions in 2008 and 2009. All students and some of their teachers received two days of training on GIS use provided by Esri South Asia. SLA and its organizing partners also provide GIS data and attend to technical queries from students throughout the entire duration of the competition. Such community-based initiatives have resulted in popularity of GIS use by teachers and students.

However, despite the various efforts in exposing GIS technologies to students in Singapore, GIS is still not incorporated into the secondary school curricula and taught as explicitly in Singapore. Only in the geography curriculum is GIS able to be relevantly used in some schools. In Singapore schools, GIS is not used in other subject areas such as mathematics and sciences.

24.3 Case Study: Raffles Institution

Raffles Institution (RI) is a prestigious school in Singapore with a long tradition of excellence and scholarship. It offers a unique six-year integrated program called the Raffles Program for exceptionally talented students, culminating in the Singapore-Cambridge GCE "A" Levels Examinations. The rigorous yet stimulating Raffles Program prepares students for the demands of a fast-changing world where they are expected to be responsible risk-takers when leading and serving the community and nation. Therefore, it allows more space in the curriculum for teachers to introduce advanced content and higher order process-and-product work and overarching themes and concepts for students to hone valuable skills they need to surge far ahead in life. One of its unique programs is for their students to learn about and with GIS technology, where their teachers had engaged faculty from NIE to offer customized training courses for both the teachers and students.

More recently, researchers from the Education and Outreach Office of the Earth Observatory of Singapore (EOS) teamed up with NIE to develop a curriculum module on volcano science and volcanic hazards for secondary school students, with the aim to spark students' interest in earth science and to stretch students' minds to think like geologists in four dimensions across timeframes and detectives piecing incomplete data together (Earth Observatory of Singapore, 2010). GIS was introduced in the module to teach students the basic concepts of data collection and processing, analysis, interpretation, and prediction. The volcano education module was piloted at Raffles Institution in October 2009. The geography teacher, Mrs Yak Siew Yang, volunteered to coordinate with the module development team in implementing the module with her students from Secondary 2 and Secondary 3 Grades.

24.3.1 Participants

Twenty-three students from Secondary 2 and five from Secondary 3 volunteered as early adopters to pioneer the volcano education module. These students were at above-average level and already studied earthquake and volcano units prior to signing up for the module.

24.3.2 Learning Design

The activities were focused on the use of GIS technology using Esri's ArcGIS 9 software program to investigate volcano distribution and analyze the extent of impacts that volcano hazards may pose to human societies using authentic data from various sources. In addition to learning and reinforcing the content knowledge on volcano and volcano hazards, students also learned ArcGIS as a mapping tool to develop their spatial thinking and analytical skills. An outline of the learning activities is listed in Table 24.1.

24.3.3 Students' Learning Experiences

The learning experiences of participants in the volcano module using GIS were evaluated through a Likert-scale based questionnaire survey and analyzed statistically. Questions in the survey cover both the perceived usefulness of GIS as well as the perceived ease-of-use of the ArcGIS software program. The perceived usefulness of GIS is reflected through three sub-categories, including the subject knowledge of geography, geographic skills, and motivation, attitude, and self-efficacy. The score students can give to each question ranges from 1 to 5, with 1 being not helpful at all or strongly disagree and 5 being very helpful or strongly agree. Students who participated in the module highlighted many of the instructional strengths of GIS that enabled them to become more competent spatial thinkers and problem solvers as well as self-directed learners. They also pointed out some of the challenges and difficulties they encountered in their interactions within the GIS environment, which educators need to bear in mind in the development and enactment of GIS enabled pedagogy. Table 24.2 provides a summary of the results from the survey.

Most of the students reported that GIS is a useful and promising technology that offered educational potential in enhancing their learning process. Students viewed GIS as an effective technology that could be leveraged by educators in improving and innovating students' learning of core competency-based geographic knowledge and skills. As one student wrote in his reflection report,

I felt that this course was really interesting, allowing us to pin-point the places of all the volcanoes around the world as well as able to zoom into a specific region to analyze it. What struck me most is the ability of the program to calculate the number of people that will be affected by the volcano if it erupts, giving the eruptions expected pattern of outflow. This will really help scientists and politicians to decide the area of evacuation in cases of emergencies.

The majority of students reported that using GIS to investigate a range of relevant volcano topics was helpful. Since the computerized GIS system has enormous digital information capacity that allows users to rapidly access, analyze, and process vast amounts of spatially referenced data, students could analytically explore a myriad of issues and engage in decision-making. This was especially evident in their last activity using GIS to analyze the impact of the volcano hazard at the Mt Merapi area in Indonesia.

Table 24.1 Volcano learning activities using GIS

Time		Outline of GIS activities
First session (9:30–11:30)	5 min	Introduce the module on GIS activities and specify the learning objectives of the module
	35 min	Create a world hazard map • Explore volcano data collected from Smithsonian Institute website (www.volcano.si.edu) using Microsoft Excel; • Get familiar with ArcGIS software program; • Map the volcano data using ArcGIS; • Explore the data through various attribute querying activities
	45 min	Extract volcano data to produce a volcano distribution map in South East Asia • Learn to query volcano data based on geographical locations; • Extract data from large database to generate dataset for SE Asia; • Generate map layout; • Answer questions
	35 min	Explore Volcanic explosivity index (VEI) using GIS • Understand the concept of VEI and the relationship between VEI and volcano eruption scale; • Learn the produce interactive graph/chart to analyze the VEI data using ArcGIS; • Answer questions
Break (15 min)		
Second session (11:45–12:45)	50 min	Map and analyze volcanic hazard using Mt Merapi as a case study • Produce a hazard map at Mt Merapi using data provided; • Produce population and population density maps; • Analyze the data to identify how many people in each of the hazard zones were affected by the volcano eruption; and • Learn to produce charts/graphs to support data analysis and presentation
	10 min	Wrap up the course

Table 24.2   Students' perceived learning experience using GIS

Items	Mean	SD
Geographic knowledge – GIS helped me to		
• Analyze data patterns, linkages, and trends	4.44	0.70
• Examine spatial organization of different places	4.37	0.74
• Investigate different topics of concern	3.89	0.64
• Establish interdisciplinary learning (geography, math, history, etc.)	3.67	0.83
Geographic skill – GIS helped me to		
• Enhance my ability to ask my own questions and seek answers	3.70	0.83
• Use maps to process and analyze geographic information	4.59	0.57
• Explore issues from different angles of analysis	4.07	0.73
• Learn to interpret and examine data better	4.44	0.64
• Make decisions using the most recent information	4.00	0.73
• Learn to facilitate and collaborate with my fellow students	3.96	0.71
Motivation, attitude, and self-efficacy		
• GIS helped me to engage in problem solving	4.11	0.75
• GIS has boosted my motivation for learning geography	4.19	0.79
• Using GIS is an effective method to improving learning	4.15	0.82
• GIS helped me to develop scientific inquiry skill	3.85	0.82
• GIS helped me to be a better learner	4.15	0.82
Perceived ease-of-use of GIS		
• I feel comfortable using GIS	3.93	0.83
• I prefer using GIS in my learning as compared to other methods	4.07	0.87
• I find it easy to use ArcGIS software program	1.69	0.42

Overall, the pedagogical productivity and promise of a GIS instructional orientation to curricular improvement has been evidently well-established in this survey. Students were satisfied with the scaffolding role of GIS as a tool for mapping and spatial analysis and felt motivated in wanting to use it to deepen their applied understanding of theoretical concepts. Students expressed a favorable stance toward embedding GIS in curricular practices to foster a paradigmatic shift from didactic geography classrooms to student-centered active learning ecologies.

However, although students were generally comfortable with the GIS program and associated tools, they reported a very low score of 1.69 on the perceived ease-of-use of ArcGIS. The major difficulty faced by students was to familiarize themselves with the ArcGIS software interface, which is catered to the needs of the professional rather than educational community. As one student commented, "the technical aspect of operating the program was quite difficult and frankly I got quite lost at times. But luckily, the friendly staff came over to help me." Appropriately addressing or mitigating these challenges during the planning and execution phases of a GIS educational module may ensure successful advocacy and implementation of GIS technology as a rigorous curricular model.

24.4 Opportunities and Challenges

The implementation of the IT Masterplan II offers various opportunities for teachers and students to engage IT, including GIS technology in their teaching and learning practice in Singapore's schools. School-based interventions using GIS technologies also demonstrate the pedagogical benefits of GIS use in developing higher order learning skills (Liu et al., 2010; Liu & Laxman, 2009). Currently the Ministry of Education's Curriculum Planning and Development Division is planning to develop a nationwide framework to introduce GIS into the geography curriculum from lower secondary and upper secondary schools to junior colleges. The integration of GIS into the school curriculum will not only motivate students in learning geography but also empower them to become spatial thinkers and problem solvers.

However, many factors exist that discourage the use of GIS in schools. Some of the barriers identified by Kerski (2000) are also shared by Singapore teachers. These include: lack of time for developing GIS lessons in addition to the standard curriculum; little support for training; lack of training geared toward educators; and the perceived complexity of GIS software. The following section identifies some significant challenges that need to be addressed by educators and educational management for effective enforcement of GIS use in schools.

24.4.1 Increasing Teachers' Competency of GIS Knowledge and Skills

One of the biggest challenges of implementing GIS in schools is the initial time investment that teachers require in learning GIS software before they feel comfortable introducing it into their classroom for students to use. Learning a full GIS software program such as ArcGIS can be overwhelming to any novice user because of its vast range of functionalities. For teachers, more time is required learning the software and then developing GIS lessons to enhance a curriculum topic (Baker, 2005; Bower, 2005; Kerski, 2000; O'Dea, 2002).

There is also the need to train and provide ongoing professional development for teachers in the proficient usage of GIS with educational intent. This support can take two different forms: (1) internal support in establishing collaborative networks or communities of practice within individual schools or school clusters to tap the expertise of experienced GIS teachers; and (2) external support comes from training vendors or agencies in disseminating information on technical development of the GIS software or in raising GIS-oriented pedagogical proficiencies.

24.4.2 Developing GIS-Based Curriculum Resources and Datasets for Easy Adoption by Teachers

The effective use of GIS in schools requires support of curriculum-specific instructional materials as well as relevant data and other resources for easy adoption by teachers. This can be achieved through the joint effort of teachers, GIS experts, as well as curriculum specialists. Teachers who received formal GIS training or who have gained experience from implementing GIS in their schools can be identified as leaders in this initiative to develop GIS-related instructional materials that are well-integrated within existing curriculum.

In addition, although many geographical datasets are available for free public access, there is still a general lack of purposely-built geographical data available for educational use, especially at regional and local scales. This suggests the need for a centrally placed geographical data repository in Singapore for teachers and students to share and use in their classes.

24.4.3 Developing GIS Software Program for Educational Use

Another big challenge to the successful implementation of GIS in schools is the availability of appropriate GIS software program for educational use (Liu & Laxman, 2009; Liu & Zhu, 2008). ArcGIS is industry oriented and has widespread usage in many different fields. However, the use of standard ArcGIS program as an instructional tool in schools does come with its costs, that is, the sophisticated interface of the program is not user-friendly for beginners, and many ArcGIS functionalities are not applicable or relevant within the secondary school geography learning context. This suggests the need for a GIS software program purposefully built for educational use.

References

Baker, T. R. (2005). Internet-based GIS mapping in support of K-12 education. *The Professional Geographers, 57*(1), 44–50.

Bower, P. A. (2005). *Using an Internet map server and coastal remote sensing for education.* Unpublished master thesis, Oregon State University.

Deng, Z. Y., & Gopinathan, S. (2005). The information technology masterplan. In J. Tan, & P. T. Ng (Eds.), *Shaping Singapore's future: Thinking schools, learning nation* (pp. 22–40). Singapore: Pearson Prentice Hall.

Earth Observatory of Singapore. (2010). Volcano science moves into the classroom. Accessed August 14, 2010, http://www.earthobservatory.sg/media/news-and-features/50-volcano-science-moves-into-the-classroom.html

Goh, C. T. (1997). Shaping our future: 'Thinking Schools' and a 'Learning Nation'. *Speeches (Singapore), 21*(3), 12–20.

Goh, C. T. (2001). Shaping lives, moulding nation. *Speeches (Singapore), 25*(4), 11–24.

Kerski, J. J. (2000). *The implementation and effectiveness of Geographic Information Systems Technology and methods in secondary education.* Unpublished Ph.D. dissertation, University of Colorado, Boulder.

Liu, S. X., & Zhu, X. (2008). Designing a structured and interactive learning environment based on GIS for secondary geography education. *Journal of Geography, 107*(1), 12–19.

Liu, Y., Bui, E. N., Chang, C.-H., & Lossman, H. (2010). PBL-GIS in secondary geography education: Does it result in higher-order learning outcomes? *Journal of Geography, 109*(4), 150–158.

Liu, Y., & Laxman, K. (2009). GIS-enabled PBL pedagogy: The effects on students' learning in the classroom. *I-manager's Journal on School Educational Technology, 5*(2), 15–27.

Ministry of Education. (1997). *Masterplan for information technology in education: A summary.* Singapore: Ministry of Education.

Ministry of Education. (1998). *Curriculum review report*. Singapore: Ministry of Education.

Ng, P. T. (2008). Teach less, learn more: Seeking curricular and pedagogical innovation. In J. Tan & P. T. Ng (Eds.), *Thinking schools, learning nation: Contemporary issues and challenges* (pp. 61–71). Singapore: Pearson Prentice Hall.

O'Dea, E. K. (2002). *Integrating geographic information systems and community mapping into secondary science education: A Web GIS approach.* Unpublished master's thesis, Oregon State University, Oregon.

Tan, J. (2002). Education in the early 21st century: Challenges and dilemmas. In D. da Cunha (Ed.), *Singapore in the New Millennium* (pp. 154–186). Singapore: Institute of Southeast Asian Studies.

Yap, L. Y., Tan, G. C. I., Zhu, X., & Wettasinghe, M. C. (2008). An assessment of the use of geographical information systems (GIS) in teaching geography in Singapore schools. *Journal of Geography, 107*(2), 52–60.

Chapter 25
South Africa: Teaching Geography with GIS Across Diverse Technological Contexts

Sanet Eksteen, Erika Pretorius, and Gregory Breetzke

25.1 Introduction

Geographic Information Systems was phased into the curriculum of South African schools from 2006 to 2008 as part of the National Curriculum Statement (NCS) for geography for Grades 10–12. Since its introduction, GIS education in secondary schools across the country has been met with a number of challenges and an array of different responses from schools and teachers. This chapter outlines these challenges and responses by describing a number of efforts to implement GIS within three different categories of schools in the country: private, public, and previously disadvantaged. While the pedagogical approach of each school to elucidate the basic concepts of GIS to students varies considerably, the chapter demonstrates how the basic geographic skills and concepts inherent in GIS can be successfully learnt regardless of the approach employed.

25.2 The South African Schooling System

Every year 13 million students attend one of the roughly 27,000 schools in South Africa. Approximately 5% of these schools are private, with 95% classified as government-funded public schools. In addition, roughly 43% of public schools are also classified as 'previously disadvantaged' schools – a term used to indicate schools that were previously disenfranchised under apartheid (School Realities, 2009). School life in South Africa typically spans 13 years or grades, from Grade R (preschool for 5 and 6 year olds) through to Grade 12 or 'matric' – the final year of school. Under the South African Schools Act of 1996, education is compulsory for all South Africans aged 7 (Grade 1) to 15, or the completion of Grade 9. South African students attend a school based on their grade: primary (Grades R–7), secondary (Grades 8–12), or combined (normally Grades R–12), with most secondary

S. Eksteen (✉)
University of Pretoria, Pretoria, South Africa
e-mail: sanet.eksteen@up.ac.za

A.J. Milson et al. (eds.), *International Perspectives on Teaching and Learning with GIS in Secondary Schools*, DOI 10.1007/978-94-007-2120-3_25,
© Springer Science+Business Media B.V. 2012

schools in South Africa offering a matric certificate. After successful completion of Grade 12, students can further their education by enrolling at an array of universities, technical colleges, community colleges, or private colleges in the country. Each school in South Africa is located within a so-called feeder zone, which is used to allocate students to their closest school. Entrance to any particular school is typically given to students within a school's feeder zone, although applications of students residing outside the feeder zone of a given school are also considered, but admission is not guaranteed.

In order to become a teacher in South Africa an individual must obtain an approved teaching qualification from a recognized tertiary institution. A teaching degree usually takes three or four years to complete. Further training and professional development (PD) of teachers in South Africa is governed by the continuing professional training and development (CPTD) system established in 2007 by the South African Department of Education (DoE). The CPTD is a management system for teachers' continuing professional development and seeks to encourage teachers to enhance their professional competence and performance. Teachers can earn professional development points over successive rolling three-year cycles by engaging in endorsed professional development activities such as in-service training and attending curriculum development workshops. According to the National Policy Framework for Teacher Education and Development it is the teachers, together with their professional body – the South African Council for Educators (SACE) – who are responsible for their own self-development and further education (South African Department of Education, 2006).

The DoE is responsible for education in South Africa through the development of national curriculum frameworks. A number of frameworks have been forthcoming postapartheid, including Curriculum 2005 (C2005) followed by the National Curriculum Statement, and, more recently, the Revised National Curriculum Statement (RNCS). Each curriculum statement sets out progressively more complex, deeper, and broader knowledge, skills, and attitudes for students to acquire from grade to grade within each subject area. Each of the nine provinces in the country has its own education departments, which are responsible for implementing the national curriculum frameworks. So while the central government provides a national framework for school policy, administrative responsibility lies within the provinces (School Realities, 2009). Within the schools themselves administrative power is further devolved to the grassroots level via elected school governing bodies, which have a significant say in the running of their schools.

25.3 History of GIS in Schools in South Africa

The rapid growth of GIS in both the public and private sectors of South Africa prompted the inclusion of GIS into the geography curriculum of secondary schools throughout the country. The curriculum for GIS was phased in incrementally and systematically over three years: Grade 10 in 2006, Grade 11 in 2007, and Grade

12 in 2008. At the Grade 10 level, the curriculum stipulates an understanding of general GIS concepts and related geographic concepts such as entity types, scale (large versus small), and resolution (spectral and spatial). At the Grade 11 level, the curriculum incorporates the functional elements of a GIS including data acquisition, satellite remote sensing as a digital data source, preprocessing, and data processing. At the Grade 12 level, which is the final level of schooling, the student is then required to obtain an understanding of the additional functional elements of a GIS, including data management, data manipulation and analysis, and spatial analysis, product generation, and application. At the completion of schooling, the student is expected to be competent in geographic numeracy through "applying GIS procedures and spatial statistics" (South African Department of Education, 2003). Despite being mandated to develop and implement this GIS curriculum into secondary schools throughout the country, the DoE has failed to provide clear guidelines regarding its actual implementation among the diverse range of South African schools. The lack of appropriate curriculum guidelines, teaching resources, and instructional manuals has further hindered its introduction, particularly into resource-poor schools in the country. As a result, private companies such as Esri South Africa, Intergraph, and Naperian Technologies have developed educational material and conducted a number of GIS in-service training courses aimed at teaching curriculum advisors and teachers. Only a limited number of these workshops have taken place, however, and they have been restricted to the training of a handful of curriculum advisors and teachers to assist and guide the implementation of GIS to schools. The educational material developed includes paper-based GIS exercises, as well as computer-based exercises at an affordable price. The South African Department of Land Affairs has also been involved in a number of initiatives to promote map awareness and currently assists the DoE in the introduction of GIS in schools and the promotion of geography as a subject.

25.4 GIS in Schools in South Africa: 3 Case Studies

To illustrate the diversity of implementation strategies employed in South African secondary schools, three teachers involved in the implementation of GIS at a private, public, and previously disadvantaged school were interviewed; their respective success stories are described in the following section.

25.4.1 Case Study 1: Private School

Bridget Fleming is a teacher at St David's Marist Inanda secondary school, which is an English private school located in Sandton, an affluent suburb of Johannesburg. Bridget holds a bachelor's degree in science and has majored in mathematics. The teacher–student ratio in Bridget's geography class is kept relatively low with not more than 25 students per class. There are approximately 110 students in geography per year from Grades 10–12. St David's has more than adequate technological and financial resources (i.e., hardware, software, Internet access), and, as a result, the geography classes are well equipped. In our interview, Bridget discusses the implementation and introduction of GIS in her school with notable enthusiasm. She introduced GIS to her students in the late 1990s when she had personally heard and learned about GIS. Since then she has attended and presented various short courses in GIS. Bridget introduces the theory of GIS to the Grade 8 and 9 students, while her Grade 10 students are then introduced to GIS software as an extension of map work described in the school curriculum. By the time the students reach Grade 12, they have the ability to use GIS in a research project. She teaches the theory and practice of GIS in her classroom by way of interactive lessons that also integrates GIS with other curriculum content. Bridget developed most of these lessons by using South African examples so as to ensure the involvement of the students. As Bridget's privileged students mostly come from technologically advanced backgrounds, she sums up the success of the implementation of GIS by saying: 'I taught them (the students) geography and they taught me GIS!' Since the introduction of GIS, the number of geography students in St. David's Marist Inanda secondary school has been increasing steadily every year, further emphasizing the success with which GIS has been implemented and subsequently taught (Fig. 25.1).

25.4.2 Case Study 2: Public School

Afrikaans Hoër Seunskool is an Afrikaans public school located in Clydesdale, a central neighborhood/suburb of the Greater Metropolitan City of Tshwane. The geography teacher, Billy Brown, has been teaching geography for the past 31 years and is well qualified, with his most recent degree being a social sciences honors

Fig. 25.1 Students of St David's Marist Inanda Secondary School working on a GIS project

degree in geography. The number of geography students per class in Afrikaans Hoër Seunskool is kept relatively small with not more than 25 students in each class. This low teacher–student ratio is unusual for public schools in South Africa and is achieved by using the school's self-generated funds to pay for extra teachers. Billy typically uses a computer to illustrate GIS terminology to the students in his class. The computer lab at the school, while well stocked and operational, is not available for teaching GIS. After Billy's computerized demonstration, the students in his class do hands-on paper-based GIS exercises that have been developed by Billy himself. These paper-based GIS exercises make use of manual techniques to teach the basic concepts of GIS to students. In addition, Billy also takes his students on various GIS-related field trips, including an annual excursion to the Satellite Application Centre (SAC) at Hartebeesthoek outside Johannesburg. Billy also invites a number of guest GIS experts from the GIS industry to give lectures to his Grade 12 students. These lectures are aimed at stimulating the students' interest in GIS in various applied contexts. Students are given the opportunity to interact with the presenter and ask relevant questions afterward. These additional learning experiences have proven to be very important in generating interest in GIS among geography students at the school, but are not part of the formal curriculum. The inclusion of GIS as part of the geography curriculum for schools in 2006 was well received at Afrikaans Hoër Seunskool. Billy has also found that the inclusion of GIS in his lessons, as well as the field trips and lectures given by guest experts working with real projects, have taught his students to think more spatially and analytically. It has also led to more students expressing an interest in studying GIS at a tertiary educational institute. The school administration has also noticed the increased interest in GIS among students and has made money available to purchase resources necessary to teach GIS through computers to students.

25.4.3 Case Study 3: Previously Disadvantaged School

Reginald Jacobs is the principal of Hoërskool Sanctor, a previously disadvantaged Afrikaans school in Port Elizabeth. Reginald has been teaching for 35 years and holds a master's degree in education. The number of students in Reginald's geography classes is high, with an average 40 students in Grade 10, 35 in Grade 11, and 25 in Grade 12. The reason for the decline in class numbers as the grades progress is due mainly to students transferring from geography to tourism, which is perceived as a less challenging subject. Being a previously disadvantaged school, Hoërskool Sanctor has low school fees (roughly R800 per year; US$100), yet not all the parents can afford to pay the amount. As a result, the school has very few resources and no access to computers or the Internet for teaching purposes. The school currently owns only one computer, which is used predominantly for administrative purposes. Internet time is also limited and considered to be too expensive to use extensively. Reginald has refused to bow to his technologically restrictive circumstances and currently teaches GIS in his classroom using a manual method.

Fig. 25.2 Mr. Reginald
Jacobs and his students
gathering to spatial
relationships using paper
overlays

Reginald starts the lesson with a paper map indicating the different provinces in South Africa. He then overlays transparencies indicating the rivers, dams, rainfall, and climate of South Africa. He finally overlays a number of agricultural activities and other socio-demographics (Fig. 25.2). The students must scrutinize and comment on the spatial relationships between the different zones and activities. Reginald has observed that their spatial thinking and literacy have improved remarkably by presenting the GIS lesson in this way. Although the students have never seen a computerized GIS, they have found it easy to understand basic GIS concepts like layers and overlays. Reginald is in the process of obtaining the necessary finances for the construction of a computer lab that he feels will benefit not only the geography students but also the broader school community.

25.5 The GIS Education Road Ahead in South Africa: Challenges and Opportunities

Based on our brief review of GIS education in South Africa, it is clear that the country faces several impediments to the successful integration of GIS within all its secondary schools. The current vagueness with which GIS is described in the school curriculum may be a fundamental contributing factor. The implementation of new curriculum content like GIS in schools in a diverse country like South Africa requires specific instruction regarding the method of implementation and the level of skill that students should obtain. This has been less than forthcoming from the DoE.

The very nature of GIS facilitates the perception that the introduction of the GIS content is reliant on substantial computer resources, retraining of teachers, and the development of new study materials. Subsequently, the lack of money, time, and effective support structures is often perceived as the biggest impediments to the successful implementation of GIS in secondary schools in South Africa. The introduction of hands-on computerized GIS in any schooling system requires considerable financial input in terms of purchasing the necessary software and hardware. The time required for teachers to attend training workshops to learn the necessary GIS software and develop or modify instructional materials supported by GIS, as well as the time required to effectively educate students about the technology, is seen as a further obstacle. The final challenge to the successful infusion of GIS into secondary schools throughout the country is adequate and continuous administrative and technological support. Ideally, three levels of support are required: (1) support from school leadership and the school community; (2) support from local tertiary institutions offering teacher education programs; and (3) support from government and industry. While many secondary schools have received some support at all three levels for its GIS education initiatives, it is increasingly evident that a lot more needs to be done to match those levels of support obtained in more developed world contexts.

The mounting challenges of implementing GIS education in 'technologically restrictive' contexts such as South Africa resulted in a number of initiatives to facilitate its successful introduction including the creation of a paper-based GIS educational package designed specifically for previously disadvantaged schools in South Africa. The educational material contained within the package include a 1:50,000 topographic map, a 1:10,000 orthophotograph, tracing paper, a ruler, colored crayons, adhesive, an exercise book for students, and a handbook for teachers. In terms of content, the exercise book consists of several practical lessons to be completed by each student. The practical lessons guide the student through the processes of defining a geographic question, collecting data using several methods, analysing data, and presenting final results. There has also been a drive from various provincial education departments to provide computers to all schools under their jurisdiction.

25.6 Conclusion

Internationally, GIS has been long regarded as an important part of geography education, both on its own and in association with other subjects such as information technology and environmental studies (Green, 2001). South Africa has recently taken measures to introduce the technology into its school syllabus, with the technology now part of the geography curriculum in Grades 10–12. GIS education in South Africa has proven to be fraught with difficulties arising from a lack of money, time, and support from the broader school community to facilitate the integration of GIS within classrooms, particularly among the country's poorer schools. Despite these, and other challenges, this chapter has shown that while there may be disparities in terms of the amount of resources available, the devotion of geography teachers across the socioeconomic secondary schooling spectrum serves as an example of what can be done at all levels to teach GIS.

References

Green, D. R. (2001). GIS in school education: You don't necessarily need a computer. In D. R. Green (Ed.), *GIS: A sourcebook for schools* (pp. 34–61). New York: Taylor & Francis.

School Realities. (2009). *South Africa.info, the official gateway*. Department of Education. Accessed October 2009, www.southafrica.info

South African Department of Education. (2003). *National curriculum statement grades 10–12 (general) geography*. Pretoria: Government Printer.

South African Department of Education. (2006). *The national policy framework for teacher education and development in South Africa*. Pretoria: Government Printer.

Chapter 26
South Korea: GIS Implementation Profiles Among Secondary Geography Teachers

Minsung Kim and Sang-Il Lee

26.1 Introduction

Recently, there has been dramatic development in Geographic Information Systems (GIS) and increasing applications of GIS to geography education. Researchers have considered GIS as a teaching tool because of its reported benefits in students' learning (e.g., Baker & White, 2003; Bednarz and van der Schee, 2006; Patterson, Reeve, & Page, 2003). Educators in South Korea also have become interested in incorporating GIS in education (e.g., Jung & Kim, 2006; Kim, 2007; Lee, Kim, & Ban, 2008). Since GIS-related content is discussed in geography textbooks and GIS in Korean society is becoming widespread, geography educators accordingly have turned their attention to GIS. However, the incorporation of GIS in education, especially at the secondary level, is at an incipient stage in South Korea. Only a few innovative teachers are beginning to consider the use of GIS in the classroom.

This chapter consists of three parts. First, contexts of education in South Korea are described through three aspects: (1) the position of geography in secondary education and the portion of GIS content in the geography curriculum, (2) geography teacher certification procedures, and (3) research trends associated with GIS in secondary education. Second, the preparedness of South Korean schools to use GIS in their classrooms is explored. Finally, opportunities and challenges in implementing GIS into secondary education are discussed.

26.2 Educational Contexts of South Korea

26.2.1 The Position of Geography in Secondary Education and GIS Content in the Geography Curriculum

The educational system in South Korea consists of pre-school, primary, secondary, and higher education. This section introduces the geography portions taught in

M. Kim (✉)
Texas A&M University, College Station, TX, USA
e-mail: minsungkim@neo.tamu.edu

A.J. Milson et al. (eds.), *International Perspectives on Teaching and Learning with GIS in Secondary Schools*, DOI 10.1007/978-94-007-2120-3_26,
© Springer Science+Business Media B.V. 2012

the secondary education curriculum: middle school (Grades 7–9) and high school (Grades 10–12), and discusses the GIS content in the geography curriculum. For the social studies curriculum, of which geography is a part, South Korea has a centralized compulsory national curriculum until Grade 10, and all textbooks are based on that curriculum. During Grades 11 and 12, students select subjects from among several electives. Even though students choose a few subjects, each elective must follow the nationally imposed curriculum. Before the current curriculum was introduced, social studies consisted of geography, history, and civics. However, in the current curriculum, history is an independent subject, and geography and civics constitute the subjects of social studies. Table 26.1 presents the sequence of history and social studies in the middle school curriculum. Geography is taught in Grades 7 and 9. A variety of geographic content is taught in the middle school, but there is no specific attention to GIS. There is a recommendation to use ICT (Information & Communication Technology), but the middle school curriculum does not explicitly guide teachers to incorporate GIS in the classroom. Table 26.2 provides the curriculum of high school. Students in Grade 10 learn the mandatory subject, social studies, while higher grade level students select from among seven elective subjects. Three geography electives – Korean geography, world geography, and economic geography – are included. In terms of GIS content, the social studies curriculum in Grade

Table 26.1 Sequence of history and social studies in the middle school curriculum (Ministry of Education, Science and Technology, 2008)

Classification	Grade 7	Grade 8	Grade 9
History		Korean History World History	Korean History World History
Social Studies	Social Studies (Geography + Civics)		Social Studies (Geography + Civics)

Table 26.2 Sequence of history and social studies in the high school curriculum (Ministry of Education, Science and Technology, 2009b)

Classification	Grade 10	Grades 11–12
History	History (Korean History + World History)	History of Korean Culture Understanding of World History History of Eastern Asia
Social Studies	Social Studies (Geography + Civics)	Korean Geography World Geography Economic Geography Law and Society Politics Economics Society and Culture

10 does not include GIS. However, most textbooks briefly introduce what GIS is and how GIS can be used. Among the three electives of geography, the world geography curriculum requires teaching GIS. However, close scrutiny of world geography textbooks reveals that the GIS content in them is almost the same as the social studies textbooks even though the world geography curriculum asks for the incorporation of GIS. Therefore, students in South Korea do not learn about GIS in high school in depth. Also, activities or exercises using GIS software are not required.

The description above reflects the currently publicized geography curriculum for South Korea known as the "2007 Revised Curriculum." However, the Ministry of Education, Science and Technology in South Korea (2009a) recently announced that a new curriculum called "2009 Revised Curriculum" will begin to be partially implemented in 2011. A detailed description of the 2009 curriculum goes beyond the scope of this chapter. While the specific content of the curriculum has not been decided yet, it is known that among the three geography electives, economic geography will be dropped. However, we do not expect that the content of Korean geography and world geography will change dramatically.

26.2.2 Procedures to Become a Geography Teacher

The pathways to become a geography teacher in public schools in South Korea can be explained in two steps: obtaining eligibility and passing an exam. First, to be eligible for a teacher certificate exam, in principle, it is necessary to graduate from a department of geography education. In South Korea, the departments of geography education specialize in educating preservice teachers. The graduates of geography education departments are eligible for teacher certification. There is an alternative way to be eligible for the exam. If an undergraduate or a graduate student fulfills requirements by attending courses offered by a department of geography education, those students can also take the teacher certification exam. However, this alternative route is restricted and very competitive. Once a person is eligible for the teacher certification exam, the person can take the exam, which is administered only once a year. It is very difficult to pass the exam because the teaching profession is popular in South Korea and only a limited number are allowed to pass.

26.2.3 Research Trends in GIS Education

Up until now, GIS has not been widely implemented in classrooms in South Korea. Nonetheless, geography educators are becoming more interested in GIS for education. This trend is reflected in the number of published articles in geography education journals. Kim (2007) categorized the research trends into three types: introduction of the status of GIS education, lessons learned from foreign countries' GIS education, and concrete teaching plans using GIS.

First, the status of GIS education in South Korea has been the subject of research. For example, Hwang and Lee (1996) analyzed the GIS component of geography textbooks and reported how geography teachers understood GIS. According to Hwang and Lee, most teachers did not have a concrete idea of what GIS is. Jung (1997) investigated the development of GIS and discussed what aspects of GIS could be applied to geography education. Oh and Seong (2003) postulated that GIS had not been implemented effectively into the classroom even though interest in and demands for GIS education had been increasing.

Second, some researchers have explored GIS education in other countries and sought to learn lessons. Kim (2002) investigated how GIS was incorporated into education in the UK to gain insight into how to effectively incorporate it into secondary education in South Korea. Similarly, GIS implementation in the USA was benchmarked. For instance, Kim (2005) discussed how GIS was used in geography education in the USA as a basis to explore its potential in the South Korean context. Jung (2005) examined GIS curricula for undergraduates in the USA. Based on that exploration, he suggested establishing an intensive GIS course and a certificate program for South Korea.

Third, research on concrete lesson plans using GIS is increasing. Hwang (1998) emphasized that the overlay function in GIS is one of the most appropriate strategies that can be used at the secondary level because the function can show the relationships between and among regions even without complicated procedures. Shin, Jeong, and Joo (2002) devised a GIS courseware that aims to enhance students' visual and spatial learning. Yang (2004) developed a GIS learning module using the Internet and demonstrated the effectiveness of the module. Kwon (2004) provided an example application of GIS in education. Kwon claimed that his study can be a model for geography teachers because the study is based on basic geographic concepts and GIS functions that are understandable to students. Jung and Kim (2006) suggested a lesson plan using GIS. Grounded in teachers' opinions, Jung and Kim exemplified how diverse spatial topics, such as plate tectonics, earthquakes, and the relationship between annual precipitation and agriculture, can be taught using GIS to high school students. Lee, Kim, Lee, and Jo (2006) analyzed geography textbooks and developed learning materials regarding geomorphology. These publications suggest that researchers in South Korea are broadening their interest in GIS education at the secondary level.

26.3 GIS Implementation Profiles

This section investigates how prepared schools are in South Korea to use GIS in the classroom. To determine preparedness, we modified and expanded the GIS implementation model suggested by Audet and Paris (1997). Nine high schools in South Korea were recruited for this study, and their GIS implementation profiles were constructed, including GIS software acquisition, hardware/equipment acquisition, data development, professional development activity, and educational context for GIS education. This exploration of nine schools provides an informative picture of the current status of GIS education at the secondary level in South

Table 26.3 A survey questionnaire for constructing GIS implementation profiles (Modified from Audet & Paris, 1997)

Please select one response for each item below	Comments				
GIS software acquisition					
I have GIS software available if I want to incorporate GIS in my teaching	SD	D	U	A	SA
I have sufficient funds to acquire GIS software	SD	D	U	A	SA
Hardware/equipment acquisition					
I have a sufficient number of computers to be used for GIS education	SD	D	U	A	SA
Computers in my school are adequate (e.g., CPU, RAM, memory) for supporting GIS applications	SD	D	U	A	SA
I have sufficient funds to acquire hardware and equipment for GIS education	SD	D	U	A	SA
Data acquisition					
GIS data is readily available	SD	D	U	A	SA
I know where to access sources of GIS (digital) data	SD	D	U	A	SA
I have sufficient funds to acquire digital data	SD	D	U	A	SA
Professional development activity					
My school and/or district provide staff development support for GIS education	SD	D	U	A	SA
I take advantage of professional development opportunities related to GIS education	SD	D	U	A	SA
I believe that professional development should be an ongoing process	SD	D	U	A	SA
School administration support for GIS in-service education is easy to obtain	SD	D	U	A	SA
I have time to learn GIS-related skills	SD	D	U	A	SA
Educational context development					
I see how GIS can be used to enhance my curriculum	SD	D	U	A	SA
GIS is an educational tool for exploring spatial concepts	SD	D	U	A	SA
I can think of interdisciplinary applications of GIS	SD	D	U	A	SA
GIS can enhance students' problem-solving ability	SD	D	U	A	SA
Project-based activities are a good way for students to apply GIS	SD	D	U	A	SA
I can envision real world applications of GIS technology	SD	D	U	A	SA

Note: SD: Strongly Disagree, D: Disagree, U: Undecided, A: Agree, SA: Strongly Agree

Korea. However, due to its limited sample size, this description is a preliminary analysis.

The survey was conducted using the questionnaire presented in Table 26.3. All geography teachers in the nine schools provided their opinions of the status of GIS in their schools; a synthesis of their opinions was used to construct each school's profile. The number of teachers in each school varied from 1 to 4, and in total, 23 teachers took part in the survey. The participants indicated their opinions via a Likert scale from 1 to 5 (1 – strongly disagree, 2 – disagree, 3 – undecided,

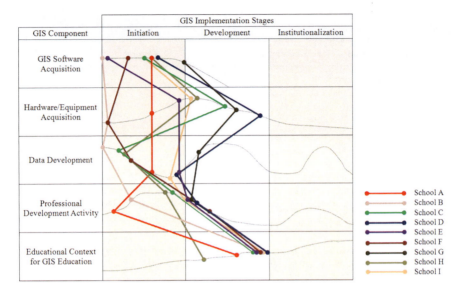

Fig. 26.1 GIS implementation profiles of schools in South Korea (Kim, 2010)

4 – agree, 5 – strongly agree) and mean scores of each school and category were calculated. Based on those mean scores, GIS implementation profiles were constructed
as shown in Fig. 26.1. For the purpose of this study, the initiation stage was defined
as an incipient phase where GIS sources are lacking, while institutionalization sig-
nifies the stage in which GIS sources are commonly shared and GIS practices are
regularly exercised. The development stage is located between the two. Considering
that the Likert scale of 2 is "disagree," 3 is "undecided," and 4 is "agree," the stages
were determined as follows: the range of mean scores of 1.0–2.5 was determined as
"initiation," 2.5–4.0 was the development stage, and more than 4.0 was deemed the
institutionalization stage. A more detailed description of this survey is found in the
article published by one of the current authors (Kim, 2010).

The GIS implementation profiles indicate that schools in South Korea are not
well-equipped to implement GIS in the classroom. Hardware, software, and data
need to be provided for teachers to incorporate GIS into their teaching, even though
there is a relatively wide variation in hardware acquisition. Teachers also do not
believe that they have adequate professional development opportunities. Moreover,
when given opportunities, teachers do not feel that they take advantage of them
effectively. One promising aspect is that the educational context for GIS education
has relatively improved; the mean score of the educational context development
category was the highest. These results suggest that teachers have positive attitudes
about the utility of GIS in educational contexts. In summary, teachers in South Korea
seem to believe in the potential of GIS as an educational tool, but a wide range of
obstacles prevent teachers from incorporating GIS into the classroom. Further study
with wider samples would confirm the results presented here.

26.4 Opportunities and Challenges

Even though GIS has not been widely incorporated into the classrooms of South Korea, we expect that the use of GIS in education can be increased in the future. Most preservice education programs for geography teachers have included GIS-related courses in their curriculum. Telephone interviews were conducted to investigate GIS courses in preservice curriculum (Fig. 26.2). We discovered that, among all eighteen geography education-related departments in South Korea, sixteen departments include GIS courses in their undergraduate curriculum. The two remaining departments offer GIS courses via the form of special lectures or other routes without explicit inclusion of a GIS course in the curriculum. Furthermore, many students actually take GIS courses. In ten out of the sixteen departments that include GIS courses in their curriculum, GIS-related courses are mandatory or most students attend the courses even if they are not mandatory. In five out of the remaining six departments, more than half of the students take the courses, and in one department, one third attend GIS-related courses. In the two departments where GIS-related courses are not explicitly included in the curriculum, whenever there is a special lecture or any other opportunity, most students take advantage of them. Therefore, today, most preservice geography teachers in South Korea are exposed to GIS-related content before they become teachers. Considering that experience with GIS is important in making geography teachers become interested in GIS for their teaching (Bednarz & Audet, 1999; Kim, Bednarz, & Lee, 2009), the current situation is favorable at least at the university level. Moreover, most teachers in South Korea appear to believe in the potential of GIS in geography education (Kim, 2010; Kim et al., 2009). The questionnaire survey of this study also suggests that teachers' awareness of GIS as an educational tool is positive. Even though teachers in South

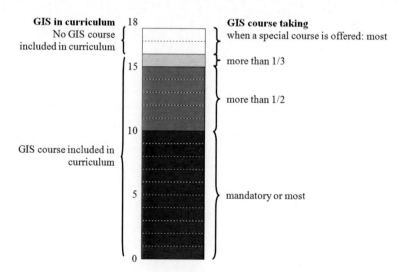

Fig. 26.2 GIS education in preservice programs

Korea think there are considerable obstacles to using GIS in the classroom, they showed positive attitudes about the educational context for GIS education, which is an indicator of teachers' evaluation of GIS as a teaching tool. Therefore, we think GIS has the potential to play a role at the secondary education level if adequate support is offered to teachers.

However, the opportunities given to GIS will be realized only when challenges are overcome. It is encouraging that most preservice teachers have the opportunity to learn about GIS during their preservice education. Nevertheless, educators must pay more attention to improving GIS preservice programs. As was identified by many researchers (e.g., Sui, 1995), teachers should have the ability to teach with GIS as well as to teach about GIS. We do not think teachers are automatically well prepared to teach with GIS without having used it themselves. Thus, teacher educators must develop effective curriculum to provide preservice teachers with experience in teaching with GIS. Moreover, a wide range of problems associated with software, hardware, data, and professional development, as were revealed in the previous section, must be solved. The current situation indicates that, even though teachers want to incorporate GIS in their teaching, obstacles prevent it. Last but not least, GIS content needs to be emphasized in the geography curriculum. In South Korea where the nationalized curriculum is imposed on all geography teachers at the secondary level, teachers have little autonomy in selecting topics and content. High-stakes exams are also based on the national curriculum. Therefore, if GIS is to be used in the classroom, the curriculum explicitly should ask geography teachers to use it.

We believe that GIS education in South Korea is at a critical point. If adequate support is provided, GIS could widen its realm in education. If the previously noted problems are not resolved, GIS may not find its way into secondary education. It is promising that teachers have relatively positive perspectives on the pedagogical role of GIS. However, the positive perspectives can either be heightened further or disappear completely depending on support given to teachers.

Acknowledgements The authors wish to thank Dr. Sarah Bednarz and Sandra Metoyer for their stimulating advice.

References

Audet, R. H., & Paris, J. (1997). GIS implementation model for schools: Assessing the critical concerns. *Journal of Geography, 96*, 293–300.

Baker, T. R., & White, S. H. (2003). The effects of GIS on students' attitudes, self-efficacy, and achievement in middle school science classrooms. *Journal of Geography, 102*, 243–254.

Bednarz, S. W., & Audet, R. H. (1999). The status of GIS technology in teacher preparation programs. *Journal of Geography, 98*, 60–67.

Bednarz, S. W., & van der Schee, J. (2006). Europe and the United States: The implementation of geographic information systems in secondary education in two contexts. *Technology, Pedagogy and Education, 15*, 191–205.

Hwang, M. (1998). Applications of geographic information system technology in geography education. *Journal of Geography Education, 40*, 1–12.

Hwang, S., & Lee, K. (1996). The present status and prospect of GIS learning in teaching geography of high school. *Journal of the Korean Association of Regional Geographers, 2*, 219–231.

Jung, A. (1997). Constructions of instruction documents about GIS in high school geography education. *Journal of the Korean Association of Geographic and Environmental Education, 5*(2), 61–73.

Jung, I. (2005). Undergraduate GIS curricula of department of geography in U.S. *Journal of the Korean Association of Geographic and Environmental Education, 13*, 225–234.

Jung, I., & Kim, J. (2006). Development of GIS teaching plans in high school geography classrooms. *Journal of the Korean Association of Geographic and Environmental Education, 14*, 251–262.

Kim, C. (2005). A study of the GIS education in the geography education: In the case of the USA. *Journal of the Korean Association of Geographic Information Studies, 8*(4), 176–190.

Kim, M. (2007). Spatial thinking and the investigation of GIS for potential application in education. *Journal of the Korean Association of Geographic and Environmental Education, 15*, 233–245.

Kim, M. (2010). The current status of GIS in the classroom and factors to consider for increasing the use of GIS. *Journal of the Korean Association of Geographic and Environmental Education, 18*, 173–184.

Kim, M., Bednarz, R., & Lee, S.-I. (2009). *Why do some teachers participate in GIS education?* Paper presented at the Annual Meeting of the National Council for Geographic Education, San Juan, Puerto Rico.

Kim, Y. (2002). The introduction and development of GIS curriculum in the UK geography education. *Journal of the Korean Association of Regional Geographers, 8*, 380–395.

Kwon, S. (2004). A GIS class exercise examples focused on the basic concepts and functions. *Journal of the Korean Association of Geographic and Environmental Education, 12*, 313–325.

Lee, M., Kim, N., & Ban, S. (2008). A study on the development of geography e-learning material using webGIS for the social studies of middle school. *Journal of the Korean Association of Geographic and Environmental Education, 16*, 17–26.

Lee, M., Kim, N., Lee, S., & Jo, E. (2006). Development of learning materials for Korean geography in high school curriculum using GIS and remote sensing: Focusing on the landform chapter. *Journal of the Korean Association of Geographic and Environmental Education, 14*, 191–200.

Ministry of Education, Science and Technology. (2008). *Explanation of middle school curriculum 2: Korean, ethics, social studies.* Seoul: Mirae N (Daehan Textbook).

Ministry of Education, Science and Technology. (2009a). *2009 revised curriculum.* Seoul: Ministry of Education, Science and Technology.

Ministry of Education, Science and Technology. (2009b). *Explanation of high school curriculum 4: Social studies (history).* Seoul: Mirae N (Daehan Textbook).

Oh, C., & Seong, C. (2003). A study of GIS education in secondary school: A case study of high school. *Journal of the Korean Geographic Information System, 11*, 89–100.

Patterson, M. W., Reeve, K., & Page, D. (2003). Integrating geographic information systems into the secondary curricula. *Journal of Geography, 102*, 275–281.

Shin, C., Jeong, Y., & Joo, S. (2002). Computational education: A development of a geographic learning courseware based on GIS. *Journal of the Korea Information Processing Society, 9-A*(1), 105–112.

Sui, D. Z. (1995). A pedagogic framework to link GIS to the intellectual core of geography. *Journal of Geography, 94*, 578–591.

Yang, S. (2004). *A study on developing geographic learning tool using Internet GIS.* Unpublished master's thesis, Seoul National University, Seoul.

Chapter 27
Spain: Institutional Initiatives for Improving Geography Teaching with GIS

Alfredo del Campo, Concepción Romera, Joan Capdevila, José Antonio Nieto, and María Luisa de Lázaro

27.1 ICT in the Spanish Educational System

The Spanish educational system has an official national curriculum called *Ley Orgánica de Educación* (LOE) (Education Organic Law, approved on 6 April 2006). One major component of the recent changes from previous curriculums is the competencies, including knowledge (minds on), skills (hands on) and behavior (hearts on). The key competencies for lifelong learning were established by the European Union (EU) in 2006 and adopted by the Spanish educational system. The competencies are not only for students: Teachers also need spatial and digital competencies that involve confident and critical use of information society technology (IST) and thus basic skills in information and communication technology (ICT). This critical thinking, pedagogical knowledge, spatial awareness, and a little knowledge about ICT were included in UNESCO's (2008) ICT Competencies Standards for Teachers. GIS could help in some ways to fulfill these competencies (Table 27.1).

There is a digital competence in the secondary school national curriculum (Table 27.2) that is defined as skills or abilities for searching, collecting, processing, and communicating information to obtain knowledge. Digital competence is necessary to access, select, and transmit information by different media. The use of information technologies and communication is essential to learn and communicate. The acquisition of this competence is based partly on using technology resources in order to solve problems efficiently and partly on having a critical and reflective attitude in the assessment of available information. The Council of Ministers approved the "Escuela 2.0" programme for students of the last two years of primary school and the first two years of secondary school (ESO) in order to have the necessary technology resources at schools to reach digital competence. In spite of the existence of the digital competence in the curriculum, there is no reference to learn with

M.L. Lázaro (✉)
Universidad Complutense de Madrid (UCM), Madrid, Spain
e-mail: mllazaro@ghis.ucm.es

A.J. Milson et al. (eds.), *International Perspectives on Teaching and Learning with GIS in Secondary Schools*, DOI 10.1007/978-94-007-2120-3_27,
© Springer Science+Business Media B.V. 2012

Table 27.1 EU key competencies (2006) and GIS. Drawn up by M.L. Lázaro

Key competencies	GIS aims	Teaching and learning process
Communication in the mother tongue and communication in foreign languages	Change the language of the interface	Express concepts and ideas with maps
Mathematical competence and basic competencies in science and technology	Make measurements through GIS tools	Necessary for reflection about specific places
Digital competence	GIS as a tool	Learning GIS ICT
Learning to learn	GIS for learning	Problem-based learning in which students can work with recent data on real-world problems, share them, and organize their own learning. Critical use of information and as a basic skill for making the tool useful
Social and civic competencies	Share our results	Encourage discussion and enable participation and engagement. Working individually and collectively in an interpersonal and intercultural environment
Sense of initiative and entrepreneurship	GIS professionals and analyst	Turning ideas into action that come from the new spatial knowledge (spatial changes, location, relations, models) make us act in different ways with new objectives
Cultural awareness and expression	GIS in Europeization (Koutsopoulos, 2008)	Allows us to express spatial ideas, experiences, and emotions about different places on the Earth/Europe

Table 27.2 The Spanish educational system. Drawn up by M.L. Lázaro using information from Ministry of Education

Ages (years)	Stages			
16–18	Secondary School	Bachillerato (high school) (2 years)	Formación Professional (Vocational studies)	Non-compulsory
		Degree in ESO		
12–16		Enseñanza Secundaria Obligatoria (ESO) (4 years)		Compulsory
5–12	Primary School	Educación Primaria (EP) (6 years)		
3–5	Kindergarten/ Child school	Educación Infantil (EI) (3 years)		

GIS in any legal document about the curriculum. Yet there are some subjects that deal with spatial components such as geography and earth sciences. This might be the reason why secondary schools do not use GIS despite the increasing use of GIS in university classrooms since the 1990s.

27.2 Tentative Applications of GIS in Schools

The first tentative applications of GIS in secondary schools began under the *Ley Orgánica General del Sistema Educativo* (LOGSE) approved in 1990 (Organic General Law for the Educative System). Some of the initiatives in using GIS in the classroom come from the members of the work group of Geography Teachers of the Spanish Geographers Association such as Miguel and Allende (1996). Other instances occurred in Catalonia, such as a doctoral thesis about Girona Town (Comas, 1995) in Barcelona Autonomous University for making a useful GIS for students. The *"Portal Educativo en SIG"* (PESIG Project, an Educational Web Portal) (Boix & Olivella, 2007) tried to improve GIS at schools. It has been implemented by Girona University and the *"Departament d'Ensenyament de la Generalitat de Catalunya"* (Teaching Department of the Government of Catalonia).

There is no doubt that GIS in Spain has influenced textbooks providing better maps than before GIS tools became available. Some of these textbooks include activities or exercises using virtual globes, such as Google Earth or ArcGIS Explorer (Calle, 2009; Lázaro et al., 2008a) in the same way that the students work in USA (Schultz, Kerski, & Patterson, 2008). Many authors consider virtual globes to be very suitable for fostering spatial competencies in secondary schools (Donert & Wall, 2008).

We can find also some reflections about GIS utility, usefulness, and advantages in schools (Lázaro & González, 2005, 2006, 2007a, 2007b; Linn, Kerski, & Wither, 2005; Zerger, Bishop, Escobar, & Hunter, 2002), but there are no specific books with applied GIS "step-by-step" as there are in other countries (Attard, 2005; Donert, 2007; Donert, Ari, Attard, O'Reilly, & Schmeinck, 2009; Kerski, 2003; Malone, Palmer, & Voigt, 2005). Many Spanish teachers write about the general advantages of websites for geography teaching and learning (Lázaro, 2000, 2003; Martín & García, 2009; Moreno, 1996), but they do not use GIS tools in secondary school. Other teachers create or gather geography materials in a Web page (Buzo, 2010; Morcillo, 2010; Ruiz, 2010; Vera, 2003) or in a blog for their students or for other geography teachers. The same happens with some education institutions, publishers, or geography associations such as *Asociación de Geógrafos Españoles* (2010), and so on; but they do not work specifically with GIS for secondary school. GIS are used in courses for training teachers, such as at the Ciudad Real University (García, 2003, 2006). The result of these teacher training courses is that when teachers teach, they will be able to use GIS in the schools. Perhaps a new opportunity for introducing GIS at secondary school is the newly created Masters degree for future secondary school teachers. Some of them have a subject in ICT, although it is not specifically about GIS.

We can see that an increased tendency of using geoinformation (GI) to teach (Jekel, 2007; Lázaro et al., 2008b) would be the reason of the great need for GIS teaching materials. To address this need, some institutions are creating and putting materials on the Internet for the national curriculum in secondary schools. We will see in the case study that two Spanish Institutions are concerned with creating a useful tool for using geoinformation: the *Instituto Geográfico Nacional* (IGN) (National Geographical Institute) and the Institute of Cartography of Andalusia (ICA). The first one concerned the SIANE project, based on the National Atlas cartography, and the Spanish spatial data infrastructure, and the second one concerned with the Didact-ICA project or the digital map of Andalusia. We hope that all of these initiatives may be the first steps to spread GIS into secondary and high schools in Spain.

27.3 Case Study: Instituto Geográfico Nacional (National Geographical Institute)

The National Geographic Institute of Spain is a Directorate General dependent on the Spanish Ministry of Public Works and Transportation (*Ministerio de Fomento*). Since its founding in 1870, the IGN has been engaged in scientific investigations and activities in the field of mapping. It is also engaged in activities related to astronomy, geodesy, geophysics, photogrammetry, remote sensing, mapping, geographic information systems, spatial data infrastructures, and administrative boundary lines. Among its commitments is its public service organization, named the National Geographic Institute, as well as its autonomous commercial body, the National Center of Geographic Information (CNIG). Both are actively involved in training activities and dissemination of geographic knowledge. Along these lines, the IGN has designed and developed several e-learning courses, such as geography for ESO (secondary school) and Thematic Cartography and Geographical Information Systems, as well as two educational projects on the National Atlas of Spain website: '*España a través de los mapas*' (Spain in maps) made in collaboration with the Spanish Geographers Association (www.ign.es/espmap) and '*La Población en España*' (Population in Spain) made in collaboration with the University of Zaragoza (www.ign.es/pobesp/). These educational resources of the National Atlas of Spain (ANE or SNA) aim to promote the usefulness of the contents of the SNA in the teaching field (Romera, Del Campo, & Sánchez, 2009). These resources from SNA are classified according to the different educational levels in Spain: primary school, secondary school, and university, and they provide educational tools with which to search through the catalog on the Web. Following the line of cartographic resources, the National Center of Geographic Information provides in its website blank maps and interactive puzzles (http://www.ign.es/ign/layout/cartografiaEnsenanza.do).

Advances in new Information and Communications Technologies (ICT) directly affect the education field. ICT for education offers a great number of educational resources for use in the classroom, some of which require teacher training to know and discover their potential. In this context, the National Geographic Institute of Spain created the Spanish National Atlas Information System (SIANE) and leads the spatial data infrastructure (SDI) (*Infraestructura de Datos Espaciales de España*, IDEE). The SIANE is an innovative project that is concerned with a new definition, development, and maintenance of the National Atlas of Spain. This geographical information system has been conceived in a way that continuous content updating and publishing is possible for work maintenance, using the Internet as the main media and without having to wait for all contents of a certain theme (in which the resource is classified) to be created in order to publish them. Specific SIANE's applications include an Excel macro for preparing data before entering the system (if needed), a content management system (CMS), a map editor, and a Web application (siANEweb) in which all ANE contents and metadata are published, making the maps as the main resource.

The IDEE is a distributed, multilingual, Internet-accessible system in which existing spatial data infrastructures in Spain are used and whose tools represent an important educational resource (www.idee.es). This project is run under the coordination of the National Geographic High Council, a governmental body, whose technical secretariat is the National Geographic Institute. The Observatory IDEE working group was launched in 2006 in order to disseminate and increase the use of spatial data infrastructures in Spain. Considering the growth perspective of SDI and its establishment as a paradigm for sharing geographic information in the context of a society that demands information, there is a need to contribute to the further spread of the SDI with regard to the potential use of them in both the public and private sectors. In this field, the use of spatial data infrastructure as an educational resource in secondary education (ESO) for dealing with subjects relevant to geographic information and the use of ICT has great potential (González, Capdevila, & Soteres, 2008).

In 2007, it became acutely apparent that the country needed to know the opinion and to have the collaboration of educational professionals and potential users to use the SDI as an educational resource aimed at bridging the gap between the world of SDI and education. Different projects were established to advance this work, including conducting workshops with teachers from the secondary school, the diffusion of SDI in forums concerned with the use of new ICTs in the classroom, and others. To the above-mentioned work, the proposal of implementing e-learning courses to teacher training in secondary school was added. Thus, the e-learning course, "E-learning training for teachers of Secondary School to use SDI as an educational resource," has been designed to use the SDI as an educational resource in the social sciences, natural sciences, and technology. This project is the result of a collaboration agreement between the National Geographic Institute, the National Centre of Geographic Information, and the Polytechnic University of Madrid.

27.4 Case Study: The Andalusia experience: The Instituto Cartográfico de Andalucía (ICA) (The Andalusia Institute of Cartography)

The Institute of Cartography of Andalusia, created by the Decree 116 of 7 September 1993 and assigned to the Public Works and Housing Cabinet of the Andalusian Government, is the organization responsible for programming the basic and derived cartography of the Andalusian Region, as well as the coordination of the thematic cartography and of the geographical data bases of this territory. This organization considers it a high-priority to offer better products and services to the citizens, facilitate an egalitarian access to those products and use more suitable technologies. In this sense, the Institute of Cartography of Andalusia has lately been working in the production and diffusion of didactic and educative material, in order to spread and to encourage the use of cartography with students (Nieto & Fajardo, 2007). This has generated an important demand that has not been always well assisted. This diffusion of cartography and geographic information in the educative community leans in three fundamental policies: (1) the free distribution of cartographic material (atlas, maps, etc.) in the schools; (2) the production of didactic and educative material related to cartography, geographic information, and maps; and (3) the programming of exhibitions, presentations, working days, competitions, and other activities oriented to the totality of the population and especially to the school community. In order to give more visibility to this work with the school community, the Institute of Cartography of Andalusia is working in Didact-ICA, the Internet portal (Fig. 27.1) where it is possible to access the educative material already made (blank maps, interactive games, digital atlases), the different didactic proposals periodically programmed, and multiple links and resources related to the cartography and the world of maps.

The use of GIS in the education community is largely confined to Andalusia. The Institute of Cartography of Andalusia works in this direction to develop initiatives where we tried to take care of the teacher's needs. When designing didactic material, the Institute of Cartography of Andalusia always has direct connection with the professors. The examples of this collaboration are many, but the best is the school version of the "Digital map of Andalusia." This is a simple application, but it is useful to facilitate the teaching–learning process of the Andalusian territory. On an interactive map of the Andalusian territory (Martín de Agar & Nieto, 2009; Nieto, 2009, 2010), the program helps students understand aspects of the physical and human geography of the region. It also makes a detailed approach to the locality of residence of the user. This allows students to analyze the human–environment relationship in the geographic area where the student resides. The interface of the application allows an intuitive use of the different tools that allow one to interact on the map. This map can be consulted on different scales by the use of a zoom. In addition, complementary graphs (evolution of the population, pyramids, and more) can be viewed. We can also make measurements of distances and surfaces, or to consult toponymy. As you can observe, although the application is not a GIS proper, it allows some functionalities that are similar to GIS.

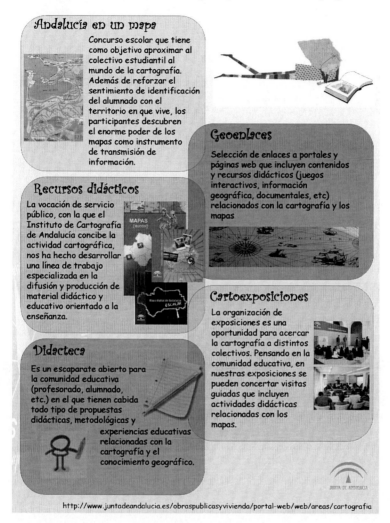

Didáct-ICA Una ventana abierta al mundo de la Cartografía y la Información Geográfica

Andalucía en un mapa

Concurso escolar que tiene como objetivo aproximar al colectivo estudiantil al mundo de la cartografía. Además de reforzar el sentimiento de identificación del alumnado con el territorio en que vive, los participantes descubren el enorme poder de los mapas como instrumento de transmisión de información.

Geoenlaces

Selección de enlaces a portales y páginas web que incluyen contenidos y recursos didácticos (juegos interactivos, información geográfica, documentales, etc) relacionados con la cartografía y los mapas

Recursos didácticos

La vocación de servicio público, con la que el Instituto de Cartografía de Andalucía concibe la actividad cartográfica, nos ha hecho desarrollar una línea de trabajo especializada en la difusión y producción de material didáctico y educativo orientado a la enseñanza.

Cartoexposiciones

La organización de exposiciones es una oportunidad para acercar la cartografía a distintos colectivos. Pensando en la comunidad educativa, en nuestras exposiciones se pueden concertar visitas guiadas que incluyen actividades didácticas relacionadas con los mapas.

Didacteca

Es un escaparate abierto para la comunidad educativa (profesorado, alumnado, etc.) en el que tienen cabida todo tipo de propuestas didácticas, metodológicas y experiencias educativas relacionadas con la cartografía y el conocimiento geográfico.

http://www.juntadeandalucia.es/obraspublicasyvivienda/portal-web/web/areas/cartografia

Fig. 27.1 "Didact-ICA" portal http://www.juntadeandalucia.es/obraspublicasyvivienda/portal-web/web/areas/cartografia/texto/d84f2420-22f2-11e0-ab22-659935a88cb0

From the school version of the "Digital map of Andalusia," the Council of Education of the Andalusian government has developed, in Averroes, its educative portal, "*Andalucía a tu alcance*" (Andalusia in your hand) (Fig. 27.2). This Web resource, conceived as a small GIS, includes cartography that incorporates mashups with services of Google Maps and IDEAndalucia and integrated toponymy and statistical data, designed with a friendly interface for a didactic use. This link allows

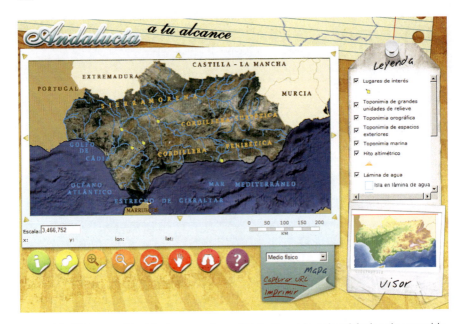

Fig. 27.2 Web page "Andalucía a tu alcance" http://www.juntadeandalucia.es/averroes/sig_educativo/proyectosig/entrada.htm

the consultation of distributed geographic information in different layers that the user can activate or deactivate. The user can capture and print maps that are of interest, interact with the map, and incorporate their own information on a place in the form of photos, videos, or text. Both resources allow professors and students to manage geographic information effectively, with the advantages of GIS and without the disadvantage of their difficult handling.

From the Institute of Cartography of Andalusia, a special interest exists in which geographic information arrives at Andalusian schools where professors and students know to use it. For this purpose, we are planning a series of activities directed to diffusing IDEAndalucia and the spatial data infrastructure of our region to schools, introducing more people to the use of GIS, and making the most of the advantages of using the free software. In this sense, through our virtual platform, an e-learning course is predicted to allow for the acquisition of basic and elementary knowledge for the accomplishment of thematic cartography.

References

Asociación de Geógrafos Españoles (AGE). (2010). (Spanish Geographers Association). Accessed September 1, 2010, http://age.ieg.csic.es/recur_didacticos/index.htm Grupo de trabajo de Didáctica de la Geografía (Working Group of Teaching Geography) http://age.ieg.csic.es/didactica/

Attard, M. (2005). Developing undergraduate GIS study-units – The experience of Malta. In K. Donert & P. Charzynski (Eds.), *Changing horizons in geography education* (pp. 97–101). Toruń: HERODOT Network Publication. Accessed September 1, 2010, http://www.herodot.net/conferences/torun2005/Changing%20Horizons%20book.pdf

Buzo, I. (2010). Recursos de Ciencias Sociales, Geografia e Historia. Accessed September 1, 2010, http://www.isaacbuzo.com

Boix, G., & Olivella, R. (2007). Los sistemas de información geográfica (SIG) aplicados a la educación. El proyecto PESIG (Portal Educativo en SIG). In M. J. Marrón, J. Salom, & J. M. Souto (Eds.), *Las competencias geográficas para la educación ciudadana* (pp. 23–32). Valencia: Grupo de Didáctica de la Geografía. Asociación de Geógrafos Españoles. Accessed September 1, 2010, more information about PESIG Project in http://www.sigte.udg.edu/pesig/

Calle, M. (2009). Aplicación de Google Earth en la formación del profesorado de educación infantil para el conocimiento geográfico. In Associaçao de Professores de Geografia et al. *A Inteligência Geográfica na Educação do Século XXI. IV Congreso Ibérico de Didáctica da Geografia* (pp. 152–157). Lisboa: Universidade de Lisboa.

Comas, D. (1995). *Urbamedia, un sistema d'informació geogràfica per analitzar la ciutat de Girona en conceptes clau. Experimentació a les ciències socials de l'ensenyament secundari obligatori.* Tesis doctoral, Universitat Autònoma de Barcelona, Barcelona.

Donert, K. (2007). *Teaching geography in Europe using GIS.* Papers and presentations from the Education track of the Esri User conference 2006. Stockholm: HERODOT, Esri Inc.

Donert, K., & Wall, K. (Eds.). (2008). *Future prospects in geography* (pp. 511–517). Liverpool: Liverpool Hope University.

Donert, K., Ari, Y., Attard, M., O'Reilly, G., & Schmeinck, D. (Eds.). (2009). *Geographical diversity.* Proceedings HERODOT conference, 29–31 May 2009. Ayvalik Turkey: HERODOT Network. (GIS chapters in pp. 264–329). Accessed September 1, 2010, http://www.herodot. net/conferences/Ayvalik/papers/manuscript-v1.pdf. Accessed September 1, 2010

European Union. (2006). European Parliament and the Council Recommendation of the European Parliament and of the Council on key competences for lifelong learning of 18 December 2006 (2006/962/EC). Official Journal of the European Union, 30.12.2006.

García, F. M. (2003). La enseñanza de las nuevas tecnologías en la Universidad. Sistemas de Información Geográfica. In M. J. Marrón, C. Moraleda, & H. Rodríguez (Eds.), *La enseñanza de la Geografía ante las nuevas demandas sociales* (pp. 180–186). Toledo: Grupo de Didáctica de la Geografía (AGE). Escuela Universitaria del Profesorado, Universidad de Castilla-La Mancha.

García, F. M. (2006). Consideraciones didácticas con SIG. Modelos medio ambientales susceptibles de ser desarrollados en el aula. In M. J. Marrón, & L. Sánchez (Eds.), *Cultura geográfica y educación ciudadana* (pp. 297–308). Murcia: Grupo de Didáctica de la Geografía. Asociación de Geógrafos Españoles. Associaçao de Professores de Geografia de Portugal. Universidad de Castilla-La Mancha.

González, M. E., Capdevila, J., & Soteres, C. (2008). Las Infraestructura de Datos Espaciales como recurso educativo para el profesorado de la Educación Secundaria Obligatoria. Una propuesta innovativa de formación e-learning. In IX Encuentro Internacional Virtual Educa Zaragoza 2008. Zaragoza, 14–18 de julio, 2008. Accessed September 1, 2010, http://www.virtualeduca. info/ponencias/246/EDUCA_ZARAGOZA_IDE-EDU-ESO.doc

Jekel, T. (2007). What you all want is GIS 2.0! In A. Car, G. Griesebner, & J. Strobl (Eds.), *Collaborative GI based learning environments for spatial planning and education* (pp. 84–89). GI-Crossroads@GI-Forum. Heidelberg: Wichmann.

Kerski, J. J. (2003). The implementation and effectiveness of geographic information systems technology and methods in secondary education. *Journal of Geography, 102*, 128–137.

Koutsopoulos, K. C. (2008). What's European about European geography? The case of geoinformatics in europeanization. *Journal of Geography in Higher Education, 32*(1), 1–14.

Lázaro, M. L. (2003). Nuevas Tecnologías en la enseñanza-aprendizaje de la Geografía. In M. J. Marrón, C. Moraleda, & H. Rodríguez (Eds.), *La enseñanza de la Geografía ante las nuevas demandas sociales* (pp. 141–167). Toledo: Grupo de Didáctica de la Geografía (AGE). Escuela Universitaria del Profesorado, Universidad de Castilla-La Mancha.

Lázaro, M. L. (2000). La utilización de internet en el aula para la enseñanza de la Geografía: ventajas e inconvenientes. In J. L. González & M. J. Marrón (Eds.), *Geografía, profesorado y sociedad. Teoría y práctica de la Geografía en la enseñanza* (pp. 211–218). Murcia: Grupo de Didáctica de la Geografía. Asociación de Geógrafos Españoles-Universidad de Murcia.

Lázaro, M. L., & González, M. J. (2005). La utilidad de los Sistemas de Información Geográfica para la enseñanza de la Geografía. *Didáctica Geográfica, 7*, 105–122.

Lázaro, M. L., & González, M. J. (2006). La utilidad de los SIG existentes en internet para el conocimiento territorial. In M. J. Marrón & L. Sánchez (Eds.), *Cultura geográfica y educación ciudadana* (pp. 443–452). Murcia: Grupo de Didáctica de la Geografía. Asociación de Geógrafos Españoles. Assicuaçao de Professores de Geografia de Portugal. Universidad de Castilla-La Mancha.

Lázaro, M. L., & González, M. J. (2007a). Spain on the web: A GIS way of teaching. In K. Donert, P. Charzynsky, & Z. Podgorski (Eds.), *Teaching in and about Europe. Geography in European higher education* (pp. 36–43). Toruń: University of Toruń and HERODOT network.

Lázaro, M. L., & González, M. J. (2007b). Learning about Spain through GIS Webs. In K. Donert (Ed.), *EUC'07 HERODOT Proceedings. Esri European User Conference 2007*: Stockholm. Esri-HERODOT Publications. Accessed September 1, 2010, http://www.herodot. net/conferences/stockholm/esri/Lazaro_Gonzalez(paper).pdf

Lázaro, M. L., González, M. J., & Lozano, M. J. (2008a). Google Earth and ArcGIS explorer in geographical education. In T. Jekel, A. Koller, & K. Donert (Eds.), *Learning with Geoinformation III – Lernen mit Geoinformation III* (pp. 95–105). Munich: Wickmann.

Lázaro, M. L., González, M. J., & Lozano, M. J. (2008b). Learning about immigration in Spain through Geoinformation on the internet. In K. Donert & G. Wall (Eds.), *Future prospects in geography* (pp. 439–445). Liverpool: Liverpool Hope University.

Linn, S, Kerski, J., & Wither, S. (2005). Development of evaluation tools for GIS: How does GIS affect student learning? *International Research in Geographical and Environmental Education, 14*(3), 217–224.

Malone, L., Palmer, A., & Voigt, C. (2005). *Mapping our world: GIS lessons for educators, ArcGIS desktop edition*. Redlands, CA: Esri.

Martín de Agar, R., & Nieto, J. A. (2009). *Andalusia in a map. Assessment of an educational experience*. XXIV International Cartographic Conference. Santiago de Chile, Chile. Accessed September 1, 2010, http://icaci.org/documents/ICC_proceedings/ICC2009/html/nonref/29_5. pdf

Martín, C., & García, F. (2009). Algunos recursos en internet para mejorar la enseñanza de la Geografía. Ar@cne. Revista electrónica de recursos en internet sobre Geografía y Ciencias Sociales, 118. Accessed September 1, 2010, http://www.raco.cat/index.php/Aracne/article/ view/130175/179613

Miguel, I. de, & Allende, F. (1996). *Uso de una aplicación S.I.G. como recurso didáctico en la enseñanza secundaria. III Jornadas de didáctica de la Geografía* (pp. 77–85). Madrid: Grupo de Didáctica de la Geografía (AGE). Departamento de Didáctica de las Ciencias Sociales de la Universidad Complutense de Madrid, Spain.

Morcillo, J. M. (2010). *Portal de Ciencias Experimentales, UCM (Devoted to physical geography)*. Accessed September 1, 2010, http://www.ucm.es/info/diciex/programas/

Moreno, A. (1996). Internet y sus recursos para enseñar Geografía. Didáctica Geográfica, 1, Segunda época, 95–102.

Nieto, J. A., & Fajardo, A. (2007). *The Andalusian government policy of diffusion of maps in the educative community*. XXIII International Cartographic Conference. Moscow, Russia. Accessed September 1, 2010, http://icaci.org/documents/ICC_proceedings/ICC2007/ documents/doc/THEME%2022/Oral%201/THE%20ANDALUSIAN%20GOVERNMENT% 20POLICY%20OF%20DIFFUSION%20OF%20MAPS%20IN%20THE.doc

Nieto, J. A. (2009). *Andalucía en un mapa. Valoración de una experiencia educativa. A Inteligência Geográfica na Educação do Século XXI. IV Congresso Ibérico de Didáctica da Geografía* (pp. 116–111). Lisboa: Universidade de Lisboa.

Nieto, J. A. (2010). *La información geográfica al servicio de la comunidad educativa andaluza. Didáctica Geográfica 11, Madrid, Grupo de didáctica de la Geografía (AGE), Real Sociedad Geográfica (RSG) (Royal Geographical Society of Spain)*. Accessed September 1, 2010, http://www.didacticageografica.es/index.php/didactica/article/view/13/21

Romera, C., Del Campo, A., & Sánchez, J. (2009). *Educational resources of the national atlas of Spain*. Accessed September 1, 2010, http://cartography.tuwien.ac.at/ica/documents/ICC_proceedings/ICC2009/html/nonref/14_10.pdf

Ruiz, F. (2010). *A useful way of making maps of Spain with the last statistics*. Accessed December 8, 2010, http://alarcos.esi.uclm.es/per/fruiz/pobesp/nueva/

Schultz, R. B., Kerski, J. J., & Patterson, T. (2008). The use of virtual globes as a spatial teaching tool with suggestions for metadata standards. *Journal of Geography, 107*, 27–34.

UNESCO. (2008). *ICT competencies standards for teachers*. Accessed September 1, 2010, http://cst.unesco-ci.org/sites/projects/cst/The%20Standards/ICT-CST-Implementation%20Guidelines.pdf

Vera, A. L. (Geohistoria, 2003). (It works interesting links for the National Curriculum) 275 http://www.geohistoria.net/index2.asp. Geography of Spain: http://www.geohistoria.net/paginas/2bgeo.htm, 276 General geography: http://www.geohistoria.net/paginas/3eso.htm. Accessed September 1, 2010.

Zerger, A., Bishop, I., Escobar, F., & Hunter, G. (2002). A self-learning approach for enriching GIS education. *Journal of Geography in Higher Education, 26*(1), 67–80.

Chapter 28
Switzerland: Introducing Geo-Sensor Technologies and Cartographic Concepts Through the Map Your World Project

Hans-Jörg Stark and Carmen Treuthardt

28.1 Structure of the Training System in Switzerland

Schooling in Switzerland consists of five levels: pre-school, primary (Grades 1–6), lower secondary (Grades 7–9), upper secondary, and tertiary. Education is compulsory through the lower secondary level and is the responsibility of the 26 cantons (provinces) of Switzerland. This decentralized structure for compulsory education provides for local governance based on regional traditions and cultural and linguistic differences. Upper secondary schooling is not compulsory and provides different options for students, including vocational training or Matura (university preparatory) schools.

According to the Swiss Conference of Cantonal Ministers of Education

> The cantons and their local municipalities finance about 87% (2005) of public educational expenditures. Most students in Switzerland (95%) complete pre-school and compulsory schooling at state schools in the municipality in which they live. Ninety percent of young people in Switzerland complete upper secondary education at the age of 18 or 19 which allows them to start working, to switch to a college of higher vocational training or – with a matura/baccalaureate – to continue their education at a university or a university of applied sciences. (http://www.edk.ch/dyn/16342.php)

28.2 GIS in Secondary Schools in the Country

Geographic inquiry and geographic information systems (GIS) technology are important tools that help educators, students, and their institutions answer personal and community questions with local to global implications. Today, more and more schools are including GIS in their curricula to help their students gain valuable

H.-J. Stark (✉)
University of Applied Sciences Northwestern Switzerland, Muttenz, Switzerland
e-mail: hansjoerg.stark@fhnw.ch

A.J. Milson et al. (eds.), *International Perspectives on Teaching and Learning with GIS in Secondary Schools*, DOI 10.1007/978-94-007-2120-3_28,
© Springer Science+Business Media B.V. 2012

background knowledge and skills with which to face global challenges (for example, climate change, poverty in the world, or the economic aspects of tourism). In Switzerland, 44 schools are using GIS; this is one third of all secondary schools.

28.2.1 ArcGIS

Within the past five years, some engaged teachers promoted GIS in Switzerland. More and more schools have been using ArcGIS 9.x in the form of a school license with the 3D-Analyst and Spatial Analyst extensions. In 2006, the first teaching materials were published. "Geographical Information Systems (GIS) – Bases And Exercises For The Secondary School" (Treier, Treuthardt Bieri, & Wüthrich, 2006) is the only material available in German-speaking countries for high schools. In the summer of 2007, the authors received the Special Achievement in GIS Award. In courses with ArcGIS, teachers teach other teachers. A goal of these courses is to encourage teachers to begin teaching GIS in schools as much as possible.

28.2.2 Map Your World

Map your World (MyW) is a conception on how students get familiar with both modern geo-sensor technologies and concepts of contemporary cartography (www.map-your-world.ch). It not only provides lectures as templates, but also guidelines and tutorials for teachers. MyW uses modern technologies and devices such as the Global Positioning System (GPS), Internet, Web 2.0, Personal Digital Assistants (PDAs), and others to introduce the collection and processing of spatial data to secondary school students. These data can further be used and applied within GIS. The project idea was developed at the University of Applied Sciences Northwestern Switzerland, School of Architecture, Civil Engineering and Geomatics at the Institute of Geomatics Engineering (IGE).

While not too many years ago large analogue wall maps were used in school subjects such as history or geography, today, thanks to fast Internet connections and the widespread availability of personal computers in developed countries, dynamic maps in two or two and a half dimensions are freely accessible on the Internet. These digital and almost freely scalable maps, such as virtual globes (earth.google.com or worldwind.arc.nasa.gov) or dynamic maps (www.bing.com/maps, maps.yahoo.com or maps.google.com), belong to the essential ingredients of the Web and the modern classroom. Students use these tools to find out more about the locations and surroundings of events in the daily news or school subjects (Bartoschek & Schöning, 2008). They mostly take this kind of spatial information and the technology to deploy the geo-data for granted. GIS on the other hand is still a tool that is mainly used by specialists in either administrative bodies or engineering and planning companies. Thus, one goal of MyW is to familiarize students at the secondary school level with the idea and use of GIS.

28.3 Cases

28.3.1 Working with ArcGIS

The following two examples for learning GIS in schools have been used several times at the Matura school Lucerne.

Which large cities worldwide are at risk from volcanic eruption? (Treier et al., 2006, p. 81).

A point data record can be provided from Internet data records, which shows volcanic eruptions worldwide. This can be classified according to volcanic type or outbreak year. With a buffer around selected volcanoes, the potential danger zones are as shown in Fig. 28.1. With a link to large cities, it becomes clear where at-risk cities lie. As a result of the topicality and the processing of the large data set, an increase in value arises from the use of GIS in instruction.

Which cities in Bangladesh will have to evacuate their population if the sea level increases around 1–5 m?

This question can be simulated with GIS, as the coastal lines change when the sea level rises (Fig. 28.2). Combined with population density and the situation of the

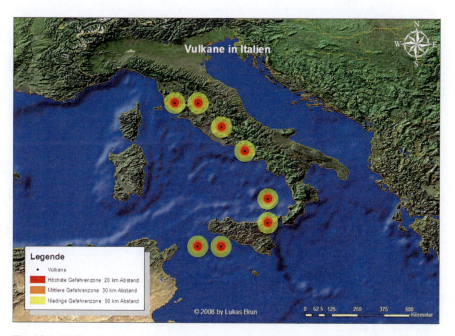

Fig. 28.1 Buffer zones around volcanoes in Italy, with a distance from 20 km, 30 km, and 50 km (Lukas Brun and a student of the Matura School Lucerne 2008)

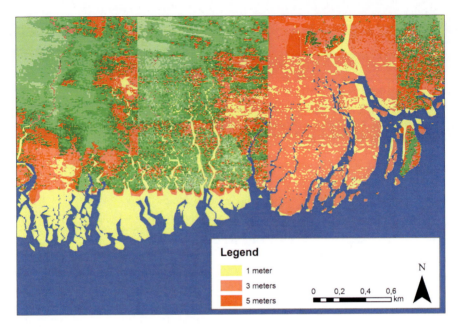

Fig. 28.2 GIS map of the sea level rise of 1 m, 3 m, and 5 m

cities in Bangladesh, it becomes clearly visible how the population is endangered in certain coastal states. Well-prepared data records on this topic can be downloaded from www.lehrer-online.de and either be looked at in ArcExplorer only or processed in ArcView. Students are very interested in problems of climate change. They know a great deal because of the IPCC Fourth Assessment Report of Climate Change 2007 and from Al Gore's film "An Inconvenient Truth."

28.3.2 Working with Map Your World

The intention of Map your World is to reveal the complex and demanding task of map-making to secondary school students. Besides this insight into a mapmaker's profession, they are also introduced to the concept of volunteered geographic information (Goodchild, 2008a, 2008b). This paradigm is very similar to social networks and other platforms that young people participate in today and is also called crowd sourcing (Ramm & Stark, 2008), participatory mapping (Aditya & Gadjah, 2008), mass collaboration (Hof, 2005), or collaborative mapping (Fischer, 2008). Basically this paradigm uses volunteers – laypeople and professionals – to collect data and store them in a central and predefined structure that continues to evolve along with the information that it contains.

28.3.3 Collaborative Mapping Projects

In the context of spatial information, OpenStreetMap (OSM) (http://www.openstreetmap.org) is perhaps the most prominent project. But there are others as well, such as OpenAddresses (OA) (http://www.openaddresses.org) and People's Map (PM) (http://www.peoplesmap.com) to name just a few. OSM aims for the collection of any relevant spatial feature to be displayed in a topographic map and is very well documented (Ramm & Topf, 2008). OA focuses on the collection of geocoded addresses (Bähler & Stark, 2008). They both share the same paradigm of open content, which means that the data are freely accessible and available under a Creative Commons License (http://creativecommons.org). In essence, this means that the data may be used by anyone as long as the license is respected, which means that if the data are published, the source must be mentioned and the processed data must be available under the same conditions/license as the source data. Both projects offer tools and interfaces for data upload, processing, and further use.

PM, on the other hand, also takes advantage of volunteers to collect data and also offers tools to collect the data, but it does not follow the same licensing paradigm. PM's intention is to serve a commercial aspect more closely and, therefore, stresses the aspect of quality assurance of the collected data. For private and noncommercial use, the data are still free of charge.

28.3.4 The Path of Spatial Data Collection

When considering the production of a map, there are several questions to be answered. The spatial extent and the contents are two of the crucial ones. If these are answered and the mapmaker starts the collection in the field, they face the next decision to make: Which of the real-world phenomena to choose and how should they be modeled to be represented on the final map? The next aspect, after having collected the raw spatial data, is to bring them into a semantic model and finally to apply design rules to the data when they are portrayed on the map. With the last aspect, it is important that not only the mapmaker be able to read the abstraction of the real world that he created, but also that an independent map reader be able to both read and understand properly what is on the map.

28.3.5 Map Your World in Action

The basic equipment used for Map your World in outdoor work is a PDA, a GPS device that is connected to the PDA, and special software package for the spatial data collection that runs on the PDA. Some of this software was developed at IGE. For the indoor work, the collected data are transferred to a PC that is connected to the Internet. The transferred raw data are then uploaded via Web applications that are provided by OSM and OA into their individual data repositories. The raw data must be post-processed by the students: Collected GPS tracks of linear or polygonal features must be adapted to an underlying reference map and the geometries along with their attributes must be entered into the project-specific data models. This part of the work takes roughly as much time as the data collection in the field. When this stage has been finished, the uploaded data are available on the Internet to anyone. Thus, it can be downloaded and applied to other school projects. The advantage of MyW is that secondary school students do not merely read about mapmaking but they also act as mapmakers – both in the field and at their desks or computers, respectively. Experiences have been very positive and documented by one of the teachers (Opferkuch, 2009).

28.3.6 Tourism Project Lucerne

A cooperation between the Matura School of Lucerne and the IGE has shown that students are very active and inspired to work in their own town. When exploring the introduced project "Map your World," students collected spatial data in Lucerne and produced their own tourist maps. The data they collected were also entered into OSM and OA. The students were technically supported by experts of the IGE and geographically supported by their teachers (Figs. 28.3 and 28.4).

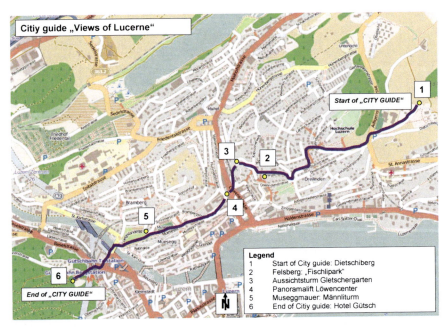

Fig. 28.3 Map of Lucerne: "City guide for tourists from Asia" (Daniela Ettlin, Alessia Gössi, and students of the Matura School Lucerne 2009)

Fig. 28.4 Map of Lucerne: City guide "Views around Lake Lucerne" (Nico Rast, Tschuensho Lam, Dominic Bühler, and students of the Matura School Lucerne 2009)

28.4 Prospects

The employment of GIS requires much commitment and initiative from a teacher. Its use in the classroom requires different instruction than is conventionally used. With GIS, teachers have a new instrument, which should be similar to using an atlas only in digital form and much more interactive way. So the lessons can be close-to-reality, up-to-date, and specialized. To be well prepared before the first lesson takes a long time for the teacher. In reality, students will know more functions faster than the teacher. Then students can teach other students. The teacher becomes a coach. In discussions with classes, this aspect is felt again and again as particularly enriching for the social climate in a classroom.

A good possibility results where synergies can be used. Thus, the cooperation between high schools and University of Applied Sciences functions very well. At present, these offers are however still too little used and are limited in each case to the activity of the individual teacher and professor. There is a high need that teachers and professors get connected and interact more closely to establish relation based on which students are able to see more clearly what further studies in the area of geography could look like. MyW offers the advantage of cooperation with the IGE. This means another quantity potential and is profitable for all sides.

A technology such as GIS, which in the future will gain ever more significance, should also be embedded in the curricula in geography lessons. Instruction with GIS is still in the pioneer phase and still depends strongly on the commitment and will of the individual educator. The stabilization of geography, in particular by the increase in value of GIS in instruction, is a central point. In a time of radical change and uncertainty, the position of geography in the school curriculum could be strengthened, especially through application and integration of GIS. With GIS, the many topics and problem areas that humankind is currently confronted with are more easily understood, and the connections and interdependences highlighted. Still, much convincing is needed before GIS will be used naturally as a work method by all teachers of geography in secondary schools in Switzerland.

References

Aditya, T., & Gadjah, M. (2008). Participatory mapping. *GIM International September*, pp. 41–43.
Bähler, L., & Stark H. J. (2008). Open Geodata – am beispiel von OpenAddresses.ch. Angewandte Geoinformatik.
Bartoschek, T., & Schöning, J. (2008). Trends und potenziale von virtuellen globen in schule, Lehramtsausbildung und Wissenschaft. *Geo Science, 4*, 28–31.
Fischer, F. (2008). Collaborative mapping – How Wikinomics is Manifest in the Geo-information Economy. *GEO Informatics März*, 28–31.
Goodchild, M. (2008a). Bürger als Sensoren / Citizens as Sensors. *GIS Trends Markets, 6*, 27–31.
Goodchild, M. (2008b). Volunteered geographic Information. *GEOconnexion International Magazine*, Oktober 08, 46–47.
Hof, R. (2005). *The power of us. Mass collaboration on the Internet is shaking up business*. BusinessWeek [online http://www.businessweek.com/magazine/content/05_25/b3938601.htm].
Opferkuch, D. (2009). Ein GIS-projekt: Schüler erfassen geodaten. *GEG-INFO*, 2-2009 April.
Ramm, F., & Stark, H. J. (2008). Crowdsourcing Geodata. *Geomatik Schweiz*, Ausgabe 06/2008, 315–319.
Ramm, F., & Topf, J. (2008). OpenStreetMap, Die frei Weltkarte nutzen und mitgestalten.
Treier, R., Treuthardt Bieri, C., & Wüthrich, M. (2006). Geografische informationssysteme (GIS) – Grundlagen und Übungsaufgaben für die Sekundarstufe II. *Bern*, 1–150.

Chapter 29
Taiwan: The Seed of GIS Falls onto Good Ground

Che-Ming Chen

29.1 Introduction

Taiwan's education system stayed highly centralized until the late 1980s. It used to serve to strengthen patriotic ideology and provide necessary human resources for economic growth. On 10 April 1994, over 30,000 marchers consisting of 95 civil groups and 57 offices of local councilors requested the modernization of education. Since then, Taiwan's education system has undergone decentralization to meet the demand for higher education and a high-quality workforce. A series of educational reforms have been made since 1994, including the multiple textbooks policy, flexible approaches for school entrance, and nine-year (Grade 1–9) integrated curriculum (Wu, 2006; Wu & Kao, 2007). Despite the central government responses to the society positively and persistently, Taiwan's education system remains centrally oriented.

As a small island country with a lack of natural resources, Taiwan stays competitive in the context of globalization by depending on its firmly established IT industry (Executive Yuan, 2001). Taiwan's government truly understands that an effective education system must provide high-quality human resources to ensure a favorable environment for the nation's development. The fact that college graduates make up more than 40% of the nation's workforce reveals this vision (National Statistics, 2009). The centralized education system and the knowledge-intensive economy make Taiwan's schools good ground to receive the seed of information and communication technologies. GIS is recognized as one of the core technologies for the nation's development so that its implementation, especially in the senior high school level, is supported by the Ministry of Education (MOE).

C.-M. Chen (✉)
National Taiwan Normal University, Taipei, Taiwan
e-mail: jeremy@ntnu.edu.tw

A.J. Milson et al. (eds.), *International Perspectives on Teaching and Learning with GIS in Secondary Schools*, DOI 10.1007/978-94-007-2120-3_29,
© Springer Science+Business Media B.V. 2012

29.2 General Structure of Secondary Education

At present, 9 years of compulsory education in Taiwan extends from primary school to junior high school. Children ages 6–11 are required to attend primary schooling. After six years of primary education, students enter junior high school for another 3 years (ages 12–14) without taking an entrance exam. There are two options following the junior high school level for students ages 15–17: senior high school or senior vocational school, both of which require students to take a standardized test to enter. Senior vocational schools emphasize career-oriented training in areas such as agriculture, industry, business, maritime studies, marine products, medicine, nursing, home economics, drama, and art (MOE, 2009).

The MOE of the Executive Yuan is responsible for formulating educational policies and managing public schools throughout Taiwan. MOE supervises the Bureaus of Education (BOE) in each city or county governments, which are in charge of local educational administrative affairs. In addition, BOEs recruit senior teachers by discipline from primary to secondary schools as teacher support teams to assist in the implementation of educational policies. For the past several decades, students in primary and secondary education were required to use only the textbooks published by the National Institute for Compilation and Translation (NICT). Since 1999, the MOE adopted the "one standard, multiple textbooks" policy. The curriculum standard is still prepared by the MOE, but the textbooks are no longer published only by the NICT. Teachers are free to choose their favorite textbooks from various publishers (Wu & Kao, 2007). With the hierarchical education authority and the "one standard" policy, the production of textbooks, the development of school entrance exams, and the implementation of educational policies are quite efficient.

To become a secondary school teacher, senior high school graduates can apply for teacher training programs in the normal university or other public or private universities. Graduate students are also eligible for applying for the same programs. All programs include common courses, discipline courses, education specialization courses and a half-year of teaching practicum. Those who complete the program have to pass a qualification exam administered by the MOE before they apply for a teaching job (MOE, 2009). Nowadays it is highly competitive to get a teaching job in secondary schools due to the significant drop in Taiwan's birth rate, the lowest of the world in 2009 (Population Reference Bureau, 2009).

29.3 The Status of GIS in Senior High Schools

Geography is a core subject in the senior high schools. It is a required course for Grade 10 to Grade 11 and is optional for Grade 12. GIS has been a part of the national curriculum standard of senior high school geography since 1995. The standard was later revised in 2006 and 2010. The revisions of the standard show that GIS is becoming more and more important in teaching and learning geography. In the 1995 standard, GIS was introduced at Grade 12 in the applied geography course, required only for students majoring in social science. It covered the introduction of GIS and emphasized the skills of geographic data input, management, and output. In the 2006 standard, there were two major revisions. The first revision was moving the introduction of GIS from Grade 12 to Grade 10, which meant that it became required for all senior high school students. In addition to the introduction of GIS, the standard of Grade 10 included the applications of GIS for everyday life. For example, students should be able to collect field data with GPS and use GIS to query and demonstrate geographic information about environmental issues in their daily lives. The class hours for GIS were raised from 3 to 6 for this expansion. The second major revision involved intensively introducing GIS to the applied geography course at Grade 12. Students should be able to understand how GIS is applied for national land planning, disease monitoring, flood forecasting, and debris flow monitoring. There were about 4 class hours introducing GIS in these topics. The 2010 standard basically follows the same contents as the 1996 standard. As one-fourth of senior high schools in Taiwan already had desktop GIS software (i.e., ArcGIS) by the end of 2009, the 2010 standard suggests that students learn how to use GIS software such as ArcGIS, Google Earth, or any Web-based GIS for 1–2 class hours. The standards of 2006 and 2010 show the trend from learning about GIS to learning with GIS.

The 1995 national curriculum was announced in 1995 but not put into practice until 2001. Geography teachers of senior high schools were not confident about teaching GIS at that time (Wang, 2001). In 2002, the MOE proposed two projects to prepare teachers for teaching GIS. Both projects were conducted by the geography department of National Taiwan University. The first project invited geography teachers of senior high schools to develop 16 GIS lesson plans covering physical geography and human geography. The results were shareable through a website and demonstrated at 3 workshops (Lay & Yu, 2004). There were 200 senior high school geography teachers, about one-seventh of the total in Taiwan, who attended these workshops. The second project, as part of the National Socio-Economic Development Plan, selected 11 "seed schools" to promote GIS in senior high schools. The MOE provided each school GIS training programs, one package of GIS software (i.e., ArcView), and spatial data CDs. The geography teachers of these schools were obligated to attend teacher workshops (40 hours), assign students to attend a GIS camp (3 days, 5 students for each school), organize a GIS task force and a student GIS club, conduct a campus or community field survey to collect geographic data, and help students to join the GIS national competition (Lay & Chiu, 2003). Taking one of the seed schools as an example, National Taichung Girls' Senior High School, successfully organized a GIS task force consisting of 13 teachers from 8 schools to develop GIS lesson plans. They shared their results including 8 lesson plans and 258 maps through a website (Wang, 2009). Since 2002, the MOE continues the project of GIS seed schools on an annual basis. Every year more than 300 teachers attend the GIS workshops. The workshops were attended by more than 1,000 teachers by 2009, about two-thirds of all geography teachers of senior high schools in the country. To encourage all geography teachers to teach with GIS in Grade 10, the MOE has invested in more resources since 2006. For example, each seed school was provided with a GIS computer lab including 40 PCs and GIS software (i.e., the lab kit licenses of ArcGIS). The GIS national competition that started in 2005 drew the attention of geography teachers and students. It includes a GIS lesson plan contest for teachers and a GIS mapping/thesis contest for students. Taking the 2009 GIS national competition as an example, the contestants included 146 student teams and 11 teachers, and 262 total attendees. All of the work of the contestants is downloadable via the official website (http://gisedu.tw/).

In addition to the MOE, there are other proponents making the senior high schools the breeding ground of GIS education. First, several universities or research institutes with GIS computer labs, such as National Taiwan Normal University (NTNU), Feng Chia University (FCU), and Academia Sinica offer GIS or spatial technology training programs for preservice and in-service high school teachers. For instance, since 2003 the geography department of NTNU has held more than 60 workshops for training secondary teachers on how to apply Google Earth and GPS to their geography lessons. These workshops were attended by over 1,000 teachers by 2010 (http://sites.google.com/site/jeremychenwww/academic_service). Second, the National Taichung Girls' Senior High School has been assigned by MOE as the Center for Geography Discipline to help with the implementation of the 2006 curriculum standard. This center not only provides teacher workshops but also coordinates the teacher support teams from each city and county to develop numerous teaching materials such as test items, animations, lesson plans, and others. These teaching materials, including many GIS tutorials and lesson plans, are free of charge for any senior high school. Third, nationwide geospatial data archives, such as National Geographical Information System (NGIS) and Taiwan e-Learning and Digital Archives Program (TELDAP) have provided digital spatial data available over the Internet since the late 1990s. Fourth, the private vendors, such as Interactive Digital Technologies Inc. (IDT, the Esri official distributor) and SuperGeo Technologies Inc. (SuperGeo, a local GIS software company) also provide GIS tutorials, training programs, and teaching materials to geography teachers. For example, when IDT sells ArcGIS software to high schools, the package includes bonus materials such as 1:25,000 digital geographical data for the whole nation, the satellite imagery of the campus (24 km × 24 km), GIS tutorials, and remote sensing tutorials. Fifth, GIS maintains an important topic in the standardized college entrance exams. The statistical data from 2002 to 2005 show that 8.3% of the test items in General Scholastic Ability Test and 16.7% of the test items in the Department Required Test are related to GIS or spatial information. The significant portion of GIS in the standardized test attracted the attention of students, teachers, parents, educational administrators, and even the general public (Lin & Lay, 2006).

There were 1,507 senior high school geography teachers in the country by 2009. Chen and Wang (2009) conducted a nationwide survey of 362 geography teachers around the country in 2009 (Table 29.1). The samples of this survey included 27 teachers from the GIS seed schools and 335 teachers from the regular schools in proportion to their numbers. They found that almost all of the geography teachers (approximately 100%) feel very confident about teaching about GIS. As for teaching with GIS, 74% of seed school teachers and 48% of non-seed schools teachers are

Table 29.1   A national survey of senior high school geography teachers on GIS education (Chen & Wang, 2009)

	Regular schools (%)	Seed schools (%)
I can teach about GIS	98	100
I can teach with GIS	48	74
I think GIS is essential for teaching geography	80	74
I can assist students to conduct GIS projects	40	74
My students are required to use GIS software	26	78
What are the barriers of teaching GIS (Top 3)	1. Limited class hours (92%) 2. Costly software (86%) 3. Lack of digital data (75%)	1. Limited class hours (91%) 2. Costly software (78%) 3. Difficult to learn software (57%)

able to use GIS to teach geography. With more software and hardware resources, over 70% of the seed school teachers ask students to use GIS software, and they can train students to conduct GIS research projects. However, only 26% of the non-seed teachers ask their students to use GIS software, and 40% of them are able to assist students in GIS projects. The common barriers to teaching GIS in the geography classrooms are insufficient class hours and the high cost of GIS software. Even though the seed schools are funded by MOE to buy the lab kit licenses of ArcGIS, it is still costly for the schools to keep upgrading the software themselves when the funded project ends after one year.

It is worth mentioning that GPS is another spatial technology popular in the senior high schools. It has been introduced in the national curriculum standard of senior high school geography since 2006 as a data collection tool for fieldwork. Although GIS is promoted by MOE in a top–down approach, GPS is gaining traction among geography teachers nationwide in a bottom–up approach. In fact, most GPS-related teacher workshops were requested by geography teachers themselves. The easy use of GPS, especially its function of navigation and perfect compatibility with Google Earth, favored fieldwork of senior high school geography. When the 2007 Asia-Pacific Regional Geography Olympiad (APRGEO) and the 2010 International Geography Olympiad (IGEO) were held in Taiwan, GPS was especially introduced into the fieldwork test and social activities. These international competitions held in Taiwan reflected local geography teachers' interest in GPS.

29.4 Case Study: National Yilan Senior High School

The National Yilan Senior High School (YLSH) is located at Yilan City in northeast Taiwan. It was selected as one of the MOE's seed schools in 2005. This school shows a model of how the seed of GIS is falling on good ground and gradually producing plentiful crops. In 2005, Shih-Yao Chou and Ming-Hung Hsu, the geography teachers of YLSH, developed 5 GIS lesson plans including 3D mapping of campus, mapping of the urban heat island, a GPS survey for community resources, a viewshed analysis for a local landscape pavilion, and the simulation of sea level rise around Yilan County. All these lessons addressed local environmental issues with the support of spatial technologies. They also led two student teams to join the 2005 GIS national competition. The first team assessed the vulnerability of every school in Yilan County to the tsunami and optimized the escape routes for each school. They won a gold medal in the student thesis contest. The second team drew the presidential election maps of 1996, 2000, and 2004 showing the power shift of two major parties of Taiwan. This team won an excellence award in the map contest. Their success stories were published in a national newspaper and inspired many students to conduct GIS research projects for the national competition. The YLSH received 9 student awards from 2005 to 2009 in the GIS national competition and the national Geography Olympiad. In 2008, Shih-Yao Chou and Ming-Hung Hsu also won an excellence award in the GIS lesson plan contest. In their lesson plan, they prepared students for both learning about and learning with GIS in the field. After all students learned basic GIS skills in the classroom, the teachers invited 18 volunteer students from Grade 10 to 11 to join a GIS summer program (12 hours). This program engaged students in a local environmental issue that is transforming the land use around a local lake from agriculture to recreation. Students were grouped into 6 teams, and each team was required to digitize land covers from aerial photos, calculate hill slope using a DEM, test water quality parameters using a water quality meter and GPS, create water quality maps and a landslide prediction map, and interview local residents (Figs. 29.1 and 29.2). Each

Fig. 29.1 The YLSH teacher briefed students on the use of water quality meter and GPS before the fieldwork

Fig. 29.2 YLSH students demonstrated the output maps of their GIS analysis

team had to reach a conclusion on either approval or disapproval of the development plan based on their investigation. Finally, each team acted as one of the stakeholders, such as government officials, tourists, local residents, fishermen, environmentalists, and vendors in a simulated public hearing to lead controversial opinions among stakeholders to a general agreement. In this program, GIS helped students to participate in environmental planning and empowered them to make spatial decisions effectively.

29.5 The Challenge Ahead

Since the education authority of the central government recognized the need of GIS in national development, GIS has been required in the curriculum standards, textbooks, and college entrance exams. It seems that the seed of GIS has taken root in the geography discipline of senior high schools as a whole. However, the uptake of GIS in secondary schools is significantly unequal. It is largely concentrated at the senior high school level and mostly on the discipline of geography. Furthermore, the MOE pools most resources in those seed schools and consequently creates a gap between regular schools and seed schools. Without a doubt, the central-oriented strategy such as the project of seed schools is very efficient for promoting GIS in secondary schools in the early stages. The challenge in the next few years is to spread the seeds of GIS to all senior high schools, other disciplines, and junior high schools.

It is likely that in this current decade GIS in the secondary schools of Taiwan will be implemented in two directions. First, GIS will be vertically introduced into junior high school geography and horizontally introduced into the other disciplines of social studies at the senior high school level as a regular tool. For example, history teachers will apply historical GIS to their teaching and teachers of civic education will take advantage of public participation GIS (PPGIS) for their issue discussion. Secondly, the MOE may remove one of the major barriers that complicates the implemention of GIS in secondary schools, which is the unaffordable cost of maintaining desktop GIS software year after year. By offering "Cloud GIS," the MOE can provide ready-to-use data, imagery, maps, tutorials, lesson plans, and GIS tools to secondary schools. When GIS data and software become accessible to all secondary teachers through a Web browser and free of charge, the diffusion of GIS in secondary schools will speed up.

References

Chen, C. M., & Wang, Y. H. (2009). *GIS education in the senior high schools of Taiwan: results of a national survey of geography teachers*, Unpublished manuscript.

Executive, Y. (2001). *Plan to develop knowledge-based economy in Taiwan, Council for Economic Planning and Development*. Accessed January 22, 2010, Web site: http://www.cepd.gov.tw/att/dot/ppt.gif

Lay, J. G., & Chiu, H. C. (2003). *E-generation education and GIS skills*. The Proceedings of 2003 National Geographical Conference of Taiwan, pp. 296–306.

Lay, J. G., & Yu, C. C. (2004). Cases study on GIS proficiency of high school teachers. *Bulletin of the Geographical Society of China, 33*, 21–47.

Lin, F. Y., & Lay, J. G. (2006). A study on the geographic information exam questions of the college entrance examination. *Journal of Cartography, 16*, 167–190.

MOE. (2009). *An education overview of Taiwan*. Accessed January 22, 2010, MOE Web site: http://english.moe.gov.tw/ct.asp?xItem=4133&CtNode=2003&mp=1

National Statistics. (2009). *Statistical yearbook of the Republic of China 2008*. Accessed January 26, 2010, Statistical Bureau Web site http://www.stat.gov.tw/public/data/dgbas03/bs2/yearbook_eng/y028I.pdf

Population Reference Bureau. (2009). World population data sheet. Accessed January 22, 2010, Population Reference Bureau Web site: http://www.prb.org/pdf09/09wpds_eng.pdf

Wang, S. F. (2001). *The research in the geographic information ability of high school teacher*, Unpublished master's thesis, National Taiwan University, Taipei.

Wang, Z. F. (2009). *Teaching with GIS in 2006 high school geography curriculum standard*, Unpublished master's thesis, National Taiwan University, Taipei.

Wu, C. S. (2006). Examination and improvement on education reform in Taiwan: 1994–2006. *Bulletin of National Institute of Education Resources and Research, 32*, 1–21.

Wu, C. S., & Kao, C. P. (2007). The analysis of the reform of secondary education in Taiwan: 1994–2007. *Bulletin of National Institute of Education Resources and Research, 34*, 1–24.

Chapter 30
Turkey: GIS for Teachers and the Advancement of GIS in Geography Education

Ali Demirci

30.1 Introduction

The education system in Turkey has undergone many changes over the last decade. Traditionally, investment in education was perceived to be mainly a matter of construction of new schools and classrooms. However, there is a growing emphasis on matters of purpose, methods, and curriculum in schooling. In other words, Turkish educators are asking why and how to teach what in schools. Making radical changes in school curricula, providing teachers with more opportunities to receive more relevant and continuous in-service education, restructuring of educational faculties to equip prospective teachers with necessary knowledge and skills in teaching their subjects, developing new teaching and learning methods and generalizing them in schools, and providing each school with Information Technology (IT) classrooms along with other necessary computer and Internet infrastructure are among the most significant indicators of this change in perceptions in Turkey. In the midst of these fundamental changes in schooling, various methods and techniques are being investigated, developed, and tested in schools for creating active and applied learning environments. One of the technologies that has gained attention recently in Turkish schools as an efficient teaching and learning tool is Geographic Information Systems (GIS). How much has Turkey achieved in understanding and benefiting from GIS in secondary schools? This and other similar questions are answered in this chapter after a short section describing the general characteristics of the education system in Turkey with an emphasis on secondary education.

A. Demirci (✉)
Fatih University, Istanbul, Turkey
e-mail: ademirci@fatih.edu.tr

A.J. Milson et al. (eds.), *International Perspectives on Teaching and Learning with GIS in Secondary Schools*, DOI 10.1007/978-94-007-2120-3_30,
© Springer Science+Business Media B.V. 2012

30.2 General Structure and Characteristics of Secondary Education

Turkey has a population of 74 million (TÜİK, 2010). In the 2008–2009 education year, 21% of this population were enrolled in primary and secondary schools (NES, 2009). Educating all the population at school age with professional teachers, adequate resources, and modern curriculums are the key issues to be achieved in today's Turkey.

Formal education in Turkey consists of pre-school, primary, secondary, and higher education institutions. Pre-school, primary, and secondary education are controlled and supervised by the Ministry of National Education (NES, 2009). Primary education is compulsory for all citizens and provides eight years of uninterrupted education to children between ages 6–14. Secondary education which lasts four years is optional for children between ages 15–18. Secondary education comprises different categories of educational institutions including General High Schools, Anatolian High Schools, Science High Schools, and Vocational High Schools. Anatolian and Science High Schools provide lessons in a number of foreign languages with a special emphasis on science education. Average grades of students from primary education and scores of students from an exam that is conducted in Grade 8 are used together to determine which students will enroll in Anatolian and Science High Schools. No nationwide exam is required for students who wish to pursue secondary education in General or Vocational High Schools.

All curricula for primary and secondary education are prepared by the Ministry of National Education in Turkey and their implementation in public and private schools is mandatory. The subjects taught in schools vary from general high schools to vocational-technical high schools. The courses taken by a Grade 9 student in a high school can be generally divided into two groups: general common cultural courses and elective courses. Turkish language and literature, religions, culture and ethics, history, geography, mathematics, biology, physics, chemistry, hygiene, foreign language, and physical education are general education requirements that are taken by all students in this Grade compulsorily. A variety of elective courses are offered to students in the same Grade from drawing, music, traffic to computer, foreign language, and environment. Students who complete Grade 12 take a university entrance exam and, if a high enough score is achieved, are placed in a university in order to continue higher education.

At the university level, education faculties are responsible for preparing prospective teachers in all subjects taught in secondary schools. The students who graduate from university education programs after five years of education including a year-long teaching internship or from art and science programs by taking necessary pedagogical formation courses can start teaching in high schools. However, they must take the Public Personnel Selection Exam (KPSS), which contains questions from three different areas, namely general culture, general skills, and pedagogic formation in order to be able to work in public schools. In-service education organized by the Ministry of National Education is the number one method for teachers' professional development in the country, although it is not legally compulsory and available to all teachers. The annual workshops organized in each province by teachers who teach the same subjects in high schools provide teachers with another opportunity to share and gain knowledge, skills, and experiences in their profession.

30.3 The Advent and Diffusion of GIS in Secondary Schools

The diffusion of GIS to secondary education did not take place quickly in Turkey. During the 1980s, GIS began to be used in the Turkish public and private sectors. The first private GIS company was established in 1981 (Demirci, 2009) and the General Command of Mapping was the first public institution where GIS began to be used in 1986 for military purposes (Yomralıoğlu, 2002). GIS spread to a larger base of users in different public and private sectors especially in the 1990s, and the first national conference on GIS was held in 1994 in Turkey (Yomralıoğlu, 2002). The use of GIS in higher education started in the late 1990s and 2000s in a number of departments; namely, geodesy and photogrammetry, geography, urban planning, geology, forest sciences, and soil sciences. The number of departments offering GIS education in universities increased almost 15-fold between 1991 and 2004 (Olgen, 2005). Although the number of disciplines and academicians utilizing GIS has increased very rapidly, the diffusion of GIS throughout higher education has not been completely achieved yet in Turkey because there are still many departments such as geography, environmental engineering, and geology without a single GIS course in their programs.

The diffusion of GIS into secondary education started in the 2000s, mainly with geography courses. Geography is a compulsory course in secondary schools for students of Grades 9–12 and it contains a broad range of topics in physical and human geography, which are taught at global and regional scales. Demirci et al. (2007) conducted a survey of geography teachers from 36 high schools from around the country in 2006 and found that none of the 46 teachers used GIS in their schools. Demirci (2009) also conducted a more detailed study of 79 teachers from 55 high schools in 33 provinces in 2008. He found that one-third of the teachers did not know what GIS was and 82% of the teachers did not know how to use GIS in their lessons. Only seven teachers indicated that they used GIS in their geography lessons, yet none of the high schools had GIS software.

The most significant step toward integrating GIS into secondary schools was taken with the new geography curriculum in 2005. Prior to the new curriculum, GIS was only discussed at a very rudimentary level in some textbooks (Demirci, 2008a). The 2005 curriculum suggests that "teachers should introduce GIS into the classroom depending on the availability of adequate hardware and software in their particular schools" (Karabağ, 2005). The inclusion of GIS in the geography curriculum was a turning point in the history of GIS in secondary schools in Turkey. The new curriculum has not only attracted teachers' interest in GIS, but it has also raised many important concerns among Turkish educators. One of the key concerns was that the curriculum neither supplied teachers with a definition of GIS nor revealed any method for implementing it in geography lessons. Another concern was whether or not teachers had enough knowledge, skill, experience, and resources to use GIS in their lessons. These concerns have triggered a surge of interest and significant research in finding ways to make GIS an effective educational tool in schools. Numerous articles, books, and doctoral dissertations with special emphasis on teaching with GIS in secondary schools have been written since 2005 (e.g., Demirci, 2006, 2007, 2008a, 2008b; Karatepe, 2007; Tuna, 2008). Recent research on teaching with GIS in secondary education focused on three main themes: (1) determining the status of GIS in schools, (2) identifying the main obstacles to incorporating GIS in lessons, and (3) developing strategies and methods to use GIS in schools (Demirci, 2008a, 2009).

The new curriculum also prompted universities, the Ministry of Education, and other institutions, such as the Turkish Geographical Society, to organize workshops and seminars to prepare teachers to incorporate GIS into their lessons and to identify the main obstacles facing them. The first "GIS for Teachers" workshop was organized in 2004 with 30 teachers participating from around the country (Demirci, 2009). The number of workshops and seminars has increased since then, yet the number of participating teachers remains small. The majority of the workshops and seminars have focused on teaching about GIS rather than teaching with GIS and have not provided teachers with sufficient knowledge, skills, and resources, such as digital data, Turkish-language software, and lesson plans to use GIS in their lessons. These problems were the main motivation for the publication of a book entitled *GIS for Teachers* (Demirci, 2008b), that provided teachers with lesson plans, digital data, and software in the Turkish language, along with information about GIS and its utilization in education.

30.4 Learning Geography with GIS-Based Exercises

Although the potential benefits of GIS for teachers and students have been documented well in the country (Demirci, 2008a, 2008b, 2009, Demirci & Karaburun, 2009), developing various methodologies to use GIS in different classroom or laboratory settings still remains a challenge. Yet, a limited number of cases can provide examples of how GIS is used in secondary schools in Turkey. These cases can be categorized into two distinct formats: implementing GIS-based exercises and conducting GIS-based projects in schools.

The book *Mapping Our World* (Malone, Palmer, & Voigt, 2003) was a significant inspiration for developing and testing the implementation of GIS-based exercises in secondary schools in Turkey. The first comprehensive study was conducted in 2006 to test the applicability and effectiveness of a GIS-based exercise in Grade 9 geography lesson (Demirci, 2008a). In this study, a GIS-based exercise was prepared along with digital data and student handouts to teach students the relationship between plates, earthquakes, and volcanoes as part of the geography curriculum. In the second step of the study, fourteen teachers from nine high schools participated in a workshop in which they were introduced to GIS and given the necessary materials to implement the GIS-based exercise in their schools. At the end of the study, twelve teachers could not implement the exercise successfully in their schools due to a variety of reasons ranging from a lack of time to accomplish the task to inappropriate physical conditions. However, the performance of the remaining two teachers and their students revealed that GIS-based exercises contributed to students' comprehension and level of success in geography lessons. After comparing the pre and post-exercise exam scores of the students, it was found that the use of GIS increased the student's success on geography lessons by 38% and 51% in two different schools (Demirci, 2008a).

The implementation of GIS-based exercises in the secondary schools of Turkey remained very limited until the book *GIS for Teachers* was published in 2008. The book, published by Fatih University with the support of Esri and many other national and international institutions, was designed to resolve some of the obstacles to incorporating GIS in secondary schools in Turkey. These obstacles were that the teachers did not: (1) know what GIS is and why it is used in different disciplines, (2) know the importance of GIS for education and how they can utilize GIS in their lessons, (3) have GIS software prepared in the Turkish language, (4) know how to use GIS software, and (5) have educational materials such as lessons plans, digital data, and GIS-based exercises to be used in their lessons (Demirci, 2009). By combining theory and practice, the book provided teachers with ArcGIS 9.2 including a one-year free single-user license, digital data, nine GIS-based exercises, and other necessary guidance to use GIS in geography lessons. The subjects of the GIS-based exercises in the book *GIS for Teachers* were selected according to the new secondary school geography curriculum (Table 30.1). Each exercise was prepared in three successive stages: (1) preparation, (2) implementation, and (3) assessment. The exercises require students to observe data in different layers, to answer questions by identifying relationships between graphic and nongraphic data, and to conduct basic spatial analysis operations. The steps of each exercise and the types of query and analysis to answer the questions were described in the handouts so that students who did not have prior GIS knowledge and skills could implement the exercises by themselves.

The *GIS for Teachers* book has received favorable attention from a wide range of disciplines and has enabled many teachers to learn about GIS and to implement GIS-based exercises in their lessons, yet only about 500 teachers and prospective teachers had obtained the book by the end of 2010. This number is not satisfactory given that there are approximately 7,000 secondary school geography teachers in the country. Of course, supplying teachers with GIS books, software, digital data, and other educational materials is not sufficient for teachers to use GIS effectively in their lessons. As indicated in many other studies, teachers must be encouraged,

Table 30.1 The title of the nine GIS-based exercises in the book *GIS for Teachers* (Demirci, 2009)

Exercise number	Topic of the GIS-based exercises
1	Which parts of the world are the most tectonically active?
2	Why do earthquakes occur frequently in Turkey?
3	Where are the minimum and maximum temperature differences on Earth?
4	What is the relationship between distribution of precipitation and vegetation cover in the world?
5	Which provinces receive the highest and the lowest amount of rainfall in Turkey?
6	Which are the most crowded and the least crowded cities and regions in the world?
7	Which economic activity earns people the most in the world's most populated country?
8	Which Turkish province has the greatest population losses and gains due to migration?
9	Why do countries need to unite under different institutions and organizations in the world?

trained, and supported if they are to effectively utilize GIS in their lessons (Bednarz, 2004; Demirci & Karaburun, 2009; Kerski, 2003; Lloyd, 2001; Marsh et al., 2007; Meyer et al., 1999).

30.5 Conducting Projects with GIS Technology at Schools

Conducting projects is a well-known method in secondary education in Turkey. Many national and international project competitions are organized each year for secondary school students in subjects such as physics, biology, chemistry, mathematics, English language, and computer courses. The number of project competitions with special interest on social sciences has also increased in recent years. Although almost all students in secondary schools are involved in a number of projects in every academic year, GIS is not typically a technology they are aware of and use in their projects. One of the few cases of conducting GIS-based projects in secondary schools was performed in 2008 at Anatolia Teacher High School (ATHS) in Bilecik, a northwestern province of Turkey. Fifteen Grade 9 students at ATHS studied whether the current network for the seven minibus lines in the province of Bilecik is fairly distributed across the city. Privately owned minibuses are a common mode of transportation in Turkey and they carry passengers only short distances within the provincial boundaries (Tuna, 2008). Students obtained street and building data from the municipality, identified minibus lines on GIS, collected information by interviewing drivers, identified the buildings located within and outside of 200 m distance from each line by creating buffers, analyzed minibus lines and buildings together, and suggested some changes of minibus lines where many buildings were farther than 200 m away. The whole project was finalized in a semester and students gained basic GIS understanding and skills.

In order to identify the main problems before implementing GIS-based projects at schools, a research project has been initiated in 2009 by the support of The Scientific and Technological Research Council of Turkey (TÜBİTAK). The project mobilized around 300 students in three public high schools to work on GIS-based projects in their school districts. Nine GIS-based projects have been conducted in pilot high schools with the titles shown in Table 30.2.

A number of activities have been targeted for the students in each project such as planning the project, making a literature review, conducting interviews, learning about GIS, collecting and storing data, using GPS, analyzing data with GIS, and reporting and disseminating results. In one of these projects, 15 Grade 9 students at Prof. Dr. Mumtaz Turhan Social Science High School have worked together for mapping the noise pollution in the Bahcelievler district where their school is located. By using GPS and an appropriate device, students measured the level of noise from 108 locations and prepared the noise pollution map of their study area on GIS by using ArcGIS 9.2 (Fig. 30.1).

Table 30.2   The GIS-based projects being conducted at three pilot high schools in Turkey (Demirci, 2009)

Project number	The title of the GIS-based projects
1	How livable is the Sisli district for disabled people? Analyzing the pedestrian ways for people with wheelchair
2	How many cars we can park in the Sisli district? Analyzing existing parking areas to find a solution to parking problem
3	How healthy are we in the classroom? Analyzing temporal and spatial change of Carbon dioxide gas in the school building
4	Do we have enough containers for solid waste in the Bahcelievler district? Evaluating the locations of waste containers and their capacity
5	Which locations are good for the thieves in the Bahcelievler district? Preparing illumination map of the district
6	How quiet is my neighborhood? Mapping the noise pollution in the Bahcelievler district
7	Which classrooms are heated better in my school? Analyzing spatial change of temperature within the school building
8	Who pollutes the coastal zone more, dogs or people? Coastal zone planning in the Buyukcekmece district with GIS
9	Which place is more favorable for fish in the Buyukcekmece bay? Analyzing marine pollution in the Sea of Marmara

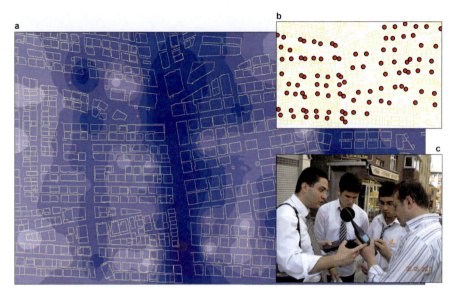

Fig. 30.1 Layout views and students from the GIS project conducted in the Prof. Dr. Mumtaz Turhan Social Science High School, **a** The students produced the noise pollution map in the study area **b** The students measured the noise level from 108 locations, **c** Students are measuring noise from the streets together with the experts from Istanbul Metropolitan Municipality

30.6 Opportunities, Challenges, and Plans to Make GIS a Common Educational Tool in Schools

A number of opportunities can be identified in Turkey to make GIS a common educational tool in secondary schools. The most significant of them is that geography has a strong place in the secondary school curriculum. There are four compulsory geography courses and GIS has been recognized officially as an important technology for teaching and learning geography. The new curriculum has created an atmosphere in the country to motivate and support teachers, school managers, governmental agencies, and academicians to take action in incorporating GIS in schools. The numbers of academic studies and opportunities for in-service education in the field of GIS have been increasing in Turkey. In one of the recent activities, a two-weeks long in-service education event was specifically organized for secondary school geography teachers in the summer of 2010. Around 150 teachers have gained knowledge and skills in various techniques used in geography lessons including GIS. Apart from the knowledge about GIS and its use in geography lessons, the teachers have used ArcGIS 9.2 with its basic tools and implemented a GIS-based exercise (Fig. 30.2). The teachers who were provided a GIS software and digital data in the in-service education have already established an e-mail group to share their experiences in using GIS in their lessons. Furthermore, the existence of resources such as the book *GIS for Teachers*, and the improvement of computer and Internet

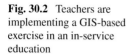

Fig. 30.2 Teachers are
implementing a GIS-based
exercise in an in-service
education

infrastructure in schools can be regarded as important opportunities in Turkey to expand the use of GIS in secondary schools. One of the most recent initiatives to provide schools with advanced ICTs in Turkey is the Fatih Project through which all classrooms in primary and secondary schools will be equipped with a computer, Internet access, and an intelligent electronic whiteboard until the end of 2013. This certainly will help teachers to use GIS in their lessons.

Despite these advances, various obstacles exist against the diffusion and efficient use of GIS in secondary schools. Many technical, systematic, and pedagogic obstacles have been identified (Demirci, 2006, 2007, 2008a, 2009). The main obstacles to using GIS in Turkish schools are related to teachers' perceptions, knowledge, skills, and experiences. Most geography teachers do not know what GIS is or how it can be used in geography lessons, and many teachers lack the time and interest to learn and use GIS. The difficulty in using GIS software is also a significant challenge for teachers who are already struggling to use computers and the Internet effectively in their lessons.

Various projects and initiatives are being planned by individual academicians or institutions such as the Ministry of Education, the Turkish Geographical Society, and universities to promote and expand the use of GIS in secondary schools throughout the country. As one of the most significant initiatives, the Ministry of Education is planning to organize more effective in-service education in cooperation with universities for not only informing teachers about GIS technology but, more importantly, for providing teachers with skills, digital data, software, and the opportunity to produce their own GIS-based lesson plans to be shared with other teachers and to be implemented in their own lessons. In addition, the organization of GIS-based project competitions is another significant initiative that should motivate and encourage school managers, teachers, and students to use GIS in their lessons. Fatih University is planning to organize such a competition, in cooperation with the Ministry of Education, in 2011. Among the other plans that are likely to be realized in the near future are launching a GIS for teachers Internet portal, publishing a journal specifically focusing on teaching with spatial technologies, encouraging master and doctoral students to work on GIS and education, offering effective GIS courses to prospective geography teachers, and starting certificate programs on teaching with GIS. As the recent developments and progress reveal, the number of teachers and schools interested in using GIS is increasing. If the outcomes of the current and planned initiatives are positive, it is likely that GIS is going to diffuse to many other schools in Turkey in the near future.

References

Bednarz, S. W. (2004). Geographic information systems: A tool to support geography and environmental education? *GeoJournal, 60*, 191–199.

Demirci, A. (2006). CBS'nin Türkiye'deki yeni coğrafya dersi öğretim programına göre coğrafya derslerinde uygulanabilirliği. In A. Demirci, M. Karakuyu, & M. A. Mcadams (Eds.), *Proceedings of the 3. GIS Days in Turkey* (pp. 241–248). Istanbul: Fatih University Publications.

Demirci, A. (2007). Coğrafi Bilgi Sistemlerinin ilk ve ortaöğretim coğrafya derslerinde bir öğretim aracı olarak kullanılması: Önem, ilke ve metotlar. *Öneri Dergisi, 28*(7), 377–388.

Demirci, A. (2008a). Evaluating the implementation and effectiveness of GIS-based application in secondary school geography lessons. *American Journal of Applied Sciences, 5*(3), 169–178.

Demirci, A. (2008b). *Öğretmenler için CBS: Coğrafi Bilgi Sistemleri*. Istanbul: Fatih University.

Demirci, A. (2009). How do teachers approach new technologies: Geography teachers' attitudes towards Geographic Information Systems (GIS). *European Journal of Educational Studies, 1*(1), 57–67.

Demirci, A., & Karaburun, A. (2009). How to make GIS a common educational tool in schools: Potentials and implications of the *GIS for Teachers* book for geography education in Turkey. *Ozean Journal of Applied Sciences, 2*(2), 205–215.

Demirci, A., Taş, H. İ., & Özel, A. (2007). Türkiye'de ortaöğretim coğrafya derslerinde teknoloji kullanımı. *Marmara Coğrafya Dergisi, 15*, 37–54.

Karabağ, S. (Ed.). (2005). *Coğrafya dersi öğretim programı. Talim ve Terbiye Kurulu Başkanlığı*. Ankara: Gazi Kitabevi.

Karatepe, A. (2007). *The use of Geographic Information Technologies in geography education*, Unpublished doctoral dissertation, Marmara University, Institute of Education Sciences, Istanbul.

Kerski, J. J. (2003). The implementation and effectiveness of geographic information systems technology and methods in secondary education. *Journal of Geography, 102*(3), 128–137.

Lloyd, W. J. (2001). Integrating GIS into the undergraduate learning environment. *Journal of Geography, 100*(5), 158–163.

Malone, L., Palmer, A. M., & Voigt, C. L. (2003). *Mapping our world: GIS lessons for educators*. Redlands, CA: Esri.

Marsh, M., Golledge, R., & Battersby, S. E. (2007). Geospatial concept understanding and recognition in G6- college students: A preliminary argument for minimal GIS. *Annals of the Association of American Geographers, 97*(4), 696–712.

Meyer, J. W., Butterick, J., Olkin, M., & Zack, G. (1999). GIS in the K-12 curriculum: A Cautionary note. *Professional Geographer, 51*(4), 571–578.

NES (2009). *National Education Statistics, Formal education 2008–2009*. Ministry of Education.

Olgen, M. K. (2005). Türkiye'de CBS eğitimi. In M. K. Ölgen (Ed.), *Ege Coğrafi Bilgi Sistemleri Sempozyumu Bildiriler Kitabı* (pp. 9–22). Izmir: Ege Üniversitesi.

TÜİK. (2010). Adrese dayalı nüfus kayıt sistemi (ADNKS), 2010 Nüfus Sayımı Sonuçları. Accessed August 2, 2011, http://tuikapp.tuik.gov.tr/adnksdagitapp/adnks.zul

Tuna, F. (2008). *Taking the advantages of Geographic Information Systems (GIS) to support the project based learning in high school geography lessons*, Unpublished doctoral dissertation, Marmara University, Institute of Education Sciences, Istanbul.

Yomralıoğlu, T. (2002). *GIS activities in Turkey*. Proceedings of international symposium on GIS, pp. 834–840, Istanbul, Turkey.

Chapter 31
Uganda: Educational Reform, the Rural–Urban Digital Divide, and the Prospects for GIS in Schools

Jim Ayorekire and Revocatus Twinomuhangi

31.1 Introduction

The education system in Uganda has gone through numerous changes over the last decade, including changes in curriculum, increased funding by the government, and the increased role of the private sector in the provision of education from primary to university education level. The broad aims of the education system in Uganda are eradication of illiteracy, promotion of scientific, technical, and cultural knowledge; promotion of national unity; and promotion of moral values. In order to improve the quality of education service delivery, the government of Uganda has identified and integrated Information Communication Technology (ICT) in the education curriculum.

31.2 General Structure and Characteristics of Education in Uganda

Formal education in Uganda consists of preprimary (nursery), primary, secondary/postprimary, and tertiary/post secondary education levels. Primary level (for pupils between 6 and 12 years) covers a period of 7 years from primary one to primary seven. The postprimary level is composed of secondary education and Business, Technical, and Vocational Education and Training (BTVET). Secondary education (for students aged between 13 and 18 years) is divided into 4 years of ordinary level ('O' level) with classes from senior one to four and two years of advanced level ('A' level) with classes from senior five to senior six. After primary school, students also have an option of joining BTVET institutions that mainly offer practical and hands-on learning experiences. There are a variety of BTVET institutions, and they include community polytechnics, farm schools, technical colleges, vocational

J. Ayorekire (✉)
Makerere University, Kampala, Uganda
e-mail: jayorekire@arts.mak.ac.ug

A.J. Milson et al. (eds.), *International Perspectives on Teaching and Learning with GIS in Secondary Schools*, DOI 10.1007/978-94-007-2120-3_31,
© Springer Science+Business Media B.V. 2012

training institutes, colleges of commerce, health training institutions, and other specialized training institutions in forestry, fisheries, wildlife, meteorology, and survey. The students are awarded certificates and diplomas after one or two years depending on the discipline.

Primary and postprimary education in Uganda are controlled and supervised by the Ministry of Education and Sports, while the National Council of Higher Education (NCHE) supervises tertiary education. Primary and secondary education (up to senior four) are compulsory for all citizens under the Universal Primary Education (UPE) and Universal Secondary Education (USE) programs. The curriculum and instructional materials for the primary and postprimary levels are developed by the National Curriculum Development Centre (NCDC) and implemented by the Ministry of Education and Sports (MoE&S), which has developed education policies, strategies, laws, and regulations. The Ministry of Education is composed of the directorates of Basic and Secondary Education, Higher Education, Education Standards, and Technical and Vocational Education, which implement policies and regulations to ensure quality education and consistency in the curriculum.

As a means of increasing the number of pupils attending primary school, the government of Uganda in 1997 introduced Universal Primary Education (UPE), which has resulted in a steady increase in pupil enrollment from 3 million in 1996 to 7.9 million in 2008. However, UPE created a demand for universal postprimary education as more pupils were completing the primary education level. In order to ensure equitable provision of quality education and training to all Ugandan pupils who successfully completed the primary education level, the government in 2007 started implementing the Universal Post Primary Education and Training (UPPET) program. As a result, the gross enrollment rate (GER) in schools is steadily increasing, with the government putting more emphasis and funding in the education sector.

31.3 Secondary Education in Uganda

Secondary school education in Uganda is guided by the National Secondary Education curriculum, most of which was inherited from the British at independence in 1962. However, the curriculum is currently undergoing a number of reforms with emphasis on altering not only the number of subjects that are taught but also 'how and why' they are taught. In the first phase of the curriculum reform in 2009, the number of subjects taught at 'O' Level was reduced from 42 to 18 in an effort to make it easier for students to easily select career-linked subject combinations. According to the National Curriculum Development Centre, the second phase of the reforms, due to begin in 2010, will include additional reforms to change the national examination system, rewrite the learning materials, and further lower the number of subjects to be taught.

Uganda has 2,908 secondary schools, with 57,158 secondary school teachers. Of the 2,908 schools, 57.3% are privately owned, 31.4% government owned, and 11% community owned. Of all the schools, 10.5% are urban based, 16.1% are peri-urban, and 37.8% are rural (MoE&S, 2008a). Before the introduction of Universal Secondary Education (USE) under the Universal Post Primary Education and Training policy in 2007, a small proportion of those who would complete Primary seven were able to join secondary education. Under the USE program, students attending secondary school drastically increased to 1.08 million in 2008 from 728,562 in 2005. The students, who were in senior one in 2007 when USE was introduced, sat for the Uganda Certificate of Education examinations in 2010. There are 803 government-owned schools and 428 privately owned secondary schools participating in the USE program, which currently caters to students in senior one to senior four. The UPPET policy was also later extended to cover the Business, Technical and Vocational Education and Training institutions. In 2009, there were 5,174 BTVET students in 46 government institutions and 1,076 students in 16 private institutions benefiting from UPPET capitalization grants (MoE&S, 2008b).

In order to be licensed by MoE&S to teach at the secondary school level in Uganda, teachers must have qualified from National Teacher Training colleges – as Grade V teachers – or they must have qualified from a university as a graduate teacher with a bachelor's degree or a postgraduate diploma in teacher education. However, due to the limited number of qualified teachers, some schools still employ unlicensed and even untrained teachers, especially in rural areas. When teachers qualify and begin teaching, there are very few professional development programs to enhance their skills and to introduce them to new knowledge and technologies. Over the last two years the Ministry of Education has carried out some in-service training, but this has mainly been limited to science and mathematics teachers from the newly created districts. A number of teachers and head teachers have also received ICT in-service training.

Although the number of students attending secondary schools is increasing, a big challenge exists in regard to the urban–rural divide in terms of the number and quality of schools, teachers, and facilities. Most rural schools tend to (1) have limited infrastructure and teaching equipment like classrooms, laboratories, electricity, and computers; (2) attract few qualified teachers as most of them are not willing to teach in hard-to-reach rural areas; and (3) rely on untrained teachers. This rural–urban divide is reflected in the national secondary school examination results, where students in urban schools tend to perform better than those in rural schools.

31.4 Integration of ICT in Secondary Schools Curriculum

The government of Uganda, along with several other developing countries, has identified Information and Communication Technology (ICT) as a key area for development, especially as countries move towards e-government and e-commerce management structures. Over the years, the government has integrated ICTs in the secondary school curriculum. The purpose is to equip students with skills that can enhance their participation in an increasingly information driven society. In order to provide a framework for mainstreaming this integration, The Republic of Uganda (2005) was drafted to rationalize and harmonize the ICT-related activities and programs that had hitherto been fragmented and uncoordinated.

The ICT policy highlights the need to revise the curriculum, train teachers to be ICT literate, and provide the required infrastructure. However, although the National Curriculum Development Centre has developed an ICT secondary school curriculum, urban schools have recorded a greater increase in the use of ICTs than rural schools (Eremu, 2009; MoE&S, 2005). This is because most rural schools are faced with challenges that constrain the use of ICT including lack of electricity, inability to afford computers, and a limited number of teachers who are ICT trained. To address the rural–urban schools ICT digital divide, a number of initiatives and projects have been rolled out by government to different levels of the education system and in different rural parts of the country. With support from a number of international organizations, various ICT initiatives have been started, and they include the following: SchoolNet Uganda, Connectivity for Educator Development (Connect-ED), CurriculumNet project, Global Teenager Program, U-Connect, among others. These initiatives focus on not only providing computers but also training teachers in using them for lesson preparation and teaching.

Despite a general increase in the use of ICT in Uganda's education system, the integration of ICT in most schools in the country is largely limited to teaching basic computer-use skills (with emphasis on word processing applications) and, to a lesser extent, the use of the Internet for accessing educational materials. Most secondary school teachers have not been equipped with skills and software to make use of ICT in the preparation of lesson plans, teaching materials, and the delivery of lessons. Based on the foregoing discussion, it is recommended that ICT integration in the education system emphasize not only merely "having ICT" but also "what can be done with ICT" to avoid the utilization of the technology even in the schools that already have adequate ICT equipment.

31.5 Advent and Integration of GIS in Secondary Education

The Government of Uganda views land and geographic information as an infrastructure that can support planning, decision making, and the country's economic growth. The public and the private sectors of Uganda are in the process of adopting GIS in their work operations. Although GIS awareness can be said to be increasing, detailed GIS knowledge is often lacking in the user environment. However, many people are increasingly realizing that adding a spatial dimension to their information may improve their understanding and work performance.

In the country's education sector, the Ministry of Education and Sports has put in place the Education Management Information System (EMIS), which is used to assess the state of the country's education system on an annual basis to ensure effective education management and planning. The primary source of data for EMIS is the annual schools census, which collects information about the schools, pupils, teachers, nonteaching staff, classrooms, textbooks, and infrastructure. To support education planning and management in a comprehensive way, GIS is integrated in EMIS so as to have a spatially oriented database that captures names, location, type, physical state, coverage, and accessibility of schools.

GIS is however a relatively new tool in the education system in Uganda, and, as a result, its usage is still limited. Its introduction and use have mainly been limited to the university level where, since the 1990s, it has been integrated in a number of disciplines such as geography, urban planning, tourism, environment management, surveying, and, recently, computing and information technology. Even then, GIS is still not widely used at all universities in the country. Out of the 29 universities in the country, it is mainly at Makerere University where GIS is widely used. The limited adoption of GIS in the university education system is mainly a result of limited number of lecturers trained to use it; limited availability of computers; high expense of the software; and limited knowledge about GIS and its capabilities.

As noted earlier, although the number of secondary schools using and teaching ICT is increasing, it is mainly for basic computer lessons. There is an absence of applied computer based tools being used in the teaching and learning process. A case in point is GIS, which is one of the tools that could be integrated in teaching subjects like geography. In addition, GIS not only has the potential for further enhancing the teaching of ICTs but can also increase students' interest, motivation, and imagination in learning ICTs (MoE&S, 2005). Despite the importance and capabilities of GIS in teaching and learning, currently there is hardly any secondary school in the country using GIS in teaching and learning or for any other purpose. More important, GIS also does not feature anywhere in secondary school curriculum (NCDC, 2008; UNEB, 2008).

Apart from the absence of GIS in Uganda's secondary school curriculum, the non use of GIS at secondary schools can also be attributed to its absence in teaching and learning in teacher training institutions. In most of the country's teacher training colleges and universities, teachers are not introduced and trained in GIS. For instance, at Makerere University (the oldest and largest public university in Uganda), although GIS is taught to geography students (which is a teaching subject), students pursuing a bachelor's degree in education do not study it. The justification put forward is that since GIS is not in the secondary school curriculum, geography teachers do not need or are not required to apply GIS skills when teaching the secondary schools. In addition, the ICT education policy does not specifically include computer-based teaching and learning tools such as GIS to be integrated in the secondary school curriculum.

Despite these limitations, a few International Secondary Schools in Uganda have started to integrate GIS into their curriculum, especially in teaching geography. Most of these schools offer curriculum following the UK- based International General Certificate of Secondary Education (IGCSE) and the General Certificate of Education (GCE). For example, at Rainbow International School in Kampala, the capital city, teachers use WebGIS tools such as Google Earth and Google Maps to teach various aspects of geography like land use, location, and human settlement.

However the schools (and others) do not have installed GIS software on their computers and as such no serious GIS is taught. Furthermore, the use of online spatial programs in teaching and learning is constrained by slow and intermittent internet connections which make images and maps take a lot of time to be downloaded.

Overall, most secondary school teachers in Uganda have not been trained in the use of GIS as a teaching and learning tool. Few teachers indicated that in 2006 they had received introductory GIS training under the Uganda GLOBE (Global Learning and Observations to Benefit the Environment) program, which was operating in 37 secondary schools across the country. However, the teachers had not used GIS in class since schools did not have GIS software installed on the school computers.

31.6 Challenges and Prospects of GIS Integration in Secondary Schools

Despite the various advantages and potential benefits of GIS to teachers and students, its integration into the secondary school curriculum in Uganda faces a number of challenges. The major challenge is the limited knowledge among education policy makers, head teachers, and other teachers about the capabilities and potential benefits that GIS can contribute to the secondary school teaching and learning environment. If this awareness can be accomplished, GIS can easily be adopted by various secondary schools in Uganda, given the existing opportunities. One reason is that the secondary school curriculum is undergoing review and more emphasis is being given to not only having ICT taught but also emphasizing how it can be applied as a tool to enhance learning and teaching. This means that introducing GIS will be more easily accepted and adopted once its capabilities and benefits are known to all stakeholders, especially those involved in curriculum design and implementation.

The prospects of popularizing GIS are high since a good number of schools already have functioning computer laboratories and will not have to incur a big cost setting up GIS laboratories. Such schools already have teachers who have the basic skills in computer use and some with basic GIS knowledge. Therefore it becomes easy to have in-house training for such teachers on how to use GIS techniques in various subjects offered in the curriculum. These teachers can then train other teachers in various subjects on how to integrate GIS in their various disciplines. Experiences from the international schools in the country that have begun integrating GIS in their curriculum, like Rainbow International School, offer good case studies that would be of great benefit when planning for the introduction of GIS in other secondary schools.

Furthermore, most teacher training colleges and universities have computer laboratories that can be used to teach GIS to teacher trainers and trainees. Staff from universities, such as Makerere, who already have competence in GIS applications and teaching can carry out such trainings. At Makerere University, GIS would easily be integrated in the curriculum of students pursuing Bachelor of Education degrees, since various disciplines are already teaching GIS. Readily available GIS teaching materials specifically developed for teacher training (such as the ones by Esri) would further facilitate the easy introduction of GIS at teacher training institutions and at various secondary school levels.

In general, the introduction of GIS in secondary schools teaching and learning is essential as it could help spur student's motivation into various science and technology careers, an aspect that the Ministry of Education and Sports strategic plan aims to achieve, among others. GIS as an educational tool would not only enrich the curriculum by creating an inquiry-based learning environment but also equip secondary school graduates with specialized skills and knowledge required to pursue tertiary education or competitively enter the workforce.

References

Connect-ED. (2003). *Training Uganda's teachers with technology (2001–2003)*. Accessed January 11, 2010, http://learnlink.aed.org/Projects/uganda.htm

Eremu, J. (2009). *ICT connectivity in schools in Uganda*. Accessed May 14, 2010, http://www.ftpiicd.org/iconnect/ICT4D_Education/ICTEducation_Uganda.pdf

GLOBE. (2009). *Global learning and observations to benefit the environment*. Accessed November 20, 2009, http://www.globe.gov

MoE&S. (2005). *Draft policy for information and communication technology in the education*. Ministry of Education and Sports, Kampala.

MoE&S. (2008a). *Education sector strategic plan 2007–2015*. Ministry of Education and Sports, Kampala.

MoE&S. (2008b). *The education and sports sector annual performance report*. Ministry of Education and Sports, Kampala.

National Curriculum Development Centre (NCDC). (2008). *Uganda certificate of education: Geography Teaching Syllabus*. Kampala, Uganda.

The Republic of Uganda. (2005). Draft Policy for information and communications in the education sector. Uganda.

UNEB. (2008). Uganda National Examination Board (UNEB). *Uganda advanced certificate of education: Regulations and syllabuses*. Kampala, Uganda.

Chapter 32
United Arab Emirates: Building Awareness of GIS in Education Through Government and University Outreach

Mohamed R. Bualhamam

32.1 Introduction

Since 1970, the United Arab Emirates (UAE) has undergone a profound transformation from an impoverished region of small desert principalities to a modern state with a high standard of living. The country offers a fully fledged educational system from primary level to university. Much has been achieved since the early 1970s, but efforts are now being made to improve the educational environment for all pupils, in line with a reevaluation of the role of government. Now that the infrastructure is in place, the educational focus is on devising and implementing a strategy that will ensure that the youth of the country are ready to meet the challenges of the twenty-first century workplace. The education strategy of the country aims to introduce the latest information technology at all school levels. Integrating Geographic Information Systems (GIS) concepts into schools is very important to teach students relevant skills in spatial analysis, reasoning, and data processing. GIS in secondary schools education is a new topic in the UAE and there are few schools and teachers who have used GIS in classrooms. This chapter reviews the current status of the secondary school system in the UAE and the prospects for learning about GIS, followed by case study that illustrates how GIS is used by one secondary school. The opportunities, challenges, and obstacles associated with incorporating GIS into UAE secondary schools are then presented.

32.2 Current Status of Secondary Education

State-funded educational opportunities in the UAE have blossomed since the establishment of the federation when only a tiny minority of the population had access to formal education. A comprehensive free education system is now available to all

M.R. Bualhamam (✉)
United Arab Emirates University, Al Ain, United Arab Emirates
e-mail: mbualhamam@uaeu.ac.ae

A.J. Milson et al. (eds.), *International Perspectives on Teaching and Learning with GIS in Secondary Schools*, DOI 10.1007/978-94-007-2120-3_32,
© Springer Science+Business Media B.V. 2012

students (Belfekih, 1993). Education at the primary and secondary level is universal and compulsory up to Grade 9 (Alhousani, 1996). At the start of the 2008–2009 academic year, 648,135 students enrolled in over 1,259 government and private schools throughout the country (Vine, Al Abed, Hellyer, & Vine, 2009). Substantial progress has also taken place in the private sector, which accounts for nearly 40% of the student population at the kindergarten, primary, and secondary levels (Carr, 2004).

The existing educational structure, which was established in the early 1970s, is a three-tier system covering twelve years of education. The Primary tier is six years in length for students of ages 6–12 years. Preparatory is three years in length for students of ages 12–15 years. Secondary is three years in length for students of ages 15–18 years. Secondary education consists of a common first year followed by specialization in science or arts. In the common first year, there is a common curriculum for all streams, which include Islamic education, Arabic language, English language, mathematics, geography, physics, chemistry, biology, geology, and computer science. After the first year, students choose either the arts or science sub-stream. They continue to study the general stream subjects, but add subjects from their chosen stream. Arts sub-stream subjects include history, geography, sociology, and economics. In the science sub-stream, subjects include physics, chemistry, biology, and geology. The certificate awarded after finishing a secondary school is a Secondary School Leaving Certificate. Also, the country has a technical secondary school system with a technical secondary diploma awarded upon completion. The Secondary cycle focuses on preparing students for university, technical or vocational training, or for joining the workforce directly (Belfekih, 1993). Most students in the UAE tend to pursue higher education at universities rather than at technical or vocational institutes (Vine et al., 2009).

Although the UAE has achieved much in the field of education there is a real awareness that constant updating of policy and continual investment in infrastructure is required to ensure that graduates are properly equipped to enter the workforce and assist in the country's development (Vine et al., 2009). To this end, the Ministry of Education has released a draft policy document outlining a strategy for educational development in the UAE up to the year 2020 based on several five-year plans (Ministry of Education, 2002). The strategy aims to introduce the latest information technology at all levels including a computer for every ten students at kindergarten, every five students at primary school, every two students at preparatory school, and a computer for every student at secondary school (Ministry of Education, 2002). The primary focus of attention will be on the needs of students, especially through the promotion of self-learning and continuous education programs. There will also be training programs for teachers since surveys have shown that although the majority of students can use computers and the Internet, their teachers are less familiar with this technology (Alhousani, 1996).

All teachers are now required to have a university degree, and pre-service and in-service programs are being redeveloped or introduced to raise the scientific and educational skills and cultural background of teachers (Belfekih, 1993). Areas of particular focus include English language skills, computing and information technology, and professional education courses. The government wants to increase the numbers of UAE nationals working as teachers and has set a target that by 2020, 90% of teaching staff will be UAE nationals (Ministry of Education, 2002).

32.3 Prospects for Learning About GIS

Formal education in the field of GIS in the UAE is very limited. There are few institutes that offer GIS courses and few universities where these courses are offered in the curriculum. Since 1995, the UAE University has offered one of the few programs available in GIS. Whilst GIS now forms part of many fields of study, an environmental theme appears to dominate the GIS courses, research, and consultancy (Yagoub, 2002). In addition, many other bodies outside the higher education system now provide a range of short courses in GIS, with the objective of raising awareness and introducing GIS theory, practice, or both. However, many universities in the UAE are still struggling to find GIS instructors, establish computer laboratories to teach it, and rearrange dense curricula to find space where they can squeeze in a GIS course (Bualhamam, 2007).

In addition, GIS technology is rapidly expanding in government and the business sectors. Today, many government organizations and private companies in the UAE utilize GIS for managing their data and making strategic decisions. Furthermore, GIS is now accepted as a mainstream technology within local government and utilities, particularly for managing infrastructure. These systems are also a key tool for many environmental and natural resource management agencies. There is great demand for GIS professionals in the UAE job market but there are not enough graduates who fulfill the high requirements of the market. Thus, an imbalance is observed in the supply sector (Bualhamam, 2007).

GIS has not yet been widely introduced to secondary school geography education in the UAE. Schools in the UAE are starting to introduce computers in their classes, which will have a direct impact on GIS education. Integrating GIS concepts into secondary education is very important to teach students relevant skills in spatial analysis, reasoning, and data processing (Alibrandi, 2003).

The new curricula of geography and geology in Grade 10 contain a broad range of topics in GIS. The textbook of geography includes: principles of aerial photography, concept of scale of images, various formats of data in computers, concept of resolution in satellite images, introduction to the Global Positioning System (GPS), and general concepts of GIS and using different layers of data for GIS analyses along with computer operations and fieldwork. The textbook for geology includes: an introduction to different sensors, electromagnetic radiation and spectrum, atmospheric effects, and spectral behavior of different objects on earth's surface. These topics are the first step to implementing GIS in schools in the UAE and teaching students fundamental information about this technology, but the Ministry of Education will need to define GIS materials and identify strategies for teaching about GIS to enhance the introductory implementation of GIS (Wiegand, 2001). Many teachers will not begin to consider using GIS software if there are not clearly defined materials available (Kerski, 2003). In addition, effective teaching strategies need to be identified by the Ministry of Education based on the operational requirements of the classroom and on the learning strategies of students using GIS.

Also, many GIS conferences in the UAE try to address the awareness and the need for GIS among the students and how it is going to help them in their day-to-day lives. Dubai Municipality organized the First GIS Students Forum in cooperation with the Ministry of Education during the Map Middle East 2006 conference, in which 60 students in the age group of 12–15 from 12 secondary schools in the UAE participated. The conference provided the participants firsthand experience with map making and data collection, in addition to presentations by international experts in the field. During the hour-long session called, "Field Survey Game" held outside the Trade Centre in Dubai, students joined a professional team from the Geodesy Unit in the Planning and Survey Department at Dubai Municipality in conducting a field survey for map making. They were shown how the data for map making is collected by GPS and processed on location, and then how it is incorporated into a map on the computer. The same GIS students participated in the Map Middle East conferences in 2007, 2008, and 2009 (http://www.mapmiddleeast.org). These forums aim at spreading awareness and setting up a sound foundation of geospatial science and technology by educating future generations.

32.4 Case Study: Al-Ain Al-Namothajia Secondary School

Over the past two years there has been an interest, particularly among GIS professors from UAE University, in introducing GIS technology into schools. One of the few cases of conducting GIS-based projects in secondary schools was performed in 2008 at Al-Ain Al-Namothajia secondary School in Al Ain city. The students in this school successfully used GIS to inquire about information and about how to make maps (Fig. 32.1). Through this training, GIS offers a scientific inquiry tool, which

Fig. 32.1 Students in a GIS training at Al-Ain Al-Namothajia Secondary School

allows the visual learners to gain particular benefit from the mapping process where they customize the properties of overlaid map layers to reflect the range of values and properties that best fit their learning and the themes of the lesson (Yagoub, 2009).

Also, the students used 2006 IKONOS satellite imagery and vector data for spatial analysis. Examples of these analyses included finding the nearest school from your home, where schools are clustered, how many people lived near your school, and to consider whether your city is in need of new schools. Students proposed new schools, considering criteria and factors such as proximity to population centers, minimum traveling distance, accessibility from roads, distance to industrial areas, and so on (Yagoub, 2009).

Because GIS requires the manipulation of graphics, charts, maps, and data, it gives students a fuller practice of the array of computer tools than spreadsheet, presentation, or desktop publishing software alone (Bednarz, 2004). In this project, students analyzed aerial photographs, field data, satellite images, attribute data, and maps in a real-world problem-solving environment. GIS provides students with an idea of the complexities of the world in which they live. Students wrestled with data relevance and data quality, identifying relationships, and drawing interpretations (Lemberg & Stoltman, 2001).

This project introduces GIS technology to the students and teachers, provides a way to address the geography standards, and matches a constructivist teaching style. The case study showed that GIS increased student motivation for geography, increased the students' ties to the surrounding community, altered communication patterns with fellow students and with teachers, stimulated students who learn visually, and reached students who are not traditional learners (Yagoub, 2009). Also, the case study showed that one of the chief constraints on GIS learning is not hardware or software, but the spatial perspective of teachers and students. Most students lacked this spatial perspective and were uncomfortable with the problem-solving style of learning, of which GIS takes advantage.

A workshop under the theme "New Trends in Geography – GIS in Schools" was organized on March 10, 2009 with the objective of increasing awareness among teachers, students, schools, and decision makers in the Ministry of Education, and called for greater support for GIS in schools. More than 200 schoolteachers attended the workshop. Five professors from the Department of Geography and Urban Planning addressed various issues related to the new wave in GIS and how it can be implemented in schools (Fig. 32.2). Similar workshops on a national level were scheduled for 2010 to create a critical mass of teachers with significant GIS exposure. The UAE University plans to continue to organize training courses, workshops, and seminars to support secondary schools to implement this technology in the schools.

Fig. 32.2 Teachers in a GIS workshop at UAE University

32.5 The Opportunities, Challenges and Obstacles of Incorporating GIS into UAE Secondary Schools

Integrating GIS technology into secondary schools education can aid students' abilities in problem solving, critical thinking, and communications (Baker & White, 2003). The education strategy of the UAE needs to prepare students in the field of GIS at the school level. The new curricula for geography and geology for secondary schools provides good opportunities and personal development for improving students' understanding of GIS concepts. Recently, many who graduated from the Geoinformatics track in the Geography Department at the UAE University joined the Ministry of Education as geography teachers. They came with a good background in GIS and that will help the Ministry of Education to introduce GIS in their classes. In addition, the universities in the UAE supported and assisted students and teachers in GIS education for secondary schools through many training courses, workshops, and seminars that increased the awareness of this new technology in schools.

Currently, the Ministry of Education does not run any GIS course in any stage. It should begin to utilize this technology at a number of levels: teacher training, curriculum integration (among a variety of disciplines in addition to the geosciences), equipment purchase, and provision of curriculum material (Bednarz & van der Schee, 2006). The continuing challenges for GIS education in secondary schools in the UAE are as follows:

1. The lack of understanding of the importance of GIS education by secondary school educators and Ministry of Education decision makers. This may be one of the most critical issues that deserves special attention.

2. The cross-disciplinary characteristics of GIS are not widely understood by the educational community and therefore GIS is usually regarded as a tool for geography (Wiegand, 2001). The incorporation of GIS into secondary schools in the UAE requires that the cross-disciplinary nature of the tool is made more tangible for the potential users.
3. The provision of teachers for areas that require some form of specialization continues to be a problem in some contexts (White & Simms, 1993). The teaching manpower requirements in GIS for secondary schools in the UAE have to be assessed. Training for geography teachers needs to be strengthened so that geography teachers feel confident that they can employ an open-ended, computerized tool in the classroom (Bednarz, 2004). Teachers have to be sure that GIS is important before they use it in the classroom (Demirci, 2009). The Ministry of Education should provide teacher training in GIS and then help teachers apply their knowledge in the classrooms.
4. Technical factors such as the availability of software and data: The price of professional desktop GIS software is usually too high, but free Web-based applications could be the starting points for introducing GIS into secondary schools (Wiegand, 2001). Also, the lack of access to data is considered as one of the obstacles to implementing GIS in secondary schools. The Ministry of Education needs to establish a spatial data library and design GIS exercises as a part of the geography curriculum (Keiper, 1999).

Current GIS efforts in the secondary schools in the UAE are small scale and often nontransferable. Despite its potential, most secondary schools in the UAE still lack the resources and know-how required to use GIS in education. In addition, the lack of understanding of the importance of GIS education by decision makers, the lack of data and GIS software, and the lack of teacher training are some of the main difficulties that secondary schools in the UAE are facing today in implementing GIS.

GIS will play a more important role in education and development in the UAE. Students are now more aware of maps and satellite imagery (Wiegand, 2001). Concentrated effort in the Ministry of Education should be devoted to integrating GIS into secondary schools education. This integration depends on several factors including teacher's preparation, availability of laboratory facilities in each school, data availability, and need for curriculum innovation (Kerski, 2003). It is hoped that this study will encourage others to pursue avenues of research and development to take advantage of GIS technology and methods to improve the quality of GIS education in the UAE.

References

Alhousani, H. A. (1996). *Secondary school administration in the United Arab Emirates: Its reality, problems and methods of development.* Unpublished doctoral dissertation. University of Wales, Cardiff.

Alibrandi, M. (2003). *GIS in the classroom: Using geographic information systems in social studies and environmental science*. Heinemann, Portsmouth, NH.

Baker, T., & White, S. (2003). The effects of GIS on students' attitudes, self-efficacy, and achievement in middle school science classrooms. *Journal of Geography, 102*(6), 243–254.

Bednarz, S. W. (2004). Geographic information systems: A tool to support geography and environmental education. *GeoJournal, 60*, 191–199.

Bednarz, S. W., & van der Schee, J. (2006). Europe and the United States: The implementation of geographic information systems in secondary education in two contexts. *Technology, Pedagogy and Education, 15*(2), 191–205.

Belfekih C. M. (1993). *Modern secondary education in the United Arab Emirates: Development, issues and perspectives*. Unpublished doctoral dissertation. Temple University Graduate Board.

Bualhamam, M. R. (2007). The preparation of students for the job market in geographical information systems at the United Arab Emirates University: Considerations and challenges. *Annual Journal of the Heidelberg Geographical Society, 21*, 191–202.

Carr, K. (2004). *An evaluation of Private schools in the UAE*. Abu Dhabi: Ministry of Education, Scientific Research Administration.

Demirci, A. (2009). How do teachers approach new technologies: Geography teachers' attitudes towards Geographic Information Systems (GIS). *European Journal of Educational Studies, 1*(1), 57–67.

Keiper, T. A. (1999). GIS for elementary students: An inquiry into a new approach to learning geography. *Journal of Geography, 98*, 47–59.

Kerski, J. J. (2003). The implementation and effectiveness of geographic information systems technology and methods in secondary education. *Journal of Geography, 102*(3), 128–137.

Lemberg, D., & Stoltman, J. P. (2001). Geography teaching and the new technologies: Opportunities and challenges. *Journal of Education, 181*(3), 63–76.

Ministry of Education. (2002). *Strategy for further educational development in the United Arab Emirates up to the year 2020*. Abu Dhabi: Scientific Research Administration.

Vine, P., Al Abed, I., Hellyer, P., & Vine, P. (2009). *United Arab Emirates 2009*. London: Trident Press Ltd.

White, K. L., & Simms, M. (1993). Geographic information systems as an educational tool. *Journal of Geography, 92*(2), 80–85.

Wiegand, P. (2001). Geographical information systems (GIS) in education. *International Research in Geographical and Environmental Education, 10*(1), 68–71.

Yagoub, M. M. (2002). *Geographical information systems (GIS) education and application in the United Arab Emirates*. UAE University. Accessed December 27, 2009, http://faculty.uaeu.ac.ae/myagoub/main_GIS.htm

Yagoub, M. M. (2009). *Geographic information systems (GIS) for Schools in the UAE*. Unpublished paper.

Chapter 33
United Kingdom: Realizing the Potential for GIS in the School Geography Curriculum

Mary Fargher and David Rayner

33.1 Introduction

As we write this chapter, school systems in the UK continue to experience a period of far-reaching and potentially very significant change. New national curricula, a degree of shift in emphases toward the central role of teachers in shaping these, and ongoing debate about creating a future-oriented, relevant schooling for young people make this an exciting time to be focusing on the potential role that GIS can play in twenty-first-century schools. Curriculum change is not the only salient characteristic of schools in the UK; the complexity of our 'overall school picture' is another. The United Kingdom of Great Britain and Northern Ireland includes England, Wales, Scotland, and Northern Ireland, each of which exhibits, to a greater or lesser degree, distinct differences in terms of its own education system. While we highlight these differences where appropriate here, our focus is on commonalities across the UK in realizing the promise of, meeting the challenges to, and articulating the prospects of teaching and learning with GIS.

33.2 UK Secondary School Education: A Synopsis

Pupils in the UK must attend school between the ages of 5 and 16. Over 90% of them are taught in the state-maintained sector, the majority of the remainder attend independent or public schools, and a small number are home schooled. Most pupils (76%) choose to stay in full-time education after 16, either in their secondary school sixth form or at further education (tertiary) colleges. State schools in England, Wales, and Northern Ireland follow a statutory National Curriculum (NC) that stipulates which subjects are to be taught, programmes of study, and

M. Fargher (✉)
Institute of Education, University of London, London, UK
e-mail: m.fargher@ioe.ac.uk

A.J. Milson et al. (eds.), *International Perspectives on Teaching and Learning with GIS in Secondary Schools*, DOI 10.1007/978-94-007-2120-3_33,
© Springer Science+Business Media B.V. 2012

the setting of standards and assessment. In Scotland, where the school curriculum is nonstatutory, responsibility for what is taught lies directly with schools and their local authorities (although national guidelines are available). Within the framework of the National Curriculum, however, schools are free to plan and organize teaching and learning in the way that best meets the needs of their particular students.

Eligibility to teach in the state-maintained sector in the UK requires qualified teacher status (QTS). Some prospective teachers both gain a Bachelor of Education degree (BEd) and carry out initial teacher training (ITT) at the same time. However, most gain a specialist subject degree first and then undertake a one-year postgraduate course. Both routes into teaching combine practical and academic training in education. At the beginning of their teaching careers, newly qualified teachers (NQTs) complete an induction year before acquiring full professional status. Teachers also follow a programme of continuing professional development (CPD), which encompasses in-house, local authority, government and optional, commercially provided training. Professional standards provide the backdrop to in-school discussions about teacher performance and individual career development. These standards define the professional attributes, knowledge, understanding, and skills for teachers at each stage of their career path.

33.3 Using GIS in UK Schools

GIS is used mainly in the support of geography education in the UK. Geography is a compulsory subject for pupils aged 5–14. At 14, students can choose to opt for geography at the General Certificate of Secondary Education (GCSE), which is examined usually after a two-year course. Students may opt to study geography further, at AS level at 17 and at A2 level at 18.

In a recent report on geography in English schools, OFSTED (Office for Standards in Education, Children's Services and Skills) stated the following:

> The use of geographical information systems is revolutionising and extending pupils' experiences in geography. Visual images from around the world bring immediacy to the learning. Satellite technology can bring landscapes to life. Data can be overlaid and used with interactive maps to interpret patterns and solve problems. Yet, while some schools are using these new opportunities very effectively, others are reluctant to take the risk or do not have the skills or resources to do so. (OFSTED, 2008)

The challenges of using GIS in classrooms continue to deter its use in many schools. In schools where GIS is in use, a range of software options are made including Esri ArcGIS, Digital Worlds, AEGIS 3 GIS for schools, and earth viewers such as Google Earth in teaching and learning geography. There is little doubt that the revised school curriculum guidelines in the UK provide plenty of potential opportunities for using a range of GIS. For example, in the Geography National Curriculum for England, the Key Stage Three programme of study (11–14-year-olds) specifically states that pupils should be given the opportunity to learn both about GIS and through GIS in terms of its value and applications in the real world. At GCSE (14–16-year-olds), subject criteria stipulate that learners at this stage must be able to demonstrate how to use GIS in carrying out geographical investigations. A-level students (16–18-year-olds) are now expected to be able to synthesize geographical information in a range of formats.

It is true that very little guidance exists for teachers wishing to develop pedagogies involving GIS. It is also clear that already overstretched teachers, including a large minority of nongeography specialists, are unlikely to embrace GIS without a considerable amount of sustained support. Some researchers suggests that difficulties experienced by some teachers confronted by the spatial querying that is an inherent aspect of using GIS is a specific area of teacher training for which more needs to be specifically provided (Bednarz & van der Schee, 2006). Other evidence also suggests that all ICT-related curriculum innovation requires teachers to develop their pedagogic strategies in considerably more complex ways than they may have done before. They need to be cognizant with a range of areas of knowledge: their own subject content knowledge, knowledge about how students think and learn, and, increasingly in the twenty-first-century classroom, knowledge about how to use technology.

The use of GIS in UK schools has to an extent become characterized by a number of these significant challenges. Initial costs and training commitments can be key causes for concern for many teachers. Access to ICT facilities remains a major stumbling block for many schools. At the secondary level in particular, the perception that using GIS requires the booking of whole ICT suites in order to be effective is discouraging many teachers from using it. Lack of user-friendly GIS data has, until recently, proven to be a real barrier for many. The search for appropriate data can be particularly daunting for the uninitiated. Even commercial GIS providers have acknowledged in the past that that UK-oriented GI can be inappropriate or difficult to manipulate in a school environment.

Prior to the 2008 National Curriculum revisions, there was little mention of GIS in National Curriculum guidance. The new curriculum highlights the use of GIS as a central element of the study of geography. Specifically GIS is highlighted as a key tool that pupils can use to collect data in inquiry learning. A growing body of literature indicates that effective use of GIS can augment inquiry-based geography education (Baker & Bednarz, 2003; Kerski, 2003). Specifically within the UK context, inquiry is identified as one of the key aspects of the National Curriculum for geography (Roberts, 2003). In conjunction with a GIS, a constructivist approach to learning allows the student to generate a range of digital GI data, which they can manipulate and adapt as they decide and require. Constructivism can be interpreted in education as a model of practice in which students create and adapt their own knowledge and skills. The teacher is seen as facilitator and student peers are often engaged in collaborative learning. Linking and layering geospatial information becomes not only part of the overall learning process but also a pivotal aspect of the pupil's own understanding (Fargher, 2006). In the two case studies that follow, we explore further the potential of using GIS in the classroom and the associated challenges for teachers.

33.4 Case Study 1: Bishop's Stortford College

Bishop's Stortford College is an independent integrated day and boarding school in Bishop's Stortford, Hertfordshire. Bishop's Stortford is an example of a secondary school using GIS in both innovative and challenging ways. While the school is extremely well resourced and this has clearly contributed to the flourishing use of GIS in the school, this case study illustrates another significant factor affecting this success: the crucial role of critical approaches by teachers to incorporating GIS.

GIS at Bishop's Stortford College has been developed through the use of Esri ArcView 9. GIS is used to support teaching and learning geography across the school with pupils ranging from Year 9 (at 14 years of age) to Year 13 (at 18 years of age). The philosophy of approach to the use of GIS in the college is very much based on the initial consideration of the underlying educational reasons for using GIS to support students' learning (O'Connor, 2006). When planning to use GIS at Bishop's Stortford, three 'conceptual levels' for learning activities are considered:

- Level 1: GIS use – presenting spatial data
- Level 2: GIS skills – processing and analyzing spatial data
- Level 3: GIS skills – data input and editing

When pupils are being introduced to GIS for the first time (Level 1), pupils are taken through the rudimentary elements of opening packages, navigating basic functions and turning layers off and on in a GIS to identify patterns of crime in England and Wales. With careful pedagogical consideration, such a simple approach can still engender access to potentially powerful GI data through which challenging geographical questions can be addressed without the need for a more complicated technological pupil competence. Once pupils are more proficient with GIS, methods of data processing and display (Level 2) are introduced. This involves rather more sophisticated teacher input: pupils operating at this level are introduced to GIS theory, principles of mapping, and a degree of statistical analysis. With this advancement, the opportunities for individuals to develop their geographical knowledge more independently through the use of GIS become more apparent. For example, access to these kinds of GIS skills can allow, for example, Year 10 students (at 15 years of age) to identify and analyze economic disparities at a global scale. At Level 3, pupils are able to apply advanced GIS techniques – this can involve them in decision making usually in association with independent projects and fieldwork (O'Connor, 2006). The use of GIS at Bishop's Stortford College is based on an emphasis on progression of pupils as they are exposed to it throughout their time at school. Key to this approach is the focus on the critical pedagogical decisions that are made in and around the use of GIS to support the teaching and learning of geography.

33.5 Case Study 2: 'Spatially Speaking': A GIS Support Project

At present, teachers are not universally prepared to integrate GIS into their continuing professional development. Training tends to be dependent on the knowledge of geography educators involved in teacher training courses and the state of GIS development within individual schools. Several successful support schemes have been implemented through national associations such as the Geographical Association (GA), the Royal Geographical Society (RGS), and the Ordnance Survey (the UK's leading map agency). For example, 'Spatially Speaking,' the GA's GIS project, ran a continuing professional development programme over a two-year period (2005–2007) aimed at developing learning and teaching approaches (in geography) with GIS. The project experienced a number of successes in developing teacher's professional skills with GIS, creating support materials and resources and disseminating findings and outcomes to geography teachers and the wider geography educator network in the UK. Findings from the project also outlined the key challenges to using GIS effectively often lie with technical issues but also the steep learning curve associated with teachers developing effective pedagogy around GIS (Spatially Speaking, 2007).

33.6 Prospects

It is perhaps not so surprising that evidence suggests that only the very best classroom practitioners are using technologies such as GIS successfully (Fargher, 2006). However, the challenges for teachers attempting to develop effective pedagogic practices around geotechnologies such as GIS can be daunting (Kankaanranta, 2005). Research suggests that connections between successful implementation of a technological innovation such as GIS and the role of teachers has not yet been fully explored or understood. Steps towards developing the use of GIS in schools can be quite readily identified, if not quite so easily implemented. For example, creating a teacher qualification in GIS use in the classroom where participants actually have to use GIS in their teaching would help to raise the profile of the significance of spatial thinking in geography. Preparing teachers to plan the integration of GIS into each key stage of the curriculum would help to develop future use of GIS. Building on well-documented empirical evidence, there is a range of GIS software available that could be used appropriately for each curriculum phase. GIS-enabled handheld GPS technology is beginning to open up another exciting world of opportunities in developing spatial thinking in school-aged learners. The latter potentialities will not be further explored here, but these innovations serve as reminders that we have to evaluate and plan for using such technology meaningfully.

When used effectively, GIS can provide comprehensive learning environments with proven potential for problem solving of real world relevance. For example, the 2010 Haiti earthquake was graphically displayed in the public media often via GIS. Such examples of applied information technology, though, illustrate to educators and learners the potential of these technologies for transferring and displaying knowledge and making powerful connections between places. Fully realizing this potential in schools will require particularly careful stewardship that embraces the momentum of current curriculum change and continued innovation in geotechnologies (McInerney, 2003).

There is growing evidence that harnessing the full potential of digital technologies such as GIS in schools may require more effective partnerships between industry and education. Increased uptake of GIS use in geography school education is likely to require development of a 'shared vision' with proven practices established for supporting innovators and beginners (Fargher, 2006). Sustained success with introduction of ICT in education depends on a complex range of internal and external factors including training, resources, and issues around organizational methods and technical support (Minaidi & Hlapanis, 2005). Integrated strategies at both a local and national level which recognize and address the complexities of these interrelated factors will be required if we are finally to fully realize the potential of GIS for geography education (Fargher, 2006).

> A classroom that uses GIS as a problem-solving tool is a classroom in which the walls are invisible and the teacher and student assume roles that are non-traditional. ... Adopting this technology is not for the fainthearted. But integrating GIS into the curriculum rewards teachers by creating intellectually challenging and demanding learning opportunities. (Audet and Ludwig, 2000).

References

Audet, R., & Ludwig, G. (2000). *GIS in schools*. Redlands: Esri Press.

Baker, T., & Bednarz, S. (2003). Lessons learned from reviewing GIS in education. *Journal of Geography, 102*, 231–233.

Bednarz, S. W., & van der Schee, J. (2006). Europe and the United States: The implementation of geographic information systems in secondary education in two contexts. *Technology, Pedagogy and Education, 15*(2), 191–205.

Fargher, M. G. (2006). *An exploration of the contribution of a 'Local Solutions' project to curriculum innovation with GIS*. Unpublished MRes dissertation, Institute of Education, University of London.

Kankaanranta, M. (2005). International perspectives on the pedagogically innovative uses of technology. *Human Technology, 1*(20), 111–116.

Kerski, J. J. (2003). The implementation and effectiveness of geographic information systems: Technology and methods in secondary education. *Journal of Geography, 102*(3), 128–137.

McInerney, M. (2003, July). *The next step with GIS in the curriculum: Approaching the question of GIS and classroom pedagogy*. Presentation to the Esri User conference, San Diego.

Minaidi, A., & Hlapanis, G. H. (2005). Pedagogical obstacles in teacher training in information and communication technology. *Technology, Pedagogy and Education, 14*(2), 241–254.

O'Connor, P. (2006). Progressive GIS. *Teaching Geography*, Autumn, 2007, 147–150.

OFSTED (2008). *Geography in schools: Changing practice*. Accessed December 8, 2010, http://www.ofsted.gov.uk

Roberts, M. (2003). *Learning through enquiry: Making sense of geography in the key stage 3 classroom*. Sheffield: Geographical Association.

Spatially Speaking. (2007). *Geographical association project*. Accessed December 8, 2010, http://www.geography.org.uk/projects/spatiallyspeaking/

Chapter 34
United States of America: Rugged Terrain and Fertile Ground for GIS in Secondary Schools

Andrew J. Milson and Joseph J. Kerski

34.1 K-12 Education in the USA

Schooling in the USA is based on a tradition of local control. From the seventeenth century Protestant Christian grammar schools of New England to the public tax-supported "Common Schools" promoted by Horace Mann during the 1840s, the formal education of children in the USA has been governed by local authorities. It was not until the late 1950s when US politicians stoked US insecurities over the USSR's launch of the Sputnik satellite into orbit that the federal government began to play a more active role in public education. The Cold War between the USSR and the USA provided US educators and scientists with the opportunity to promote legislation and federal funding for mathematics and science education. Similarly, the struggle over the racial desegregation of public schools during the 1960s required the involvement of President Lyndon B. Johnson and the US Congress. Yet, despite this increased federal influence in education over the past 50 years, governance of US public schools remains primarily a local matter.

The USA does not have an official national curriculum. Instead, each state establishes curricular guidelines and policies for achievement testing, teacher training, and high school graduation requirements. Given the differing standards and policies of each state, it is not possible to portray US education as a coherent set of practices or policies. Nonetheless, homogenizing influences in recent years have served to reduce the differences between states. For example, various national associations of educators developed voluntary subject-area standards during the 1990s (e.g., AAAS, 1993; NCSS, 1994; NGESP, 1994; NRC, 1996). These national standards have been adapted and integrated into many state curriculum guidelines. More recently, the George W. Bush administration promoted and passed through Congress the No Child Left Behind (NCLB) Act of 2001. One aim of NCLB was to introduce stronger accountability measures into public schooling. In practice, this policy

A.J. Milson (✉)
University of Texas at Arlington, Arlington, TX, USA
e-mail: milson@uta.edu

A.J. Milson et al. (eds.), *International Perspectives on Teaching and Learning with GIS in Secondary Schools*, DOI 10.1007/978-94-007-2120-3_34,
© Springer Science+Business Media B.V. 2012

has translated into greater emphasis on standardized testing in public schools and a resulting homogenization of the curriculum to focus on tested academic content (Cuban, 2008).

34.2 GIS in Secondary Education in the USA

Despite the growing standardization of US schooling at present, the integration of GIS and other geospatial technologies into US secondary classrooms remains haphazard. Milson and Roberts (2008) found that 22 states had secondary school geography standards that included reference to GIS, Global Positioning System (GPS), or the analysis of geographic data with technology, yet only five of these states had standards that promoted the application of geospatial technologies for higher-order thinking about spatial patterns. In a survey of 1,520 secondary schools that owned GIS software, Kerski (2003) found that fewer than half of the schools that owned ArcView, Idrisi, or MapInfo GIS software actually used it. Evidence suggests that this condition has improved recently as a result of the publication of curriculum materials for teaching with GIS (Baker, Palmer, & Kerski, 2009) and the development of Internet-based GIS (Baker, 2005; Kerski, 2008a; Milson & Earle, 2007). However, it remains common for US secondary teachers to work in isolation and to face numerous barriers when attempting to use GIS in their classrooms. The barriers identified in most surveys and case studies published since 1990 include insufficient access to computers, time to spend on project-based learning, support from administration, technical support, time to create or adapt curriculum, attention to GIS or spatial analysis in teacher preparation programs, and familiarity with what GIS can offer (Audet & Paris, 1997; Bednarz & Audet, 1999; Bednarz & Bednarz, 2008; Kerski, 2003; McClurg & Buss, 2007).

More important than technological, pedagogical, and administrative hurdles may be that spatial thinking and analysis lacks a home in the curriculum. If spatial analysis is perceived as a curricular add-on rather than something that is integral to instruction in tested and funded disciplines, such as language arts, science, and mathematics, then few educators will pursue it due to pressures to focus on standardized test content. Although geography could be seen as a natural home for spatial thinking and analysis, the subject has been integrated into the social studies for most of the twentieth century in the USA and is commonly given less attention than history within the social studies (Stoltman, 1990). Consequently, geography is not taught often in US secondary schools, and many geography teachers do not have the strong foundation in geographic skills and content needed to teach it with an open-ended tool such as GIS (Bednarz & Audet, 1999).

Despite these barriers, GIS has been steadily gaining ground in US secondary education due to a convergence of factors. First, the rising sense of urgency about human-caused environmental change and the need for environmental education coupled with reports on the key role of outdoor education on human health and the environment (e.g., Louv, 2005), has led to increased support of geotechnology-driven field work, even if the "field" is only the school campus. Second, the role of visualization in learning has been receiving dual support from the research community (Piburn & Reynolds, 2005) as well as from a rapid increase in the number and diversity of mapping and virtual globe tools (Schultz, Kerski, & Patterson, 2008). Third, although standardized assessment instruments tend to stifle creativity and emphasize rote memorization, the national content standards for geography and other sciences emphasize the importance of learning science through inquiry-oriented problem solving about authentic issues with real-world data. Furthermore, spatial thinking in a GIS environment was recommended to be a fundamental part of K-12 education in the USA by the National Academy of Sciences (National Research Council, 2006). Finally, US educators and policy-makers are increasingly focused on secondary school reform efforts that emphasize Career and Technical Education (CTE) and the Science, Technology, Engineering, and Mathematics (STEM) disciplines. These reform efforts are fostering interest and funding for innovative technology-driven learning and may provide the impetus needed for GIS to become a common tool for learning in US high schools.

34.3 The Landscape of Learning with and About GIS in USA Secondary Schools

Toward the end of the 1980s, a few pioneering earth and environmental science teachers in the USA began to use GIS in their secondary classrooms. These teachers perceived GIS to be an effective tool for engaging students in inquiry-oriented field-work and for connecting science content to community and societal issues (Baker & Kerski, in press). The chief proponents of GIS education at this time were indi-vidual teachers. Many of these educators spread the word of their successes with this technology at the conferences of the National Science Teachers Association (NSTA) and the International Society for Technology in Education (ISTE) and some of their stories were publicized in GIS trade magazines. By the mid-1990s, more geography and social studies teachers began to discover and use GIS in their classrooms (Keiper, 1999), but science teachers remained the primary innovators of classroom GIS use (Goodchild & Palladino, 1995; Kerski, 2003). GIS in education gained momentum by the end of the 1990s following the efforts of private vendors, such as Environmental Systems Research Institute (Esri). The outreach to educators included a set of national institutes, a repeated presence at educational and technol-ogy conferences, and a series of Educational GIS Conferences (EdGIS) championed first by Technology Education Research Center (TERC) and then by universities. Since 2000, the use of GIS in US schools has become more common due to sev-eral factors: (1) the availability of curricular materials (e.g., English & Feaster, 2002; Malone, Palmer, & Voigt, 2005), (2) funded professional development initia-tives for teachers, (3) a growing body of implementation and effectiveness research (e.g., Baker & White, 2003; Shin, 2006; West, 2003), (4) improved computers, networks, and storage media, (5) the availability of easily obtainable spatial data,

(6) national conferences such as the Esri Education Users Conference (http://www.esri.com/educ), (7) school district-wide and statewide GIS licenses and support, particularly for Esri GIS software, and (8) the increased cohesiveness of the GIS education community (Alibrandi, Milson, & Shin, 2010, Kerski, 2008b). Subjects in which GIS is most commonly used include GIS classes inside career and technical education, earth, life, and environmental science, chemistry, and geography, but subjects on the rise include history, mathematics, and English language arts.

Although there is growing momentum for using GIS in US classrooms at present, GIS is not required in the curriculum of most states (Gatrell, 2001; Milson & Roberts, 2008). Despite the absence of a curricular mandate, there are numerous teachers in the USA who act as islands of innovation by integrating GIS into teaching and learning in their classrooms. These teachers teach a variety of subject areas in both private and public schools. Some teach in small rural schools, while others work in large urban school districts. Many of these teachers have had their work profiled in journal articles and books in the past decade (e.g., Alibrandi, 2003; Alibrandi, Beal, Thompson, & Wilson, 2000; Alibrandi & Sarnoff, 2006; Audet & Ludwig, 2000; Keiper, 1999; Milson & Curtis, 2009; Milson, Gilbert, & Earle, 2007; Wiegand, 2003; Wigglesworth, 2003). Most of these examples of teaching and learning with GIS place emphasis on using the technology as a tool for learning rather than learning about the technology itself. Sui (1995) described the distinction as teaching with GIS versus teaching about GIS:

> For most geography students, GIS technology should not be an end in itself. Instead, it is a means to a higher end, to enrich geography's four grand traditions, to find new laws, and to have a more thorough understanding about human-environment interaction and various physical processes. If we fail to establish a tight bond between GIS and geography's intellectual core, GIS will remain a greatly improved means for unimproved ends. This demands that teaching about GIS should be well balanced with teaching with GIS (p. 587).

Secondary educators are much more likely to teach *with* GIS, while community college and university instructors are more likely to teach *about* GIS. A growing number of US secondary school teachers are demonstrating that the balance of the curricular and the technical – learning with GIS and learning about GIS – can occur simultaneously in an US high school. In the following case study, we will profile one such example.

34.4 Case Study: Piner High School

Piner High School (PHS) in Santa Rosa, California, USA is the site of an innovative program that we believe provides one model for the future of GIS in US secondary schools. In 2007, Kurt Kruger and Kristi Erickson, teachers at PHS, set out to establish a program that would provide students with both academic and career-oriented learning experiences. Students would learn important concepts in science, mathematics, and technology, while also gaining skills they could apply to a career in the geospatial technology industry. With the support of funding from the California Department of Education, Kruger and Erickson designed and implemented the

Geospatial Technology Pathway (GTP) program at PHS in 2008. The GTP offers a three-year course sequence for students in their sophomore, junior, and senior years of high school. US secondary students are classified as Freshman, Sophomore, Junior, and Senior during their first through fourth years of high school, respectively. Typically, Freshman students are entering their tenth year of formal schooling and are 14–15 years of age. Seniors are in their final year of secondary schooling and are 17–18 years of age. Seniors who successfully complete all course requirements mandated by their state are awarded a High School diploma.

During year one, students enroll in an introductory GIS course in which they are introduced to GIS and GPS concepts and skills while learning physical geography. The textbooks for the course include *Geosystems* by Robert W. Christopherson and the *Our World GIS Education* series from Esri press. The course emphasizes local investigations, problem-based learning, and environmental stewardship. For example, as students learn about and investigate the hydrologic system, they apply their GIS skills to analyze data from a nearby creek. Students also have the opportunity to participate in field trips (Fig. 34.1) and to meet with local professionals who use GIS in their work (Fig. 34.2). During the second year of the GTP program, students enroll in a "Principles of GIS" course concurrent with an applied trigonometry and statistics or an applied space science course. Students gain additional experience with data acquisition and analysis. In the final year of the GTP, students focus on a GIS application area and complete an internship with a local government or industry partner. Kruger and Erickson have secured an articulation agreement with Santa Rosa Junior College so that students who complete the GTP in high school may test out of college-level coursework. Junior colleges and community colleges in the USA are post-secondary institutions that focus on vocational or technical education that prepare students to enter the workforce after a two-year course of study. In addition, growing numbers of US students complete two years of lower-division university coursework at a junior college, where tuition rates are much lower, before transferring to a four-year university to pursue a baccalaureate degree.

Fig. 34.1 GTP students explore the history of mapmaking at a local museum

Fig. 34.2 A GIS professional
gives GTP students an
orientation to equipment used
in the field

34.5 Future Prospects for GIS in US Secondary Education

It is likely that the diffusion of GIS in education will remain slow due to continued challenges with IT infrastructure, educational support, and standardized testing, coupled with new and serious shortfalls in educational funding at state levels. However, new WebGIS tools that allow educators and students to easily use a Web browser for spatial analysis, construct their own maps, and share those maps with others may provide the means that will open the door for GIS to spread beyond educational innovators and early adopters (per Rogers (2003) diffusion of innovations theory) to the majority of educators. The spread of GIS and GPS to mobile devices such as smartphones coupled with the rise of citizen science such as National Geographic's FieldScope may heighten interest in mapping and analysis. The Partnerships for 21st Century Skills (http://www.p21.org) and the common core state standards initiative (http://www.corestandards.org/) are examples of national movements in the USA that are, through the efforts of educators, embedding geotechnologies.

Another factor encouraging GIS in education is that career-oriented programs are becoming increasingly common in US high schools as educators seek to provide students with a blend of academic content and practical career skills. In 2010, the US Department of Labor released a geospatial technology competency model (http://www.careeronestop.org). This model will serve as a resource for career guidance, curriculum development and evaluation, career pathway development, recruitment, professional development, certification and assessment development, apprenticeship program development, and outreach efforts to promote geospatial technology careers. The GTP program at Piner High School is one of several career pathway options provided to students and is an example of the a larger reform movement that emphasizes College and Career Readiness for students in US high schools (Tierney, Bailey, Constantine, Finkelstein, & Hurd, 2009). As GIS technologies become more versatile, as teachers and researchers discover new paths for accomplishing curricular goals using GIS, and as educational policies impact what and how educators teach, the rugged terrain of GIS in US schools may become much smoother.

References

Alibrandi, M. (2003). *GIS in the classroom: Using geographic information systems in social studies and environmental science*. Portsmouth, NH: Heinemann.

Alibrandi, M., & Sarnoff, H. (2006). Using GIS to answer the 'whys' of 'where' in social studies. *Social Education, 70*(3), 138–143.

Alibrandi, M., Beal, C., Thompson, A., & Wilson, A. (2000). Reconstructing a school's past using oral histories and GIS mapping. *Social Education, 64*(3), 134–140.

Alibrandi, M., Milson, A. J., & Shin, E. K. (2010). Where we've been; Where we are; Where we're going: Geospatial technologies and social studies. In R. Diem & M. J. Berson (Eds.), *Technology in retrospect: Social Studies' place in the information age 1984–2009* (pp. 109–132). Charlotte, NC: Information Age.

American Association for the Advancement of Science (AAAS). (1993). *Benchmarks for science literacy*. New York: Oxford University Press.

Audet, R. H., & Ludwig, G. S. (Eds.). (2000). *GIS in schools*. Redlands, CA: Esri.

Audet, R. H., & Paris, J. (1997). GIS implementation model for schools: Assessing the critical concerns. *Journal of Geography, 96*, 293–300.

Baker, T. R. (2005). Internet-based GIS mapping in support of K-12 education. *The Professional Geographer, 57*(1), 44–50.

Baker, T. R., & Kerski, J. J. (in press). Lone trailblazers: GIS in K-12 science education. In J. MaKinster (Ed.), *GIS in science inquiry*. New York: Springer.

Baker, T. R., Palmer, A., & Kerski, J. J. (2009). A national survey to examine teacher professional development and implementation of desktop GIS. *Journal of Geography, 108*, 174–185.

Baker, T. R., & White, S. H. (2003). The effects of GIS on students' attitudes, self-efficacy, and achievement in middle school classrooms. *Journal of Geography, 102*, 243–254.

Bednarz, S. W., & Audet, R. H. (1999). The status of GIS technology in teacher preparation programs. *Journal of Geography, 98*, 60–67.

Bednarz, S. W., & Bednarz, R. S. (2008). Spatial thinking: The key to success in using geospatial technologies in the social studies classroom. In A. J. Milson & M. Alibrandi (Eds.), *Digital geography: Geospatial technologies in the social studies classroom* (pp. 249–270). Charlotte, NC: Information Age.

Cuban, L. (2008). *Frogs into princes: Writings on school reform*. New York: Teachers College.

English, K. Z., & Feaster, L. S. (2002). *Community geography: GIS in action*. Redlands, CA: Esri.

Gatrell, J. D. (2001). Structural, technical, and definitional issues: The case of geography, GIS, and K-12 classrooms in the United States. *Journal of Educational Technology Systems, 29*(3), 237–249.

Goodchild, M. F., & Palladino, S. D. (1995). Geographic information systems as a tool in science and technology education. *Speculations in Science and Technology, 18*(4), 278–286.

Keiper, T. A. (1999). GIS for elementary students: An inquiry into a new approach to learning geography. *Journal of Geography, 98,* 47–59.

Kerski, J. J. (2003). The implementation and effectiveness of geographic information systems technology and methods in secondary education. *Journal of Geography, 102,* 128–137.

Kerski, J. J. (2008a). The world at the student's fingertips: Internet-based GIS education opportunities. In A. J. Milson & M. Alibrandi (Eds.), *Digital geography: Geospatial technologies in the social studies classroom* (pp. 119–134). Charlotte, NC: Information Age.

Kerski, J. J. (2008b). Toward an international geospatial education community. In A. J. Milson & M. Alibrandi (Eds.), *Digital geography: Geospatial technologies in the social studies classroom* (pp. 61–74). Charlotte, NC: Information Age.

Louv, R. (2005). *Last child in the woods: Saving our children from Nature-Deficit disorder.* Chapel Hill, NC: Algonquin.

Malone, L., Palmer, A., & Voigt, C. (2005). *Mapping our world: GIS lessons for educators, ArcGIS desktop edition.* Redlands, CA: Esri.

McClurg, P. A., & Buss, A. (2007). Professional development: Teachers use of GIS to enhance student learning. *Journal of Geography, 106,* 79–87.

Milson, A. J., & Curtis, M. (2009). Where and why there? Spatial thinking with geographic information systems. *Social Education, 73*(3), 113–118.

Milson, A. J., & Earle, B. D. (2007). Internet-based GIS in an inductive learning environment: A case study of ninth grade geography students. *Journal of Geography, 106,* 227–237.

Milson, A. J., Gilbert, K. M., & Earle, B. D. (2007). Discovering Africa through Internet-based geographic information systems: A pan-African summit simulation. *Social Education, 71*(3), 140–145.

Milson, A. J., & Roberts, J. A. (2008). The status of geospatial technologies in state geography standards in the United States. In A. J. Milson & M. Alibrandi (Eds.), *Digital geography: Geospatial technologies in the social studies classroom* (pp. 39–59). Charlotte, NC: Information Age.

National Council for the Social Studies. (1994). *Expectations of excellence: Curriculum standards for social studies.* Washington, DC: Author.

National Geography Education Standards Project. (1994). *Geography for life: National geography standards, 1994.* Washington, DC: National Geographic Research & Exploration.

National Research Council. (1996). *National science education standards.* Washington, DC: National Academy Press.

National Research Council. (2006). *Learning to think spatially—GIS as a support system in the K-12 curriculum.* Washington, DC: National Academies Press.

Piburn, M. D., & Reynolds, S. J. (2005). The role of visualization in learning from computer-based images. *International Journal of Science Education, 27*(5), 513–527.

Rogers, E. (2003). *Diffusion of innovations* (5th ed., 576p.). New York: Free Press.

Schultz, R. B., Kerski, J. J., & Patterson, T. (2008). The use of virtual globes as a spatial teaching tool with suggestions for metadata standards. *Journal of Geography, 107,* 27–34.

Shin, E. (2006). Using geographic information system (GIS) to improve fourth graders' geographic content knowledge and map skills. *Journal of Geography, 105*(3), 109–120.

Stoltman, J. P. (1990). Geography education for citizenship. Bloomington, IN: ERIC Clearinghouse for Social Studies/Social Science Education.

Sui, D. (1995). A pedagogic framework to link GIS to the intellectual core of geography. *Journal of Geography, 94,* 578–591.

Tierney, W. G., Bailey, T., Constantine, J., Finkelstein, N., & Hurd, N. F. (2009). *Helping students navigate the path to college: What high schools can do: A practice guide* (NCEE #2009-4066). Washington, DC: National Center for Education Evaluation and Regional Assistance, Institute of Education Sciences, U.S. Department of Education.

West, B. A. (2003). Student attitudes and the impact of GIS on thinking skills and motivation. *Journal of Geography, 102*, 267–274.

Wiegand, P. (2003). School students' understanding of choropleth maps: Evidence from collaborative mapmaking using GIS. *Journal of Geography, 102*, 234–242.

Wigglesworth, J. C. (2003). What is the best route? Route-finding strategies of middle school students using GIS. *Journal of Geography, 102*, 282–291.

Chapter 35
Synthesis: The Future Landscape of GIS in Secondary Education

Joseph J. Kerski, Andrew J. Milson, and Ali Demirci

35.1 Introduction

The use of GIS in secondary education around the world is a relatively recent phenomenon. Settled almost entirely since 1990, the landscape is marked by pedagogical and technological trailblazing by enterprising educators who have overcome numerous challenges to teach about GIS and teach with GIS in a wide variety of settings, to students of all ages, in lecture rooms, in computer labs, and in the field. Yet for all of this variety, a remarkably small number of themes are common. In this chapter, we begin by identifying those themes, discuss trends that will impact the future of GIS in secondary education and beyond, and make recommendations about what we believe needs to happen to advance the use of GIS in education and why it needs to happen.

35.2 The Common Landscape

By this time, every reader of this book will be able to easily identify the common themes in the landscape of GIS in secondary education, because of their frequent mention by nearly every author. Themes were common among the challenges as well as the benefits experienced by educators who are engaging students in spatial thinking and analysis. Given the frequency with which they were mentioned, we feel no need to delve into details of each challenge and benefit. However, it may be helpful to summarize the challenges as technological and societal. Technological challenges include access not only to computers that have enough internal and graphics memory, hard disk space, and the proper software to be able to handle spatial analysis but also to those computers in the school and support from the school's information technology staff. It is tempting to relegate these challenges to the early

J.J. Kerski (✉)
Esri, Broomfield, Colorado, USA
e-mail: jkerski@esri.com

A.J. Milson et al. (eds.), *International Perspectives on Teaching and Learning with GIS in Secondary Schools*, DOI 10.1007/978-94-007-2120-3_35,
© Springer Science+Business Media B.V. 2012

days of GIS in education, but they remain challenges in many regions of the world today, even in many developed countries.

It is also clear, though, that despite the hardware, and software challenges repeatedly mentioned by educators, societal issues appear to cast the greatest constraint on GIS becoming an embedded, required tool throughout education. Of major importance seems to be the lack of awareness of spatial thinking and analysis and their importance in education and society. Coupled with the segmentation of education into discrete subjects, this translates into a lack of a "home" for GIS in the curriculum and, consequently, a lack of funding and support for professional development institutes and for educators to attend these even if more were offered, for curriculum development efforts, and for research to examine the best methods of curricular implementation and assessment. A logical home for GIS lies in geography, and in places such as Turkey, Norway, Taiwan, and the UK, this is where it is most frequently taught. In Turkey, for example, geography is a compulsory course from Grades 9 through 12, and equally important is that in these courses, project-based learning is common. In the UK, geography is a compulsory subject for pupils aged 5 to 14 and offered and available after that through graduation. Sadly, however, the importance of geography education around the world is more of the exception than the rule.

In most countries, a lack of awareness exists even at the educational policy level of what geography is and what it is not. Geography is frequently buried in the social studies and, given perennial educational funding crises, faces continual pressure from education and political officials to even maintain the anchor points it has achieved. Furthermore, an overreliance on standardized tests that do not assess critical thinking skills in countries with and without a national curriculum constrains any inquiry-based method such as the use of spatial analysis through geotechnologies.

Despite these challenges, the authors of these chapters make it clear that educators throughout the world are captivated not only by the potential for GIS in education, but are making it a real, daily, and viable part of their educational practice. These instructors are using GIS effectively with their students, as evidenced in the research results that they and their colleagues are finding, and in the achievement of their instructional goals. And for every chapter and case study that we have included in this book, many other countries and stories could have been included as further evidence for the benefits of spatial analysis through geotechnologies. The stories included here illustrate well what might be thought of as "the three Cs" of linking students to their communities, to citizenship education, and to meaningful careers. Another way of categorizing the benefits is scholarship (developing skills in thinking and communicating), artisanship (gaining key skills), and citizenship (becoming thoughtful citizens). Key to all of these benefits is engagement, cited again and again by the authors—that students are interested and motivated not only to learn but also to take action. Indeed, students using geotechnologies exemplify well the geographic inquiry process of asking geographic questions, gathering geographic data, assessing geographic information, analyzing geographic information, and acting on the decisions that they make with their new-found knowledge. Even though this book's purpose was to address formal secondary education, the power of informal educational opportunities cannot be ignored, such as the work that students in the National 4-H program do with GIS and GPS in the USA.

This book makes it clear that educators and students using these tools and methods are engaging in deep and meaningful inquiry, problem-based learning, peer mentoring, and working with real-world problems, tools, and data. They are using technology in meaningful ways, as a tool to grapple with relevant, current issues, rather than using technology for technology's sake. This work is being accomplished, as was described by our Portugal colleagues' collaboration among geography, biology, geology, philosophy, and computer teachers, in an interdisciplinary manner.

35.3 Trends Affecting the Future Landscape

What will the landscape of GIS in education look like by the end of the current decade? Modifying that landscape will be technological, societal, and educational policy trends.

Technological trends having a positive impact on the use of GIS in education include increased Internet bandwidth to move large spatial data files, faster and less expensive computers, and more powerful, easier-to-use geotechnologies. While energy shortages and consequential electrical outages could have a negative effect on GIS in education, longer-lasting batteries and solar-powered computers could reduce the potential negative effect. Still, concerns about the "digital divide" will remain and GIS will be an option only for those with a decent Internet connection and computer. However, initiatives such as the worldwide "One Laptop Per Child," school programs that enable each student to have a computer, and bandwidth improvement initiatives occurring in many countries will ensure that at least the technological infrastructure is present in many more parts of the world. Another positive technological trend is that nowadays one only needs a smart phone to take GPS-tagged photographs and videos that can be geotagged to build rich field-based GIS projects, rather than a separate GPS receiver and a digital camera. Finally, voice-enabled GIS could be of particular help when the students' native language is not the same as that of the software, as well as for disabled or special needs students.

We predict that an even more powerful force already affecting GIS professionals will dramatically change the landscape of GIS in education—the use of Web-based GIS. Web-based GIS can be used entirely online with only a Web browser, negating software installation issues and extensive work with school IT staffs. WebGIS solutions in which students can store spatial data and projects on servers operating in the cloud environment represents a major shift for the largely desktop-based paradigm of GIS instruction since 1990. Online portals such as ArcGIS Online (http://www.arcgis.com) also allow a combination of Web-based and desktop tools to be used, taking advantage of the best of both worlds—the ease of the cloud, and the power of the desktop. Furthermore, the system also allows GIS projects to be easily shared. In the past, this task was always a cumbersome one that required compressing and sending a series of large files and ensuring that layers were linked to their associated data sets. Over the current decade, we anticipate that more and more analytical capabilities of desktop GIS software will migrate to the cloud. While the Web admittedly has challenges of its own, such as continued problems with bandwidth in schools in many parts of the world and the plethora of nonauthoritative and undocumented data sources, we believe that examples such as the ArcGIS Server-based projects in Colombia (see chapter 7) will become commonplace, enabling Web-based GIS use in schools to eclipse that of desktop GIS long before even the midpoint of the decade.

In terms of societal trends, GIS in education is benefitting from an increased emphasis at the educational policy level on preparing students for the workforce, particularly in science and technology. Furthermore, through programs such as GeoMentor (http://www.geomentor.org), GIS professionals are partnering with local teachers or teachers around the world. These partnerships may last for years or may exist for just one event. For example, in the UAE, the Dubai Municipality organized the First GIS Students Forum in cooperation with the Ministry of Education during a large Middle East GIS conference.

GIS is poised to take advantage of the heightened interest in preparing students for careers in green technologies and in careers that can have a positive impact on society. One of the best examples is the career-focused, yet "make a difference," vision shared by our colleague from Rwanda for "responsible and well educated citizens." "Rwanda will have young people who can explore the world, and who can make good decisions... [ ] So, with GIS in all secondary schools of Rwanda we will be on a good level to help the Rwandan society with proposals for decision makers, and students will develop a good understanding of geography and the situation of our country so that they actively participate and think about future development of our country."

A frequently mentioned constraint to the implementation of GIS is the lack of available spatial data in many countries. The trend toward open data policies will increase the number of schools that can use that data and hence make it easier to teach with it. Distance education could increase the amount of GIS taught, the standardization of GIS curricula worldwide, and the number of students exposed to GIS. An increased reliance on home schooling and after-school programs opens a doorway to GIS in these venues, as National 4-H has done in the USA. GIS is also benefitting from a trend to develop an active citizenry that understands and cares about their own community, illustrated well by the neighborhood safety project described by our colleagues in Japan.

Yet we predict that these trends will pale in comparison to the rise of geo-enabled devices and technologies, beginning with GPS receivers and spreading to smart phones and Web maps. This movement even extends to electronic sensors of all kinds and, more recently, the concept of humans as sensors. This has been accompanied by GPS-driven activities such as geocaching as well as, more recently, the enabling of volunteered geographic information or citizen science projects, which has in turn given rise to citizen demand for government transparency and accountability. As people use geographic information in their everyday lives, at work, and at play, they more fully realize the value of geographic information and may begin demanding educational reform that makes heavy use of inquiry and critical thinking such as GIS can provide.

Educational policy trends will affect the use of geotechnologies in secondary education in positive and negative ways. On the positive side, workforce-driven education with its focus on skill building and multiple competencies will certainly help, at least in terms of teaching about GIS. In terms of teaching with GIS, efforts to implement national curricula in different countries can positively affect GIS implementation if those curricula are inquiry driven and negatively if they are based on the memorization of facts. Even if the curricula are inquiry driven, if students are tested with standardized fact-based tests, or if educators are assessed on the results of those "high-stakes" tests, all inquiry in secondary schools will suffer. Budgetary constraints can also have positive and negative impacts. On the positive side, educators and policy makers who increasingly ask for the return on investment (ROI) for everything teachers are doing in the classroom may look to the mounting evidence of the value of GIS in education. At the same time, if the campuses and school districts where this occurs are also using GIS on the administrative side of the house, such as in recruitment and retention, analysis of test scores, campus safety, and school bus routing, administrators will already understand what GIS is because they are using it to make their own operations more efficient. Many authors of chapters in this book noted that GIS is now mentioned in several of their national standards. At the moment, though, these standards tend to be similar to those of Malta where no specifics are given in terms of how or what should be taught. This open-ended nature gives freedom to the teachers knowledgeable about GIS to do what they feel is best, but makes it difficult for the bulk of the teachers who are not comfortable with GIS and need more specific instructions. A good example of a rich and specific set of standards is that from India, where the syllabus for Class XII focuses on human and economic geography, data processing, and spatial information technology, with 30 hours of practical lab work including the "use of computers in data processing and mapping, tabulating and processing of data including the calculation of averages, measures of central tendency, deviation and rank correlation, representation of data including the construction of diagrams; bars, circles and flowcharts, thematic maps including the construction of dot, choropleth and isopleth maps; and introduction to GIS, hardware requirements and software modules, raster and vector data, data input, editing and topology building, data analysis, overlay and buffer."

A positive development that is only beginning to bear fruit is the recent call for the return of outdoor education. Furthermore, increased demand for citizenship education and global education could hasten the growth of GIS in secondary schools. We predict that environmental education will receive attention that has been long overdue as human impact on the carrying capacity of the Earth is no longer possible to ignore. If environmental education rises as we predict it will, GIS could stand to make significant gains if those environmental educators can use the technology. In fact, the increased importance of environmental education could be very well what puts GIS firmly on the secondary educational landscape as policy makers take steps to fund and support it.

35.4 Recommendations for Landscape Change

Rogers (1995) postulated a theory on the diffusion of innovations that may be helpful to frame the progress that GIS in secondary education has made. Central to the theory is a curve describing the adoption and diffusion of innovations that begins with innovators who are so captivated by an innovation that they do the extra work that makes it easier for those who follow. Those who will follow are first the early adopters, then the early majority, late majority, and, finally, the laggards. Have we come to the point on the innovations phase on the curve to the early adoption phase? The answer to this question has implications for how teacher professional development needs to be conducted and whether GIS in education has enough internal momentum without requiring external support. One might argue that the spread of GIS beyond higher education into secondary education represents a "tipping point" that guarantees an increased rate of diffusion into the rest of education, formal and informal. However, GIS in secondary education has entered its third decade. It is still largely ignored by the bulk of the secondary education community. It is largely perceived as irrelevant except for those who teach career and technical education or too difficult or time consuming for the bulk of educators to use. With the educational landscape of GIS confined largely to innovators and a few early adopters, the momentum from the early majority and late majority components that make up the bulk of the innovation curve may never be realized. We argue that without external support mechanisms and reforms, GIS is likely to continue to spread only slowly into secondary schools throughout the current decade. What are those mechanisms and reforms?

We believe that several key things need to happen to advance the use of GIS in education. First, we recommend that educators must publicize what they are doing for other teachers, students, and school administrators. It continually amazes us to learn about exemplary projects that almost nobody knows about besides a solitary teacher and his or her students. Plot posters and hang them on the school walls, give a presentation about GIS, have the students give presentations, participate in or organize a local GIS day event, invite the local media to observe, write articles or case studies for both popular and scholarly outlets, and use Web 2.0 tools such as blogs, Facebook, Twitter, Linked In, and others to tell your story. One never knows where it could lead. In South Africa, for example, "The school administration has also noticed the increased interest in GIS among students and has made money available to purchase resources necessary to teach GIS through computers to students." Even if no tangible benefits result, spatial analysis about twenty-first-century issues, we believe, is simply too important to hide away in one classroom at the back of the school.

Second, we call on all those involved with GIS in education to conduct teacher professional development. Face-to-face as well as Web-enabled institutes continue to bear much fruit. In Taiwan, for example, over 1,000 teachers attended GIS workshops by 2009. The teachers' participation in these workshops was a key factor in ensuring that spatial information and GIS are important topics in standardized college entrance exams and in probably a higher percentage of high schools than anywhere else in the world. The Taiwan example points to the third recommendation: that higher education take an active role in promoting and supporting GIS in education at the secondary level. This partnership must include professional development, curriculum development, the use of secondary classrooms by university professors and graduate students, and more.

Fourth, given the dozens of examples showing that the GIS industry has made a positive impact on the landscape, we recommend not only that it continue, but that the GIS industry reexamine its pricing policy. Time after time, the prohibitive cost of GIS software was mentioned as being out of reach for secondary schools.

Fifth and even more important, we recommend educational reform efforts that support inquiry-based methods throughout primary and secondary education. Yet reform cannot end with secondary school. Inquiry-based education needs to be supported throughout universities to achieve the kind of workforce that industry, nonprofit organizations, and government are demanding. Momentum from inquiry-based education can be quickly lost if at the university level students are once again relegated to memorizing and taking exams. Perhaps the best example of collaborative partnerships for reform came from the Netherlands. One of the main textbooks for secondary geography education there in 2003 included a chapter on GIS, followed by a Dutch version of ArcExplorer Java Edition for Education (AEJEE), lessons, and data sets. Around the same time, a WebGIS entitled Bosatlas online (www.bosatlasonline.nl) allowed students to create maps and charts by themselves in an online environment and an Internet portal for secondary education called EduGIS (www.edugis.nl) was developed through cooperative efforts between government agencies, research institutes, teachers, teacher training institutes, and textbook publishers. Later, EduGIS began providing data and lessons, and conducted workshops and courses. Another reason we recommend collaboration among industry, nonprofit, and government is so that societies will cease bemoaning the fact that students are graduating without the kinds of skills needed in the workforce.

As is evident by this book and other documents published in the past few years, GIS education is not the sum total of what just a few educators are championing and practicing. Rather, in a similar manner in which Geographic Information Systems evolved into Geographic Information Sciences 20 years ago, GIS in education now has its own research base, developers, and educators who advance its theory and practice through research, curriculum, and teaching. The research is anchored in geographic education but incorporates elements of experiential learning, problem-based learning, and a small amount of cognitive science. The GIS in education bibliography (on http://edcommunity.esri.com) reached 1,750 entries by 2011 with an increasing number of dissertations and other major studies produced annually. However, the GIS educational landscape remains small. Educational policy makers and administrators demand additional evidence that clearly shows how and why the use of any technology and method makes a difference in student learning. We therefore advocate as our sixth recommendation that an international GIS in education virtual research center be established to promote and support research that is desperately needed in this field. A five-year funded international GIS in education center would accomplish what the establishment of the National Center for Geographic Information Systems did in 1990 for GIScience—an established research base that develops GIS-based curricular efforts, teacher training programs, and policy advocacy documents that are well grounded in learning theory. A model for the impact that a large study could have comes from Turkey. In order to identify problems before implementing GIS-based projects at schools there, the Scientific and Technological Research Council of Turkey involves 300 students, many teachers, and three public high schools. It is projects such as these that could provide a good template for a global study of GIS in education, and the cases could include some that are featured in this book. Our Austrian colleagues in this book discuss a program where teachers are paid to conduct research. Could this model be a valuable component in such a GIS education research center?

Those who advocate the use of GIS in education tend to focus on its benefits for analyzing data and solving problems. While this has great value, our seventh recommendation is that GIS advocates take steps to illustrate how GIS can help teach content to advance GIS at the educational policy level. Content is what policy makers often focus on, and so we need to be clear that, for example, while students in Hungary are engaged in collecting data in the field, they are at the same time learning about ecoregions and watersheds, the Colombian students are learning about the cultural and physical geography of their country, and the New Zealand students are learning about plate tectonics. We also need to be clear that students are learning content in a variety of disciplines simultaneously. Students in Dr. Yaghi's class Lebanon are learning English and those at Piner High School in the USA are learning about surveying and mathematics through the use of GIS and GPS, for example. Finally, it cannot be overstated that the need to demonstrate that skills and content learned is standards based will only increase in this decade with increased accountability required of schools through transparency in government.

We recognize that for all of the discussion about the advantage of having students solve problems with real-world data, Piaget's words that "the trouble with education... is that the best teaching methods are in fact the most difficult" have special relevance here (Piaget, 1929). Perhaps the most telling understatement of this book comes from our Switzerland colleagues, who stated that, "Its use in the classroom requires different instruction than is conventionally used." Teaching with any inquiry-based method is more difficult because it is fraught with pedagogical uncertainties and inconsistencies in the data used and in the results that could be discovered. We need to instill the kind of educational culture where teachers are comfortable with this approach.

GIS has been on the secondary educational landscape for 20 years. Yet even in Canada, where the first GIS was implemented through the Canadian Land Inventory during the 1960s, GIS is not well established in education. Sadly, the Canada situation is the norm around the world: The bulk of educators still do not really know about GIS. They have increasingly heard about it, but they may confuse it with GPS or digital maps or geovisualization. Even if they know about it, they may dismiss it as "just software" or something that is only suitable to teach at the university level. Others sometimes see GIS as a computerized atlas, thus missing the analytical capabilities of the tool. We have specific recommendations to move spatial analysis beyond something that a few students may be fortunate enough to experience in school to a key component of all secondary education throughout the world.

First, we need to get past the perception of maps as mere reference documents and geography as memorization of facts. The authors from Canada referred to the "stigma" of GIS being "just maps." If GIS is "just" maps, then we are "just" using it to teach about climate, biodiversity, population, energy, water, and other key issues of our time. Rather than reducing possibilities, teaching with GIS enables students and educators, expands their horizons, and invites interdisciplinary education. In a decade where the amount of data will dwarf even the advances of the past decade, GIS will be absolutely essential for helping students sort through the plethora of information, helping them to think critically about not only the data but also the issues that the data are speaking to. To advance GIS in education, we must get beyond promoting maps, which resonate with geography educators but with few other educators. In addition, we must not focus only on geography, which will likely remain constrained around the world given budgetary, educational policy, and curriculum pressures from other disciplines. We also must not focus simply on science, technology, engineering, and mathematics (STEM) education, which only resonates in some countries and comes and goes depending on shifting educational policies and political slogans. Rather, we need to focus on something that nearly every educator, standard, and policy document state that they want to see more of—critical thinking. Positioning GIS in its rightful place as a critical thinking tool will advance it throughout secondary education and beyond.

We must also promote the fact, as this book makes clear, that GIS is an amazing unifier. In a culture where students and citizens bemoan the segmentation of education, GIS brings together content, skills, faculty, and students from different disciplines and from different educational levels, including primary, secondary, tertiary, and informal institutions and settings. Educators also seek tools and methods that can help them teach "at-risk" students and those with special needs. Spatial analysis, since it relies so much on visual learning, seems to be something that can effectively help these students learn content and skills to become effective citizens and find meaningful employment.

We need to actively work toward fundamental changes in our educational systems to raise the kind of decision makers that societies will need to grapple with the fundamental issues of our time. Each of these issues, including water, energy, biodiversity, hazards, agriculture, and economics, has a geographical component. Without large changes—such as an entire year devoted to service learning and immersion in fieldwork in secondary education—we are relegated in the meantime to a fall-back position. This fall-back position includes being diligent in keeping tabs on continued assaults on specific disciplines, inquiry, and fieldwork. One never knows when or from where these assaults will occur, and they may occur in a country or in a discipline that had a rich tradition of use in education. For example, as reported by our colleagues in Malta, "One of the proposals is for geography to be amalgamated with History and Social Studies, therefore reducing the actual time allocated for these core subjects in the first two years of secondary school." We cannot assume that the disciplines, schools, or policies that support GIS today will exist tomorrow.

If we want to see a rapid increase in the numbers of students and educators using GIS in the world, we may do well to concentrate efforts on several key countries. China and India are natural choices, even though they both suffer from the decades-long practice of locking away of spatial data beyond the reach of teachers, in terms of cost or use permissions. Russia is another fertile ground, given the host of issues that could be examined in such a vast country, from climate to resource extraction, but again, the lack of public domain spatial data is a hindrance. Of any of the large countries with a great deal of growth potential for GIS in secondary education, the one that actually could see great progress in this decade is Brazil. Even though we do not have a chapter on Brazil in this book, the IGBE in Brazil operates a rich data depository with data on census, the environment, and much more. The Esri distributor is keen on working with schools, and sustainability of urban areas and agricultural lands are issues that are at the forefront of most citizens' minds. In terms of other countries with a high potential for growth, Indonesia's government agencies have begun to share data with educators and the general public, and as we have seen from Dr. Ida and Dr. Yuda, much progress has been achieved in Japan.

We also advocate that professional development in teacher education be expanded. Professional development is needed not only in the techniques of spatial analysis but also in the pedagogical approaches to GIS. It needs to embrace the "technological pedagogical content knowledge" recommended by our colleagues in the Netherlands. This knowledge (Mishra & Koehler, 2006) captures the complex interplay among three main components of learning environments: content, pedagogy, and technology. We agree with and support the approach that the authors in France referred to as "confidence building rather than competence building" in these professional development sessions. They advocated that GIS "needs support rather than control" and "more space for experimentation." However, as these stories make painfully clear, most professional development in secondary education with geotechnologies is carried out by and for in-service teachers. We recommend a concentrated effort toward preservice teachers. While teacher training programs and courses at university colleges of education are very difficult to change, we must be diligent in our efforts, because of the propensity for all of us to teach as we have been taught. Concentrated work with associations of teacher preparation universities and individual colleges of education to incorporate spatial analysis in science, math, social studies, and other teacher education courses could bear more fruit than working exclusively with in-service educators. Common sense would dictate that as teachers who are "digital natives" are hired to replace retiring "digital immigrant" teachers, GIS may be implemented more frequently. However, in the professional development that we have conducted, we have not observed new teachers being necessarily more comfortable with teaching with technology as veteran teachers. Therefore, we do not believe that those preparing to be teachers would "naturally" use GIS without specific inclusion of it in their teacher preparation coursework. South Korea provides a good model of what could happen; it was "discovered that, among all 18 geography education-related departments in South Korea, 16 departments include GIS courses in their undergraduate curriculum," with the last two offering "GIS courses via the form of special lectures or other routes."

Whether for future teachers or current teachers, the GIS in education community must remind educators of several things through teacher education. First, we recognize that teaching a technology where they do not perceive themselves as experts is a difficult task. Educators who feel that they need to be experts in GIS to teach it most likely will never use it. Our South Africa author stated that, "I taught them (the students) geography and they taught me GIS!" Our New Zealand colleagues said, "The teacher of this class (Anne Olsen) and the students learned the required GIS skills together." When the teacher learns the tools alongside the students, an environment of collaborative learning is fostered. Some might argue that the teacher's role is diminished in such an environment. On the contrary, we need to emphasize that the teacher's role is critical to the learning process. Teachers can rely on students to become adept at the tools and provide peer mentoring, but it is the teacher who initially provides and frames the inquiry-based questions and issues. Therefore, using GIS frees teachers to do what they love to do best—engage students in meaningful issues.

We advocate the incorporation of fieldwork with GIS. If we are teaching about the Earth using GIS, but students do not ever get an opportunity to see, touch, hear, smell, and feel that real Earth, we are not doing our job as educators. As is so well illustrated by our colleagues in Hungary, collecting data in the field along with coordinates gathered from any device that provides from GPS satellites, mapping those points, and analyzing results in a GIS reveals patterns and deepens understanding. If the hurdles of getting students off site are insurmountable, we advocate the use of the school yard or campus as the field study laboratory.

We recommend any effort that builds the community of educators, in their own school or with the wider community, so that they won't feel alone and can partner with one another in planning and teaching lessons. Our colleagues in Singapore provided a good model that does exactly that. The most valuable resources will come from the teachers themselves. As is well stated by our colleagues in Norway, "Teachers urgently need teaching resources such as textbooks and ready-to-use teaching materials, but they should also be empowered with GIS skills enabling them to make such textbooks and teaching materials themselves."

The combined approach identified by our colleagues in Spain, who said that teaching with GIS needs to include knowledge (minds on), skills (hands on) and behavior (hearts on), resonates well with our own philosophy. Teaching with GIS and even teaching about GIS is never just about the software; indeed, the button pushing is only a means to the end. Even the final map is only a means to the end. The end goal is the understanding brought about by spatially analyzing an issue, and the action it may lead the student to take to make a positive difference.

The stories told in this book seem to confirm what Audet and Ludwig (2000) wrote more than a decade ago: "A classroom that uses GIS as a problem-solving tool is a classroom in which the walls are invisible and the teacher and student assume roles that are non-traditional. …Adopting this technology is not for the fainthearted. But integrating GIS into the curriculum rewards teachers by creating intellectually challenging and demanding learning opportunities." It is up to all of us who believe so firmly in the power of spatial analysis within a GIS environment to take active steps to ensure that the kind of educational and societal transformation can be achieved—for the health of the planet and for the students who will inhabit that planet.

References

Audet, R., & Ludwig, G. (2000). *GIS in schools.* Redlands, CA: Esri Press.
Mishra, P., & Koehler, M. (2006). Technological pedagogical content knowledge: A framework for teacher knowledge. *Teachers College Record, 108*(6): 1017–1054.
Piaget, J. (1929). *The child's conception of the world.* London: Routledge.
Rogers, E. (1995). *Diffusion of innovations.* New York: Free Press.

Bibliography

Aditya, T., & Gadjah, M. (2008). Participatory mapping. *GIM International September*, pp. 41–43.

Akimoto, H. (1996). GIS (chiri joho shisutemu) to koko chiri kyoiku. *Shin chiri, 44-3*, 24–32.

Alhousani, H. A. (1996). *Secondary school administration in the United Arab Emirates: Its reality, problems and methods of development*. Unpublished doctoral dissertation. University of Wales, Cardiff.

Alibrandi, M. (2003). *GIS in the classroom: Using geographic information systems in social studies and environmental science*. Portsmouth, NH: Heinemann.

Alibrandi, M., Beal, C., Thompson, A., & Wilson, A. (2000). Reconstructing a school's past using oral histories and GIS mapping. *Social Education, 64*(3), 134–140.

Alibrandi, M., Milson, A. J., & Shin, E. K. (2010). Where we've been; Where we are; Where we're going: Geospatial technologies and social studies. In R. Diem & M. J. Berson (Eds.), *Technology in retrospect: Social Studies' place in the information age 1984–2009* (pp. 109–132). Charlotte, NC: Information Age.

Alibrandi, M., & Sarnoff, H. (2006). Using GIS to answer the 'whys' of 'where' in social studies. *Social Education, 70*(3), 138–143.

Alter, N. (2002). *L'innovation ordinaire*. Paris: Presses Universitaires de france.

American Association for the Advancement of Science (AAAS). (1993). *Benchmarks for science literacy*. New York: Oxford University Press.

Anders, K. H. (2008). *Applications on the move. Development of a mobile gaming application for young people*. Location based games from students for students. http://www.sparklingscience. at/en/projects/14--applications-on-the-move. Accessed January 25, 2010.

Anders, K. H., et al. (2009). Applications on the move: Ortsbezogene spiele von schülern für schüler. In: T. Jekel, A. Koller, & K. Donert (Eds.), *Learning with GI IV* (pp. 97–102). Heidelberg: Wichmann.

Andersland, S. (2005). GIS i geografiundervisning. In R. Mikkelsen, & P. J. Sætre (Eds). *Geografididaktikk for Klasserommet. En Innføringsbok i Geografiundervisning for Studenter og Lærere* (pp. 213–229). Kristiansand: Høyskolerådet.

Andersland, S. (2006). Eg syns ArcView var eit skamtøft program. Om GIS og geodata i skule og undervisning. *Kart og Plan, 99*(3), 195–200.

Arellano-Marín, J. P. (2001). Educational reform in Chile. *Cepal Review, 73*, 81–91.

Asiedu-Akrofi, A. (1982). Education in Ghana. In B. Fafunwa & J. U Aisiku (Eds.), *Education in Africa: A comparative survey*. London: George Allen & Unwin.

Asociación de Geógrafos Españoles (AGE). (2010). (Spanish Geographers Association). Accessed September 1, 2010, http://age.ieg.csic.es/recur_didacticos/index.htm Grupo de trabajo de Didáctica de la Geografía (Working Group of Teaching Geography) http://age.ieg.csic.es/ didactica/

A.J. Milson et al. (eds.), *International Perspectives on Teaching and Learning with GIS in Secondary Schools*, DOI 10.1007/978-94-007-2120-3,
© Springer Science+Business Media B.V. 2012

Association of American Geographers. (2008). *Supporting basic math and science education in the Muslim world.* Accessed December 8, 2010, http://www.aag.org/cs/mycoe/middle-east-north-africa

Attard, M. (2005). Developing undergraduate GIS study-units – The experience of Malta. In K. Donert & P. Charzynski (Eds.), *Changing horizons in geography education* (pp. 97–101). Toruń: HERODOT Network Publication. Accessed September 1, 2010, http://www.herodot.net/conferences/torun2005/Changing%20Horizons%20book.pdf

Attard, M. (2009). Developing undergraduate GIS study-units. In K. Donert (Ed.), *Using geoinformation in European geography education* (pp. 102–111). Rome: Società Geografica Italiana, International Geographical Union.

Audet, R. H., & Ludwig, G. S. (Eds.). (2000). *GIS in schools.* Redlands, CA: Esri.

Audet, R. H., & Paris, J. (1997). GIS implementation model for schools: Assessing the critical concerns. *Journal of Geography, 96,* 293–300.

Auer, B., et al. (2008). Perspektiven österreichischer Gletscherskigebiete. *GW-Unterricht, 112,* 51–56.

Avalos, B. (2002). *Profesores para Chile. Historia de un proyecto.* Santiago: Ministerio de Educación.

Avalos, B. (2005). *Secondary teacher education in Chile: An assessment in the light of demands of the knowledge society.* Paper submitted to the Ministry of Education. Santiago.

Bahbahani, K., & Huynh, N. T. (2008). *Teaching about geographical thinking.* R. Case & B. Sharpe (Series Eds.).Vancouver, BC: The Critical Thinking Consortium and The Royal Canadian Geographical Society.

Bähler, L., & Stark H. J. (2008). Open Geodata – am beispiel von OpenAddresses.ch. Angewandte Geoinformatik.

Baker, T., & Bednarz, S. (2003). Lessons learned from reviewing GIS in education. *Journal of Geography, 102,* 231–233.

Baker, T. R. (2005). Internet-based GIS mapping in support of K-12 education. *Professional Geographer, 57*(1), 44–50.

Baker, T. R., & Kerski, J. J. (in press). Lone trailblazers: GIS in K-12 science education. In J. MaKinster (Ed.), *GIS in science inquiry.* New York: Springer.

Baker, T. R., Palmer, A., & Kerski, J. J. (2009). A national survey to examine teacher professional development and implementation of desktop GIS. *Journal of Geography, 108,* 174–185.

Baker, T. R., & White, S. H. (2003). The effects of GIS on students' attitudes, self-efficacy, and achievement in middle-school science classrooms. *Journal of Geography, 102*(6), 243–254.

Bartoschek, T., & Schöning, J. (2008). Trends und potenziale von virtuellen globen in schule, Lehramtsausbildung und Wissenschaft. *Geo Science, 4,* 28–31.

Bednarz, S. W. (2004). Geographic Information Systems: A tool to support geography and environmental education? *GeoJournal, 60,* 191–199.

Bednarz, S. W., & Audet, R. H. (1999). The status of GIS technology in teacher preparation programs. *Journal of Geography, 98,* 60–67.

Bednarz, S. W., & Bednarz, R. S. (2008). Spatial thinking: The key to success in using geospatial technologies in the social studies classroom. In A. J. Milson & M. Alibrandi (Eds.), *Digital geography: Geospatial technologies in the social studies classroom* (pp. 249–270). Charlotte, NC: Information Age.

Bednarz, S. W., & van der Schee, J. (2006). Europe and the United States: The implementation of geographic information systems in secondary education in two contexts. *Technology, Pedagogy and Education, 15,* 191–205.

Bednarz, S., Bednarz, R. S., Mansfield, T. D., Semple, S., Dorn, R., & Libbee, M. (2006). Geographical education in North America (Canada and the United States of America). In J. Lidstone & M. Williams (Eds.), *Geographical education in a changing world: Past experience, current trends, and future challenges* (pp. 107–126). Dordrecht, The Netherlands: Springer.

Belfekih C. M. (1993). *Modern secondary education in the United Arab Emirates: Development, issues and perspectives.* Unpublished doctoral dissertation. Temple University Graduate Board.

Berry, A., Loughran, J., & Van Driel, J. H. (2008). Revisiting the roots of pedagogical content knowledge. *International Journal of Science Education, 10,* 1271–1279.

Berry, B. (1968). *Spatial analysis: A reader in statistical geography.* New York: Prentice Hall.

BM:UKK (Austrian Federal Ministry for Education, Arts and Culture). (2009). *Schools and education.* Accessed December 1, 2010, http://www.bmukk.gv.at/enfr/school/index.xml

BM: WF (Austrian Federal Ministry of Sciences and Research). (2009). *SparklingScience. Science linking with school. School linking with science.* Accessed January 7, 2010, http://www.sparklingscience.at/en/

Boix, G., & Olivella, R. (2007). Los sistemas de información geográfica (SIG) aplicados a la educación. El proyecto PESIG (Portal Educativo en SIG). In M. J. Marrón, J. Salom, & J. M. Souto (Eds.), *Las competencias geográficas para la educación ciudadana* (pp. 23–32). Valencia: Grupo de Didáctica de la Geografía. Asociación de Geógrafos Españoles. Accessed September 1, 2010, more information about PESIG Project in http://www.sigte.udg.edu/pesig/

Borg, C., Camilleri, J., Mayo, P., & Xerri, T. (1995). Malta's National curriculum: A critical analysis. *International Review of Education, 41*(5), 337–356.

Bower, P. A. (2005). *Using an Internet map server and coastal remote sensing for education.* Unpublished master thesis, Oregon State University.

Brown, M. (1994). Presentation material NZFSS conference, Hamilton. Unpublished.

Bualhamam, M. R. (2007). The preparation of students for the job market in geographical information systems at the United Arab Emirates University: Considerations and challenges. *Annual Journal of the Heidelberg Geographical Society, 21,* 191–202.

Buzo, I. (2010). (A very complete page): Accessed September 1, 2010, http://personales.ya.com/isaacbuzo/geografia.html

Cabinet Office. (2007). Dai 5-kai johoka to seishonen ni kansuru ishiki chosa ni tsuite (sokuho). Accessed December 1, 2010, http://www8.cao.go.jp/youth/kenkyu/jouhou5/g.pdf

Caldeweyher, D., Zhang, J., & Pham, B. L. (2006). OpenCIS – Open source GIS-based web community information system. *International Journal of Geographical Information Science, 20*(8), 885–898.

Calle, M. (2009). Aplicación de Google Earth en la formación del profesorado de educación infantil para el conocimiento geográfico. In Associaçao de Professores de Geografia et al. *A Inteligência Geográfica na Educação do Século XXI. IV Congreso Ibérico de Didáctica da Geografia* (pp. 152–157). Lisboa: Universidade de Lisboa.

Canadian Association of Geographers (CAG). (2008). *Canadian Association of Geographers Geographic Education Study Group.* http://info.wlu.ca/%7Ewwwgeog/CAGEDU.htm. Accessed 12 September 2010.

Canadian Association of Geographers (CAG). (2009a). *Canadian Association of Geographers Homepage.* http://www.cag-acg.ca/en/index.html. Accessed 12 September 2010.

Canadian Association of Geographers (CAG). (2009b). *Geographic awareness week and geographic information systems day.* http://www.cag-acg.ca/en/geography_week.html. Accessed 12 September 2010.

Canadian Council for Geographic Education (CCGE). (2001). *Canadian National Standards for Geography: A Standards-Based Guide to K-12 Geography.* http://www.ccge.org/programs/geoliteracy/geography_standards.asp. Accessed 12 September 2010.

Canadian Council for Geographic Education (CCGE). (2005). *Projecting geography in the Public Domain in Canada.* http://www.ccge.org/programs/geoliteracy/geolit_symposium.asp. Accessed 12 September 2010.

Canadian Council for Geographic Education (CCGE). (2010). http://www.ccge.org/. Accessed 15 September 2010.

Carr, K. (2004). *An evaluation of Private schools in the UAE.* Abu Dhabi: Ministry of Education, Scientific Research Administration.

Caruana, C. M., (1992). *Education's role in the socio-economic development of Malta*. New York: Praeger Publishers.

Cassar, G. (2003). Politics, religion and education in nineteenth century Malta. *Journal of Maltese Education Research*, *1*(1), 96–118.

Cassar, G. (2009). Education and schooling: From early childhood to old age. In J. Cutajar & G. Cassar (Eds.), *Social transitions in Maltese society* (pp. 51–74). Malta: Agenda Publishers.

Central Advisory Board of Education. (2005). *Universalisation of secondary education*. New Delhi: Ministry of Human Resource Development. Report nr F.2-15/2004-PN-1.

Chen, C. M., & Wang, Y. H. (2009). *GIS education in the senior high schools of Taiwan: results of a national survey of geography teachers*, Unpublished manuscript.

Christensen, K. E., Thomas J., & Torben P. J. (2003). Det digitale atlas. MTM-Geoinformatik, Aalborg Universitet.

CNNIC (2010). *Statistical report on Internet development in China. China Internet Network Information Centre (CNNIC)*. Accessed July 2010, Website: http://www.cnnic.net.cn

CNTV (2010). *China's mobile phone users top 800 mln.* Accessed September 2010, http://english. cntv.cn/20100721/101314.shtml

Collicard, J. P., Trisson, C. M., Genevois, S., & Joliveau, T. (2005). *L'utilisation d'un Système d'Information Géographique en classe*. Bilan d'une expérimentation Paper presented at the Actes de Géoforum Lille 2005 "Savoir penser et partager l'information géographique: les SIG", Lille.

Comas, D. (1995). *Urbamedia, un sistema d'informació geogràfica per analitzar la ciutat de Girona en conceptes clau. Experimentació a les ciències socials de l'ensenyament secundari obligatori*. Tesis doctoral, Universitat Autònoma de Barcelona, Barcelona.

Connect-ED. (2003). *Training Uganda's teachers with technology (2001–2003)*. Accessed January 11, 2010, http://learnlink.aed.org/Projects/uganda.htm

Cope, M., & Elwood, S. (2009). *Qualitatife GIS. A mixed methods approach*. Thousand Oaks, CA: SAGE Publications Ltd.

Coulter, B. (2003, July). *Maximising the potential for GIS to enhance education*. Paper presented at the ESRI education user conference, San Diego, CA.

Crechiolo, A. (1997). *Teaching secondary school geography with the use of a Geographical Information System (GIS)*. Unpublished master's thesis. Wilfrid Laurier University, Waterloo, Ontario, Canada.

Cremer, P., Richter, B., & Schäfer, D. (2004). GIS im Geographieunterricht – Einführung und Überblick. *Praxis Geographie*, *34*(2), 4–7.

CSDMS. (n.d.). *Methodology for neighbourhood mapping exercise*. Accessed December 10, 2010, http://www.csdms.in/NM/project/methodology1.htm

Cuban, L. (2008). *Frogs into princes: Writings on school reform*. New York: Teachers College.

Daniel, S., & Badard, T. (2008). *Mobile geospatial augmented reality, games and education: The Geoeduc3D project*. Paper presented at the Second International Workshop on Mobile Géospatial Augmented Reality, August 28–29, Quebec.

Dankwa, W. A. (1997). *SchoolNet: A catalyst for transforming education in Ghana*. Accessed October 20, 2003, http://www.isoc.org/isoc/whatis/conferences/inet/96/proceedings/c6/c6_1. htm

De Blij, H. (2008). *The power of place: Geography, destiny, and globalization's rough landscape*. London: Oxford.

Dehmer, W., & Koller, A. (2000). Computer und internet in der Lehrerfortbildung Österreichs, Seminarmodule und Zugänge für Lehrer. In M. Flath & G. Fuchs (Eds.), *Lernen mit Neuen Medien im Geographieunterricht* (pp. 61–75). Gotha und Stuttgart: Klett-Perthes.

Delannoy, F. (2000). Education reforms in Chile, 1980–1998: A lesson in pragmatism. Country studies. *Education Reform and Management Publication Series* (Vol. 1(1), pp. 1–80). Washington, DC: The World Bank.

Demirci, A. (2006). CBS'nin Türkiye'deki yeni coğrafya dersi öğretim programına göre coğrafya derslerinde uygulanabilirliği. In A. Demirci, M. Karakuyu, & M. A. Mcadams

(Eds.), *Proceedings of the 3. GIS Days in Turkey* (pp. 241–248). Istanbul: Fatih University Publications.

Demirci, A. (2007). Coğrafi Bilgi Sistemlerinin ilk ve ortaöğretim coğrafya derslerinde bir öğretim aracı olarak kullanılması: Önem, ilke ve metotlar. *Öneri Dergisi, 28*(7), 377–388.

Demirci, A. (2008a). Evaluating the implementation and effectiveness of GIS-based application in secondary school geography lessons. *American Journal of Applied Sciences, 5*(3), 169–178.

Demirci, A. (2008b). *Öğretmenler için CBS: Coğrafi Bilgi Sistemleri.* Istanbul: Fatih University.

Demirci, A. (2009). How do teachers approach new technologies: Geography teachers' attitudes towards Geographic Information Systems (GIS). *European Journal of Educational Studies, 1*(1), 57–67.

Demirci, A., & Karaburun, A. (2009). How to make GIS a common educational tool in schools: Potentials and implications of the *GIS for Teachers* book for geography education in Turkey. *Ozean Journal of Applied Sciences, 2*(2), 205–215.

Demirci, A., Taş, H. İ., & Özel, A. (2007). Türkiye'de ortaöğretim coğrafya derslerinde teknoloji kullanımı. *Marmara Coğrafya Dergisi, 15,* 37–54.

Deng, Z. Y., & Gopinathan, S. (2005). The information technology masterplan. In J. Tan, & P. T. Ng (Eds.), *Shaping Singapore's future: Thinking schools, learning nation* (pp. 22–40). Singapore: Pearson Prentice Hall.

Deutsche Gesellschaft für Geographie. (Ed.). (2008). *Bildungsstandards im Fach Geographie für den mittleren Schulabschluss – mit Aufgabenbeispielen.* 5. Auflage: Selbstverlag Deutsche Gesellschaft für Geographie.

Deutsche Gesellschaft für Geographie. (Ed.). (2009). *Rahmenvorgaben für die Lehrerausbildung im Fach Geographie an deutschen Universitäten und Hochschulen.* Bonn: Selbstverlag Deutsche Gesellschaft für Geographie.

Diaz, N. (2010). *Prototipo de aplicación de los sistemas de información geográfica S.I.G en la enseñanza de geografía en la educación básica secundaria en Colombia.* Tesis para optar por el título de Ingeniera Catastral y Geodesta. Bogotá D.C.: Universidad Distrital Francisco José de Caldas.

Doering, A., & Veletsianos, G. (2007). An investigation of the use of real-time, authentic geospatial data in the K12 classroom. *Journal of Geography, 106,* 217–225.

Dokken, Ø., Johansen, O. I., & Øverjordet, A. H. (2004). *Geografi: Landskap, ressurser, mennesker og miljø.* Oslo: Cappelen.

Donert, K. (2007). *Teaching geography in Europe using GIS.* Papers and presentations from the Education track of the Esri User conference 2006. Stockholm: HERODOT, Esri Inc.

Donert, K., Ari, Y., Attard, M., O'Reilly, G., & Schmeinck, D. (Eds.). (2009). *Geographical diversity.* Proceedings HERODOT conference, 29–31 May 2009. Ayvalik Turkey: HERODOT Network. (GIS chapters in pp. 264–329). Accessed September 1, 2010, http://www.herodot. net/conferences/Ayvalik/papers/manuscript-v1.pdf. Accessed September 1, 2010, Escuela2000 programme in http://www.la-moncloa.es/ActualidadHome/2009-2/040409-enlace20

Donert, K., & Wall, K. (Eds.). (2008). *Future prospects in geography* (pp. 511–517). Liverpool: Liverpool Hope University.

Earth Observatory of Singapore (2010). Volcano Science Moves into the Classroom. http://www. earthobservatory.sg/media/news-and-features/50-volcano-science-moves-into-the-classroom. html. Accessed 14 August 2010.

EDUCAR CHILE. (2005). *The education portal.* Accessed November 14, 2009, http://www. educarchile.cl/Portal.herramientas/quienessomosingles/index.html

Engelhardt, R. (2009). *OHG-Schüler erzielen ersten Preis.* Accessed January 15, 2010, http://www. ohg-ka.de/

English, K. Z., & Feaster, L. S. (2002). *Community geography: GIS in action.* Redlands, CA: Esri.

ENLACES. (2005). *Centro de Educación y Tecnología.* Ministerio de Educación, Santiago, Chile.

ENLACES. (2009). *Centro de Educación y Tecnología.* Ministerio de Educación, Santiago, Chile. http://www.enlaces.cl/index.php?t=44&i=2&cc=800&tm=2. Accessed 2 November 2009.

Erebus International. (2008). A Study into the Teaching of Geography in Years 3–10. Australian Government Department of Education, Employment and Workplace Relations.

Eremu, J. (2009). *ICT connectivity in schools in Uganda*. Accessed May 14, 2010, http://www. ftpiicd.org/iconnect/ICT4D_Education/ICTEducation_Uganda.pdf

Esri. (2010). *Open source*. Accessed January 12, 2010, http://www.esri.com/technology-topics/ open-source/index.html

European Commission. (2008). The Education system in France: Directorate-General for Education and Culture.

European Commission. (2009). EURYDICE. National summary sheets on education system in Europe and ongoing reforms. Finland, August 2009. Education, Audiovisual & Culture Executive Agency. Accessed August 20, 2010, http://eacea.ec.europa.eu/education/eurydice/ documents/eurybase/national_summary_sheets/047_FI_EN.pdf

European Union. (2006). European Parliament and the Council Recommendation of the European Parliament and of the Council on key competences for lifelong learning of 18 December 2006 (2006/962/EC). Official Journal of the European Union, 30.12.2006.

Executive, Y. (2001). *Plan to develop knowledge-based economy in Taiwan, Council for Economic Planning and Development*. Accessed January 22, 2010, Web site: http://www.cepd.gov.tw/att/ dot/ppt.gif

Falch, T., Borge, L. E., Lujala, P., Nyhus, O. H., & Strøm, B. (2010). *Årsaker til og konsekvenser av manglende fullføring av videregående opplæring. SØF-rapport nr. 03/10*. Accessed September 10, 2010, http://www.sof.ntnu.no/SOF-R%2003_10.pdf

Falk, G., & Nöthen, E. (2005). *GIS in der Schule Potenziale und Grenzen*. Berlin: Mensch & Buch-Verlag.

Fargher, M. G. (2006). *An exploration of the contribution of a 'Local Solutions' project to curriculum innovation with GIS*. Unpublished MRes dissertation, Institute of Education, University of London.

Favier, T. T., & Schee, J. A. v. d. (2009). Learning geography by combining fieldwork with GIS. *International research in geographical and environmental education, 18*, 261–274.

Ferrari, J. (2006a). Naive syllabus neglects basics. *The Australian*, 25 September, 2006, p. 1.

Ferrari, J. (2006b). The geography wars. *The Australian*, 28 September 2006, http://www. theaustralian.news.com.au/story/0,20867,20487109-28737,00.html

Fischer, F. (2008). Collaborative mapping – How Wikinomics is Manifest in the Geo-information Economy. *GEO Informatics März*, 28–31.

Fitzpatrick, C., & Maguire, D. J. (2001). GIS in schools. Infrastructure, methodology and role. In D. R. Green (Ed.), *GIS: A sourcebook for schools* (pp. 62–72). London: Taylor & Francis.

Fontanieu, V., Genevois, S., & Sanchez, E. (2007). Les pratiques géomatiques en collège-lycée. D'après les résultats d'une enquête nationale sur les usages des outils géomatiques dans l'enseignement de l'Histoire-Géographie et des sciences de la vie et de la Terre. *Géomatique Expert*, 49–55.

Forster, M., & Mutsindashyaka, Th. (2008). *Experiences from Rwandan secondary schools using GIS*. Proceedings of the Esri Education User Conference 2008. Accessed February 1, 2010, http://proceedings.esri.com/library/userconf/educ08/educ/papers/pap_1119.pdf

Friedman, T. L. (2005). *The world is flat: A brief history of the 21st century*. New York: Farrar, Straus, and Giroux.

Gaffga, P., Barheier, K., Bloching, K., Engelmann, D., Felsch, M., Harms, E., et al. (2008). *Diercke GWG für Gymnasien in Baden-Württemberg 5*. Braunschweig: Westermann.

García, F. M. (2003). La enseñanza de las nuevas tecnologías en la Universidad. Sistemas de Información Geográfica. In M. J. Marrón, C. Moraleda, & H. Rodríguez (Eds.), *La enseñanza de la Geografía ante las nuevas demandas sociales* (pp. 180–186). Toledo: Grupo de Didáctica de la Geografía (AGE). Escuela Universitaria del Profesorado, Universidad de Castilla-La Mancha.

García, F. M. (2006). Consideraciones didácticas con SIG. Modelos medio ambientales susceptibles de ser desarrollados en el aula. In M. J. Marrón, & L. Sánchez (Eds.), *Cultura geográfica y educación ciudadana* (pp. 297–308). Murcia: Grupo de Didáctica de la Geografía. Asociación de Geógrafos Españoles. Associaçao de Professores de Geografia de Portugal. Universidad de Castilla-La Mancha.

Gatrell, J. D. (2001). Structural, technical, and definitional issues: The case of geography, GIS, and K-12 classrooms in the United States. *Journal of Educational Technology Systems, 29*(3), 237–249.

Genevois, S. (2008). Quand la géomatique rentre en classe. Usages cartographiques et nouvelle éducation géographique dans l'enseignement secondaire. Thèse de doctorat, Université de Saint-Etienne, UMR 5600.

Genevois, S., & Joliveau, T. (2009). Using a geoinformation-based learning environment in geography education: The GeoWebExplorer platform. In T. Jekel, A. Koller, & K. Donert (Eds.), *Learning with Geoinformation IV* (pp. 113–120). Heidelberg: Wichmann.

Genevois, S., Carlot, Y., Joliveau, T., & Collicard, J. P. (2003). Le SIG: un outil didactique innovant pour la géographie scolaire. *Dossiers de l'ingéniérie éducative* (n°44), 10–13.

GEOPORTAL. (2010). *Ministry of National Goods*. Accessed February 8, 2010, http://www.geoportal.cl/Portal/ptk

Gilson, E., De Battista, R., & Quintano, A. (2009). *Geography syllabus. Department for Curriculum Management and eLearning, Directorate for Quality and Standards in Education.* Ministry of Education, Youth and Sport, Malta. Accessed January 5, 2010, http://www.curriculum.gov.mt/docs/Geo_Gen_JL_1_5.pdf

GLOBE. (2009). *Global learning and observations to benefit the environment*. Accessed November 20, 2009, http://www.globe.gov

Goh, C. T. (1997). Shaping our future: 'Thinking Schools' and a 'Learning Nation'. *Speeches (Singapore), 21*(3), 12–20.

Goh, C. T. (2001). Shaping lives, moulding nation. *Speeches (Singapore), 25*(4), 11–24.

González, M. E., Capdevila, J., & Soteres, C. (2008). Las Infraestructura de Datos Espaciales como recurso educativo para el profesorado de la Educación Secundaria Obligatoria. Una propuesta innovativa de formación e-learning. In IX Encuentro Internacional Virtual Educa Zaragoza 2008. Zaragoza, 14–18 de julio, 2008. Accessed September 1, 2010, http://www.virtualeduca.info/ponencias/246/EDUCA_ZARAGOZA_IDE-EDU-ESO.doc

Goodchild, M. (2008a). Bürger als Sensoren / Citizens as Sensors. *GIS Trends Markets, 6*, 27–31.

Goodchild, M. (2008b). Volunteered geographic Information. *GEOconnexion International Magazine*, Oktober 08, 46–47.

Goodchild, M. F. (1992). Geographical information science. *International Journal of Geographical Information Systems, 6*, 3–45.

Goodchild, M. F. (2008). What does Google Earth mean for the social sciences? In M. Dodge, M. McDerby, & M. Turner (Eds.), *Geographic visualization. Concepts, tools and applications* (pp. 11–23). Chichester: Wiley.

Goodchild, M. F., & Palladino, S. D. (1995). Geographic information systems as a tool in science and technology education. *Speculations in Science and Technology, 18*(4), 278–286.

Government of India. (1968). *National policy on education*. New Delhi: Ministry of Human Resource Development.

Government of India. (1986). *National policy on education*. New Delhi: Ministry of Human Resource Development.

Government of India. (2009). *Annual report of the department of school education and literacy*. New Delhi: Ministry of Human Resource Development.

Graham, C. K. (1971). *The history of education in Ghana: From the earliest times to the declaration of independence*. London: F. Cass.

Green, D. R. (2001). GIS in school education: You don't necessarily need a computer. In D. R. Green (Ed.), *GIS: A sourcebook for schools* (pp. 34–61). New York: Taylor & Francis.

Grima, G., Grech, L., Mallia, C., Mizzi, B., Vassallo, P., & Ventura, F. (2008). *Transition from primary to secondary schools in Malta: A review*. Floriana, Malta: Ministry for Education, Youth and Sport.

Gryl, I., Jekel, T., & Donert, K. (2010). GI & spatial citizenship. In T. Jekel, A. Koller, K. Donert, & R. Vogler (Eds.), *Learning with GI V* (pp. 2–11). Berlin & Offenbach: Wichmann.

Hall-Wallace, M. K. (2002). *Exploring tropical cyclones.* Brooke Coles-Thomason Learning.

Han, L., & Zheng, J. (2006). Practical methods for GIS education in China – How to meet social need of high quality human resources. *IEEE International Conference on Geoscience and Remote Sensing.* pp. 999–1002. Denver, CO.

Haubrich, H. (Ed.). (2006). *Geographie unterrichten lernen, Die neue Didaktik der Geographie konkret* (2 ed.). München: Oldenbourg.

Hawkins, R. J. (2002). *Ten lessons from ICT and education in the developing world.* World link for development program, The World Bank Institute. Accessed March 23, 2010, http://www.cid.harvard.edu/archive/cr/pdf/gitrr2002_ch04.pdf

Hepp, P., Hinostroza, J. E., & Laval, E. (2004). A systematic approach to educational renewal with new technologies: Empowering learning communities in Chile. In A. Brown & N. Davis (Eds.), *World yearbook of education 2004: Digital technologies, communities and education* (pp. 299–311). London: Routledge Falmer.

Heyden, K. (2009). *Einsatz von geographischen Informationssystemen an der LG.* Accessed January 15, 2010, http://www.lg-ratzeburg.de/index.php?id=172

Hinostroza, J. E., Labbé, C., & Claro, M. (2005). ICT in Chilean schools: Student's and teacher's access to and use of ICT. *Human Technology, 1*(2), 246–264.

Hof, R. (2005). *The power of us. Mass collaboration on the Internet is shaking up business.* BusinessWeek [online http://www.businessweek.com/magazine/content/05_25/b3938601.htm].

Höhnle, S., Schubert, J. C., & Uphues, R. (2010). The frequency of GI(S) use in the geography classroom – Results of an empirical study in German secondary schools. In T. Jekel, A. Koller, K. Donert, & R. Vogler (Eds.), *Learning with Geoinformation V – Lernen mit Geoinformation V* (pp. 148–158). Berlin: Wichmann.

Holt-Jensen, A. (1981). *Geography: Its history and concepts: A student's guide.* London: Harper & Row.

Houtsonen, L. (2003). Maximising the use of communication technologies in geographical education. In R. Gerber (Ed.), *International handbook on geographical education* (pp. 47–63). Dordrecht: Kluwer Academic Publishers.

Hungarian National Core Curriculum. (2009). *Ministry of education and culture.* Accessed January 7, 2009, www.okm.gov.hu/english/hungarian-national-core

Hutchinson, N., & Pritchard, B. (2006). 'True Blue' geography. *Geography Bulletin, 38*(4), 16–18.

Huynh, N. (2009). The role of geospatial thinking and geographic skills in effective problem solving with GIS: K-16 education. Ph.D. dissertation, Wilfrid Laurier University, Canada. Retrieved November 9, 2010, from Dissertations & Theses: Full Text. (Publication No. AAT NR54258).

Hwang, M. (1998). Applications of geographic information system technology in geography education. *Journal of Geography Education, 40,* 1–12.

Hwang, S., & Lee, K. (1996). The present status and prospect of GIS learning in teaching geography of high school. *Journal of the Korean Association of Regional Geographers, 2,* 219–231.

Ida, Y. (2008). Mijika na chiiki shirabe – chiiki no tokucho wo miidasu hoho. In K. Asakura, J. Ito, & Y. Hashimoto (Eds.), *Chugakko shakai wo yori yoku rikai suru* (pp. 131–136). Nihon bunkyo shyppan.

Iglesias, J. (2002). Problem-based learning in initial teacher education. *Prospects, 32*(2), 319–332.

Ismail, M. (2002). *Readiness for the networked world: Ghana assessment.* Information Technologies Group, Center for International Development. Harvard and Digital Nations, The Media Lab, MIT.

Jekel, T. (2007). What you all want is GIS 2.0! In A. Car, G. Griesebner, & J. Strobl (Eds.), *Collaborative GI based learning environments for spatial planning and education* (pp. 84–89). GI-Crossroads@GI-Forum. Heidelberg: Wichmann.

Jekel, T., & Gryl, I. (2010). Spatial citizenship. Beiträge von Geoinformation zu einer mündigen Raumaneignung. *Geographie und Schule, 186*, 39–45.

Jekel, T., Koller, A., & Donert, K. (Eds.). (2008). *Learning with Geoinformation III – Lernen mit Geoinformation III*. Heidelberg: Wichmann.

Jekel, T., Koller, A., & Donert, K. (Eds.). (2009). *Learning with Geoinformation IV – Lernen mit Geoinformation IV*. Heidelberg: Wichmann.

Jekel, T., Koller, A., & Strobl, J. (Eds.). (2006). *Lernen mit Geoinformation*. Heidelberg: Wichmann.

Jekel, T., Koller, A., & Strobl, J. (Eds.). (2007). *Lernen mit Geoinformation II*. Heidelberg: Wichmann.

Jekel, T., Koller, A., Donert, K., & Vogler, R. (Eds.). (2010). *Learning with Geoinformation V – Lernen mit Geoinformation V*. Berlin: Wichmann.

Johansson, T. (2006). GISAS project in a nutshell. In T. Johansson (Ed.), Geographical Information Systems applications for schools–GISAS (pp. 7–21). Helsinki: Dark Oy.

Joliveau, T., Calcagni, Y., & Mayoud, R. (2005). *Geowebexplorer, un outil géomatique collaboratif au service des enseignants et des élèves*. Paper presented at the Actes de Géoforum Lille 2005 "Savoir penser et partager l'information géographique: les SIG", Lille.

Jung, A. (1997). Constructions of instruction documents about GIS in high school geography education. *Journal of the Korean Association of Geographic and Environmental Education, 5*(2), 61–73.

Jung, I. (2005). Undergraduate GIS curricula of department of geography in U.S. *Journal of the Korean Association of Geographic and Environmental Education, 13*, 225–234.

Jung, I., & Kim, J. (2006). Development of GIS teaching plans in high school geography classrooms. *Journal of the Korean Association of Geographic and Environmental Education, 14*, 251–262.

Kankaanranta, M. (2005). International perspectives on the pedagogically innovative uses of technology. *Human Technology, 1*(20), 111–116.

Karabağ, S. (Ed.). (2005). *Coğrafya dersi öğretim programı. Talim ve Terbiye Kurulu Başkanlığı*. Ankara: Gazi Kitabevi

Karatepe, A. (2007). *The use of Geographic Information Technologies in geography education*, Unpublished doctoral dissertation, Marmara University, Institute of Education Sciences, Istanbul.

Karikari, I. B., Stillwell, J. C. H., & Carver, S. (2003). Land administration and GIS: The case of Ghana. *Progress in Development Studies, 3*(3), 223–242.

Keiper, T. A. (1999). GIS for elementary students: An inquiry into a new approach to learning geography. *Journal of Geography, 98*, 47–59.

Kerski, J. (2003). The implementation and effectiveness of Geographic Information Systems Technology and methods in secondary education. *Journal of Geography, 102*(3), 128–137.

Kerski, J. J. (2000). *The implementation and effectiveness of Geographic Information Systems Technology and methods in secondary education*. Ph.D. Dissertation, University of Colorado.

Kerski, J. J. (2008a). The world at the student's fingertips: Internet-based GIS education opportunities. In A. J. Milson & M. Alibrandi (Eds.), *Digital geography: Geospatial technologies in the social studies classroom* (pp. 119–134). Charlotte, NC: Information Age.

Kerski, J. J. (2008b). Toward an international geospatial education community. In A. J. Milson & M. Alibrandi (Eds.), *Digital geography: Geospatial technologies in the social studies classroom* (pp. 61–74). Charlotte, NC: Information Age.

Kim, C. (2005). A study of the GIS education in the geography education: In the case of the USA. *Journal of the Korean Association of Geographic Information Studies, 8*(4), 176–190.

Kim, M. (2007). Spatial thinking and the investigation of GIS for potential application in education. *Journal of the Korean Association of Geographic and Environmental Education, 15*, 233–245.

Kim, M. (2010). The current status of GIS in the classroom and factors to consider for increasing the use of GIS. *Journal of the Korean Association of Geographic and Environmental Education, 18*, 173–184.

Kim, M., Bednarz, R., & Lee, S.-I. (2009). *Why do some teachers participate in GIS education?* Paper presented at the Annual Meeting of the National Council for Geographic Education, San Juan, Puerto Rico.

Kim, Y. (2002). The introduction and development of GIS curriculum in the UK geography education. *Journal of the Korean Association of Regional Geographers, 8*, 380–395.

Kinniburgh, J. C. (2008). An investigation of the impediments to using Geographical Information Systems (GIS) to enhance teaching and learning in mandatory Stage 5 geography in New South Wales. *Geographical Education, 21*, 20–38.

Klein, U. (2007). *Geomedienkompetenz. Untersuchung zur Akzeptanz und Anwendung von Geomedien im Geographieunterricht unter besonderer Berücksichtigung moderner Informations- und Kommunikationstechniken.* Dissertation, Christian-Albrechts-Universität, Kiel.

KMK. (2010). *Bundesweit geltende Bildungsstandards.* Accessed September 15, 2010, http://www.kmk.org/bildung-schule/qualitaetssicherung-in-schulen/bildungsstandards/ueberblick.html

KNAG. (2003). *Gebieden in perspectief: natuur en samenleving, nabij en veraf. Rapport Commissie Aardrijkskunde Tweede Fase.* Utrecht: Koninklijk Nederlands Aardrijkskundig Genootschap.

KNAG. (2008). *Kijk op een veranderende wereld: Voorstel voor een nieuw examenprogramma aardrijkskunde VMBO.* Utrecht: Koninklijk Nederlands Aardrijkskundig Genootschap.

Kobayashi, T. (1999). Hyo keisan sofuto niyoru chiri joho shisutemu. *Tsukuba shakaika kenkyu, 18*, 61–70.

Korevaar, W. (2003). GIS in het voortgezet onderwijs. *Vi Matrix, 11.*

Korevaar, W. (2004). Modern aardrijkskundeonderwijs met GIS op de kaart gezet. *Geografie, 13.*

Koutsopoulos, K. C. (2008). What's European about European geography? The case of geoinformatics in europeanization. *Journal of Geography in Higher Education, 32*(1), 1–14.

Kwon, S. (2004). A GIS class exercise examples focused on the basic concepts and functions. *Journal of the Korean Association of Geographic and Environmental Education, 12*, 313–325.

L94. (1994). *Læreplan for videregående opplæring. Geografi. Felles, allment fag i studieretning for allmenne, økonomiske og administrative fag.* Oslo: Kirke, utdannings- og forskningsdepartementet.

Lay, J. G., & Chiu, H. C. (2003). *E-generation education and GIS skills.* The Proceedings of 2003 National Geographical Conference of Taiwan, pp. 296–306.

Lay, J. G., & Yu, C. C. (2004). Cases study on GIS proficiency of high school teachers. *Bulletin of the Geographical Society of China, 33*, 21–47.

Lázaro, M. L. (2000). La utilización de internet en el aula para la enseñanza de la Geografía: ventajas e inconvenientes. In J. L. González & M. J. Marrón (Eds.), *Geografía, profesorado y sociedad. Teoría y práctica de la Geografía en la enseñanza* (pp. 211–218). Murcia: Grupo de Didáctica de la Geografía. Asociación de Geógrafos Españoles-Universidad de Murcia.

Lázaro, M. L. (2003). Nuevas Tecnologías en la enseñanza-aprendizaje de la Geografía. In M. J. Marrón, C. Moraleda, & H. Rodríguez (Eds.), *La enseñanza de la Geografía ante las nuevas demandas sociales* (pp. 141–167). Toledo: Grupo de Didáctica de la Geografía (AGE). Escuela Universitaria del Profesorado, Universidad de Castilla-La Mancha.

Lázaro, M. L., & González, M. J. (2005). La utilidad de los Sistemas de Información Geográfica para la enseñanza de la Geografía. *Didáctica Geográfica, 7*, 105–122.

Lázaro, M. L., & González, M. J. (2006). La utilidad de los SIG existentes en internet para el conocimiento territorial. In M. J. Marrón, & L. Sánchez (Eds.), *Cultura geográfica y educación ciudadana* (pp. 443–452). Murcia: Grupo de Didáctica de la Geografía. Asociación de Geógrafos Españoles. Assicuaçao de Professores de Geografia de Portugal. Universidad de Castilla-La Mancha.

Lázaro, M. L., & González, M. J. (2007a). Spain on the web: A GIS way of teaching. In K. Donert, P. Charzynsky, & Z. Podgorski (Eds.), *Teaching in and about Europe. Geography in European higher education* (pp. 36–43). Toruń: University of Toruń and HERODOT network.

Lázaro, M. L., & González, M. J. (2007b). Learning about Spain through GIS Webs. In K. Donert (Ed.), *EUC'07 HERODOT Proceedings*. *Esri European User Conference 2007*: Stockholm. Esri-HERODOT Publications. Accessed September 1, 2010, http://www.herodot. net/conferences/stockholm/esri/Lazaro_Gonzalez(paper).pdf

Lázaro, M. L., González, M. J., & Lozano, M. J. (2008a). Google Earth and ArcGIS explorer in geographical education. In T. Jekel, A. Koller, & K. Donert (Eds.), *Learning with Geoinformation III – Lernen mit Geoinformation III* (pp. 95–105). Munich: Wickmann.

Lázaro, M. L., González, M. J., & Lozano, M. J. (2008b). Learning about immigration in Spain through Geoinformation on the internet. In K. Donert & G. Wall (Eds.), *Future prospects in geography* (pp. 439–445). Liverpool: Liverpool Hope University.

Lee, J. W., & Bednarz, R. (2009). Effect of GIS learning on spatial thinking. *Journal of Geography in Higher Education, 33*(2), 183–198.

Lee, M., Kim, N., & Ban, S. (2008). A study on the development of geography e-learning material using webGIS for the social studies of middle school. *Journal of the Korean Association of Geographic and Environmental Education, 16*, 17–26.

Lee, M., Kim, N., Lee, S., & Jo, E. (2006). Development of learning materials for Korean geography in high school curriculum using GIS and remote sensing: Focusing on the landform chapter. *Journal of the Korean Association of Geographic and Environmental Education, 14*, 191–200.

Lemberg, D., & Stoltman, J. P. (2001). Geography teaching and the new technologies: Opportunities and challenges. *Journal of Education, 181*(3), 63–76.

Leon G. G. (2002). *La imposición de modelo pedagógicos en Colombia – Siglo XX*. *Revista de Estudios Latinoamericanos*. Pasto: Universidad de Nariño, Centro de Estudios e Investigaciones Latinoamericanas.

Lin, F. Y., & Lay, J. G. (2006). A study on the geographic information exam questions of the college entrance examination. *Journal of Cartography, 16*, 167–190.

Lindner-Fally, M. (2009). Digital:earth:at – centre for teaching and learning geography and geoinformatics. In K. Donert, Y. Ari, M. Attard, G. O'Reilly, & D. Schmeinck (Eds.), *Geographical diversity. Proceedings of the Herodot conference in Ayvalik, Turkey* (pp. 332–338). Berlin: Mensch und Buch.

Linn, S, Kerski, J., & Wither, S. (2005). Development of evaluation tools for GIS: How does GIS affect student learning? *International Research in Geographical and Environmental Education, 14*(3), 217–224.

Lisle, R. J. (2006). Google Earth: A new geological resource. *Geology Today, 22*(1), 29–32.

Liu, S. X., & Zhu, X. (2008). Designing a structured and interactive learning environment based on GIS for secondary geography education. *Journal of Geography, 107*(1), 12–19.

Liu, Y., Bui, E. N., Chang, C.-H., & Lossman, H. (2010). PBL-GIS in secondary geography education: Does it result in higher-order learning outcomes? *Journal of Geography, 109*(4), 150–158.

Liu, Y., & Laxman, K. (2009). GIS-enabled PBL pedagogy: The effects on students' learning in the classroom. *I-manager's Journal on School Educational Technology, 5*(2), 15–27.

Lloyd, W. J. (2001). Integrating GIS into the undergraduate learning environment. *Journal of Geography, 100*(5), 158–163.

Longley, P., Goodchild, M., Maguire, D., & Rhind, D. (2005). *Geographic Information Systems and science* (2nd ed.). New York: Wiley.

Louv, R. (2005). *Last child in the woods: Saving our children from Nature-Deficit disorder*. Chapel Hill, NC: Algonquin.

Mahoney, P. (2005). *National certificate of educational achievement executive summary*. New Zealand Parliamentary library.

Mallick, R. (2005). *Infusing map culture through participatory mapping*. Accessed December 10, 2010, http://www.gisdevelopment.net/magazine/years/2005/feb/infusing.htm

Malone, L., Palmer, A., & Voigt, C. (2002). *Mapping our world: GIS lessons for educators, ArcView edition*. Redlands, CA: Esri.

Malone, L., Palmer, A., & Voigt, C. (2005). *Mapping our world: GIS lessons for educators, ArcGIS desktop edition*. Redlands, CA: Esri.

Manitoba Social Science Teachers' Association. (2010). Accessed September 15, 2010, http://www.mssta.mb.ca/

Marsh, M., Golledge, R., & Battersby, S. E. (2007). Geospatial concept understanding and recognition in G6- college students: A preliminary argument for minimal GIS. *Annals of the Association of American Geographers, 97*(4), 696–712.

Martín, C., & García, F. (2009). Algunos recursos en internet para mejorar la enseñanza de la Geografía. Ar@cne. Revista electrónica de recursos en internet sobre Geografía y Ciencias Sociales, 118. Accessed September 1, 2010, http://www.raco.cat/index.php/Aracne/article/view/130175/179613

Martín de Agar, R., & Nieto, J. A. (2009). *Andalusia in a map. Assessment of an educational experience*. XXIV International Cartographic Conference. Santiago de Chile, Chile. Accessed September 1, 2010, http://icaci.org/documents/ICC_proceedings/ICC2009/html/nonref/29_5.pdf

McClurg, P. A., & Buss, A. (2007). Professional development: Teachers use of GIS to enhance student learning. *Journal of Geography, 106*, 79–87.

McHarg, I. (1995). *Design with nature* (2nd ed.). New York: Wiley.

McInerney, M. (2003, July). *The next step with GIS in the curriculum: Approaching the question of GIS and classroom pedagogy*. Presentation to the Esri User conference, San Diego.

Meyer, J. W., Butterick, J., Olkin, M., & Zack, G. (1999). GIS in the K-12 curriculum: A cautionary note. *Professional Geographer, 54*(1), 571–578.

Miguel, I. de, & Allende, F. (1996). *Uso de una aplicación S.I.G. como recurso didáctico en la enseñanza secundaria. III Jornadas de didáctica de la Geografía* (pp. 77–85). Madrid: Grupo de Didáctica de la Geografía (AGE). Departamento de Didáctica de las Ciencias Sociales de la Universidad Complutense de Madrid, Spain.

Mikkelsen, R. (2009). Geografi i K06 – læreplanprosess, utfordringer og endringer. In O. Fjær & E. Eikli (Eds.), *Geografi og Kunnskapsløftet. Rapport fra Norsk Geografisk Selskaps konferanse i Trondheim; Sted, levemåter og sårbarhet 27–28. mars 2008* (pp. 9–25). Acta Geographica – Trondheim. Serie B, No 16. Trondheim: Department of Geography.

Milson, A. J., & Curtis, M. (2009). Where and why there? Spatial thinking with geographic information systems. *Social Education, 73*(3), 113–118.

Milson, A. J., & Earle, B. D. (2007). Internet-based GIS in an inductive learning environment: A case study of ninth grade geography students. *Journal of Geography, 106*, 227–237.

Milson, A. J., & Roberts, J. A. (2008). The status of geospatial technologies in state geography standards in the United States. In A. J. Milson & M. Alibrandi (Eds.), *Digital geography: Geospatial technologies in the social studies classroom* (pp. 39–59). Charlotte, NC: Information Age.

Milson, A. J., Gilbert, K. M., & Earle, B. D. (2007). Discovering Africa through Internet-based geographic information systems: A pan-African summit simulation. *Social Education, 71*(3), 140–145.

Minaidi, A., & Hlapanis, G. H. (2005). Pedagogical obstacles in teacher training in information and communication technology. *Technology, Pedagogy and Education, 14*(2), 241–254.

Ministerio de Educación Nacional (MEN). (1988). Propuesta de programa curricular. Ciencias Sociales. Sexto Grado. Educación Básica Secundaria. Bogotá D.C.

Ministerio de Educación Nacional (MEN). (1989). Propuesta de programa curricular. Ciencias Sociales. Séptimo Grado. Educación Básica Secundaria. Bogotá D.C.

Ministerio de Educación Nacional (MEN). (1990). Propuesta de programa curricular. Ciencias Sociales. Octavo Grado. Educación Básica Secundaria. Bogotá D.C.

Ministerio de Educación Nacional (MEN). (1991). Propuesta de programa curricular. Ciencias Sociales. Noveno Grado. Educación Básica Secundaria. Bogotá D.C.

Ministerio de Educación Nacional (MEN). (1994). *Ley 115 de 1994 (8 de Febrero). Ley General de Educación.* Accessed August 10, 2010, http://www.mineducacion.gov.co/1621/w3-channel. html

Ministerio de Educación Nacional (MEN). (2004). *Estándares Básicos de Competencias en Ciencias Naturales y Ciencias Sociales.* SERIE GUÍAS No 7. Formar en Ciencias: ¡El desafió! Lo que necesitamos saber y saber hacer. Bogotá D.C.

Ministerio de Educación Nacional (MEN). (2008). *Revolución educativa: Plan Sectorial de educación 2006–2010.* Bogotá D.C.

Ministerio de Tecnologías de la Información y las Comunicaciones (MTIC). (2008). Plan Nacional de Tecnologías de la Información y las Comunicaciones 2008–2019. Bogotá D.C.

Ministerium für Bildung Wissenschaft und Kultur Mecklenburg-Vorpommern. (2002). *Rahmenplan Geographie - Regionale Schule, Verbundene Haupt- und Realschule, Hauptschule, Realschule, Integrierte Gesamtschule.* Accessed September 10, 2007, http:// www.bildungsserver-mv.de/download/rahmenplaene/rp-geografie-7-10-reg.pdf

Ministerium für Kultus Jugend und Sport Baden-Württemberg. (2004). *Bildungsplan allgemein-bildendes Gymnasium.* Accessed July 10, 2007, www.bildung-staerkt-menschen.de/service/ downloads/Bildungsplaene/Gymnasium/Gymnasium_Bildungsplan_Gesamt.pdf

Ministry of Education. (1974). *The new structure and context of education for Ghana.* Accra: Republic of Ghana.

Ministry of Education. (1997). *Masterplan for information technology in education: A summary.* Singapore: Ministry of Education.

Ministry of Education. (1998). *Curriculum review report.* Singapore: Ministry of Education.

Ministry of Education. (1999). *National minimum curriculum.* Valletta: Ministry of Education.

Ministry of Education. (1999a). *Review of education sector analysis in Ghana 1987–1998.* Accessed October 20, 2003, http://www.adeanet.org/wgesa/en/doc/Ghana/chapter_2.htm

Ministry of Education. (1999b). *Comprehensive framework on education.* Accra: Republic of Ghana.

Ministry of Education. (2002). *Strategy for further educational development in the United Arab Emirates up to the year 2020.* Abu Dhabi: Scientific Research Administration.

Ministry of Education. (2002). *The People's Republic of China. Geography curriculum standards for full-time compulsory education (experimental).* Beijing: Beijing Normal University Press (in Chinese).

Ministry of Education. (2003). *The People's Republic of China. High school geography curriculum standards (trial version) [S].* Beijing: People's Education Press (in Chinese).

Ministry of Education. (2005). *For all children to succeed.* Malta: MEYE.

Ministry of Education. (2006). ICT Professional Development Clusters – schools. Accessed June 2010, www.tki.org.nz.

Ministry of Education. (2007). *The New Zealand curriculum.* Wellington, New Zealand: Learning Media Limited.

Ministry of Education. (2008). *Supporting professional development for teachers.* Performance Audit Report.

Ministry of Education of the Republic of Rwanda. (2002a). *EFA plan of action.* Kigali: Ministry of Education.

Ministry of Education of the Republic of Rwanda. (2002b). *Annuaire statistique 2001/2002.* Kigali: Ministry of Education.

Ministry of Education of the Republic of Rwanda. (2006). *Education sector strategic plan 2006–2010 (Draft).* Kigali: Ministry of Education.

Ministry of Education of the Republic of Rwanda. (2008). *Education sector strategic plan 2008–2012.* Kigali: Ministry of Education.

Ministry of Education of the Republic of Rwanda. (2009). *Official launch of GIS textbook for secondary schools in Rwanda.* Online article. Accessed February 1, 2010, http://www.mineduc. gov.rw/spip.php?article484

Ministry of Education, Science and Technology. (2008). *Explanation of middle school curriculum 2: Korean, ethics, social studies*. Seoul: Mirae N (Daehan Textbook).

Ministry of Education, Science and Technology. (2009a). *2009 revised curriculum*. Seoul: Ministry of Education, Science and Technology.

Ministry of Education, Science and Technology. (2009b). *Explanation of high school curriculum 4: Social studies (history)*. Seoul: Mirae N (Daehan Textbook).

Ministry for Education, Youth and Sport. (2009). *Malta: A guide to education and vocational training*. Accessed January 2, 2010, http://www.education.gov.mt/edu/edu_03.htm

Ministry for Investment, Industry and Information Technology. (2008). *The smart island the national ICT strategy for Malta 2008–2010*, Malta. Accessed January 21, 2009, http://www.thesmartisland.gov.mt

Mishra, S. (2009). GIS in Indian retail industry – A strategic tool. *International Journal of Marketing Studies, 1*(1), 50–57.

Mishra, P., & Koehler, M. (2006). Technological pedagogical content knowledge: A framework for teacher knowledge. *Teachers College Record, 108*(6): 1017–1054.

MOE. (2009). *An education overview of Taiwan*. Accessed January 22, 2010, MOE Web site: http://english.moe.gov.tw/ct.asp?xItem=4133&CtNode=2003&mp=1

MoE&S. (2005). *Draft policy for information and communication technology in the education*. Ministry of Education and Sports, Kampala.

MoE&S. (2008a). *Education sector strategic plan 2007–2015*. Ministry of Education and Sports, Kampala.

MoE&S. (2008b). *The education and sports sector annual performance report*. Ministry of Education and Sports, Kampala.

Morcillo, J. M. (2010). *Portal de Ciencias Experimentales, UCM (Devoted to physical geography)*. Accessed September 1, 2010, http://www.ucm.es/info/diciex/programas/

Moreira-Riveros, G. (2004). Sistema de aprendizaje basado en la web para facilitar el desarrollo del pensamiento. In J. Sánchez-Ilabaca (Ed.), *Memorias, IX taller internacional de software educativo, TISE* (pp. 29–32). Santiago: Universidad de Chile.

Moreira-Riveros, G. (2006). Activities of learning of geography using open source and ArcReader, Paper 1951, Proceedings of the Conference of Esri Users, San Diego, California. http://proceedings.esri.com/library/userconf/educ06/abstracts/a1951.html. Accessed 26 October 2009.

Moreno, A. (1996). Internet y sus recursos para enseñar Geografía. Didáctica Geográfica, 1, Segunda época, 95–102.

Mota, M. (2005). Concepção de curricula em análise espacial para o terceiro ciclo do ensino básico, MSc thesis, 181 pp., Lisbon ISEGI-UNL, http://www.isegi.unl.pt/servicos/documentos/TSIG/TSIG0007.pdf. ISEGI-UNL. Accessed 15 May 2010.

Mukama, E., & Andersson, S. B. (2008). Coping with change in ICT-based learning environments: Newly qualified Rwandan teachers' reflections. *Journal of Computer Assisted Learning, 24*, 156–166.

Muñiz, O. (2004). School geography in Chile. In A. Kent, L. Rawling, & A. Robinson (Eds.), *Geographical education. Expanding horizons in a shrinking world. Geographical Education Commission of the International Geographical Union* (pp. 177–180). Glasgow: Scottish Association of Geography Teachers.

Muñiz-Solari, O. (2009). Geography education: 'The North' and 'The South'. In O. Muñiz-Solari & R. G. Boehm (Eds.), *Geography education. Pan American perspectives, a volume in the International Geography Education Series (IGES)* (pp. 29–32). The Gilbert M. Grosvenor Center for Geographic Education, Austin, TX: Allen Griffith.

Napoleon, E. J., & Brook, E. A. (2010). *Thinking spatially using GIS: Our world GIS education, Level 1*. Student Workbook, Esri. ISBN 9781589481848.

National Commission for Higher Education Malta. (2009). *Annual report 2008*. Valletta: National Commission for Higher Education.

National Committee for Geography. (2007). *Australian's need geography.* Canberra: Australian Academy of Science.

National Council for the Social Studies. (1994). *Expectations of excellence: Curriculum standards for social studies.* Washington, DC: Author.

National Council of Education Research and Training. (2005). Education policies and curriculum at the upper primary and secondary education levels. In *Globalization and living together* (150p.). New Delhi: Discovery Publishing House.

National Curriculum Development Centre. (2008). *Uganda certificate of education: Geography Teaching Syllabus.* Kampala, Uganda.

National Curriculum Development Centre. (2010). *Online curricular programs.* Accessed January 29, 2010, http://www.ncdc.gov.rw

National Geography Curriculum Steering Committee. (2009). *Towards a national geography curriculum for Australia – Background report,* June 2009. National Geography Committee. www.ngc.org.au/report/index.htm

National Geography Education Standards Project. (1994). *Geography for life: National geography standards, 1994.* Washington, DC: National Geographic Research & Exploration.

Natural Resources Canada (NRC). (2007). *Atlas of Canada.* Accessed September 12, 2010, http://atlas.nrcan.gc.ca/site/english/index.html

National Research Council. (1996). *National science education standards.* Washington, DC: National Academy Press.

National Research Council. (2006). *Learning to think spatially—GIS as a support system in the K-12 curriculum.* Washington, DC: National Academies Press.

National Statistics. (2009). *Statistical yearbook of the Republic of China 2008.* Accessed January 26, 2010, Statistical Bureau Web site http://www.stat.gov.tw/public/data/dgbas03/bs2/yearbook_eng/y028I.pdf

National Statistics Office. (2009). *Malta in figures 2009.* Valletta: National Statistics Office.

NES. (2009). *National Education Statistics, Formal education 2008–2009.* Ministry of Education.

Ng, P. T. (2008). Teach less, learn more: Seeking curricular and pedagogical innovation. In J. Tan & P. T. Ng (Eds.), *Thinking schools, learning nation: Contemporary issues and challenges* (pp. 61–71). Singapore: Pearson Prentice Hall.

Niedersächsisches Kultusministerium. (2008). *Kerncurriculum für die Realschule Schuljahrgänge 5-10 Erdkunde.* Hannover: Niedersächsischer Bildungsserver (NIBIS).

Nieto, J. A. (2009). *Andalucía en un mapa. Valoración de una experiencia educativa. A Inteligência Geográfica na Educação do Século XXI. IV Congresso Ibérico de Didáctica da Geografia* (pp. 116–111). Lisboa: Universidade de Lisboa.

Nieto, J. A. (2010). *La información geográfica al servicio de la comunidad educativa andaluza. Didáctica Geográfica 11, Madrid, Grupo de didáctica de la Geografía (AGE), Real Sociedad Geográfica (RSG) (Royal Geographical Society of Spain).* Accessed September 1, 2010, http://www.realsociedadgeografica.com/en/site/index.asp

Nieto, J. A., & Fajardo, A. (2007). *The Andalusian government policy of diffusion of maps in the educative community.* XXIII International Cartographic Conference. Moscow, Russia. Accessed September 1, 2010, http://icaci.org/documents/ICC_proceedings/ICC2007/documents/doc/THEME%2022/Oral%201/THE%20ANDALUSIAN%20GOVERNMENT%20POLICY%20OF%20DIFFUSION%20OF%20MAPS%20IN%20THE.doc

Norris, P. (2001). *Digital divide: Civic engagement, information poverty, and the Internet worldwide.* Cambridge: Cambridge University Press.

Nourbakhsh, I. (2006). Mapping disaster zones. *Nature, 439*(16), 787–788.

Obura, A. (2003). *Never again: Educational reconstruction in Rwanda.* Paris: UNESCO International Institute for Educational Planning.

O'Connor, P. (2006). Progressive GIS. *Teaching Geography,* Autumn, 2007, 147–150.

O'Dea, E. K. (2002). *Integrating geographic information systems and community mapping into secondary science education: A Web GIS approach.* Unpublished master's thesis, Oregon State University, Oregon.

OFSTED (2008). *Geography in schools: Changing practice.* Accessed December 8, 2010, http://
    www.ofsted.gov.uk
Oh, C., & Seong, C. (2003). A study of GIS education in secondary school: A case study of high
    school. *Journal of the Korean Geographic Information System, 11,* 89–100.
Olgen, M. K. (2005). Türkiye'de CBS egitimi. In M. K. Ölgen (Ed.), *Ege Cografi Bilgi Sistemleri
    Sempozyumu Bildiriler Kitabı* (pp. 9–22). Izmir: Ege Üniversitesi.
Olsen, A., & Eddy, S. (2010). *GISMAPED company records.*
Olsen, A., Eddy, S., Page, N., & Arthur, P. (2006). IMAGIS: Interactive mapping with GIS. (CD),
    Wellington.
Ontario Association for Geographic and Environmental Education (OAGEE). (2010). Accessed
    September 15, 2010, http://www.oagee.org/
Opferkuch, D. (2009). Ein GIS-projekt: Schüler erfassen geodaten. *GEG-INFO,* 2-2009 April.
Oppong, J. R. (2010). Transport, communication and information technologies in Sub-Saharan
    Africa: Digital bridges over spatial divides. In S. A. Attoh (Ed.), *Geography of Sub-Saharan
    Africa* (pp. 243–264). New York: Prentice Hall.
Ota, H. (2006). Chizu to GIS de sekai to kyodo wo shiro. *Chizu, 44-4,* 36–39.
Palfrey, J., & Gasser, U. (2008). *Born digital: Understanding the first generation of digital natives.*
    New York: Basic Books.
Parthemore, J. (2003). *A secondary school computer lab in rural Brong Ahafo: A case study
    reflection on the future of secondary school computer literacy and computer based distance
    education in Ghana.* Accessed March 23, 2010, http://www.wess.edu.gh/lab/reports/paper.pdf
Patterson, M. W., Reeve, K., & Page, D. (2003). Integrating geographic information systems into
    the secondary curricula. *Journal of Geography, 102,* 275–281.
Persson, A. (2007). *Digital Norway. GIM international.* Accessed September 14, 2010, http://
    www.gim-international.com/issues/articles/id816-Digital_Norway.html
Piaget, J. (1929). *The child's conception of the world.* London: Routledge.
Piaget, J. (1937). *La construction du réel chez l'enfant / The construction of reality in the child.*
    New York: Basic Books.
Piburn, M. D., & Reynolds, S. J. (2005). The role of visualization in learning from computer-based
    images. *International Journal of Science Education, 27*(5), 513–527.
Population Reference Bureau. (2009). World population data sheet. Accessed January 22, 2010,
    Population Reference Bureau Web site: http://www.prb.org/pdf09/09wpds_eng.pdf
Prenzel, M., Artelt, C., Baumert, J., Blum, W., Hammann, M., Klieme, E., et al. (2008).
    PISA 2006 in Deutschland - Die Kompetenzen der Jugendlichen im dritten Ländervergleich.
    Accessed January 5, 2010, http://www.wir-wollen-lernen.de/resources/PISA-E_2006_vollst_
    Bericht_HH.pdf
Püschel, L., Hofmann, K., & Hermann, N. (2007). *Blickpunkt WebGIS. Der Einstieg für Schulen in
    Geographische Informationssysteme (GIS).* Koblenz: Landesmedienzentrum Rheinland-Pfalz.
Ramm, F., & Stark, H. J. (2008). Crowdsourcing Geodata. *Geomatik Schweiz,* Ausgabe 06/2008,
    315–319.
Ramm, F., & Topf, J. (2008). OpenStreetMap, Die frei Weltkarte nutzen und mitgestalten.
Real Sociedad Geográfica (RSG). (2010). (Royal Geographical Society of Spain). Accessed
    September 1, 2010, http://www.realsociedadgeografica.com/en/site/index.asp.
Roberts, M. (2003). *Learning through enquiry: Making sense of geography in the key stage 3
    classroom.* Sheffield: Geographical Association.
Rød, J. K., Larsen, W., & Nilsen, E. (2010). Learning geography with GIS: Integrating GIS into
    undergraduate geography curricula. *Norwegian Journal of Geography, 64*(1), 21–34.
Rogers, E. (1995). *Diffusion of innovations.* New York: Free Press.
Rogers, E. (2003). *Diffusion of innovations* (5th ed., 576p.). New York: Free Press.
Romera, C., Del Campo, A., & Sánchez, J. (2009). *Educational resources of the national atlas
    of Spain.* Accessed September 1, 2010, http://cartography.tuwien.ac.at/ica/documents/ICC_
    proceedings/ICC2009/html/nonref/14_10.pdf

Ruiz, F. (2010). *A useful way of making maps of Spain with the last statistics.* Accessed December 8, 2010, http://alarcos.esi.uclm.es/per/fruiz/pobesp/nueva/

Sanchez, E. (2009). Innovative teaching/learning with geotechnologies in secondary education. In A. Tatnall & T. Jones (Eds.), *Education and technology for a better World* (pp. 65–74). Berlin: Springer.

Sanchez, E., Delorme, L., Jouneau-Sion, C., & Prat, A. (2010). Designing a pretend game with geotechnologies: Toward active citizenship. In T. Jekel, A. Koller, K. Donert, & R. Vogler (Eds.), *Learning with geoinformation V* (pp. 31–40). Heidelberg: Wichman.

Sanchez, E., & Jouneau-Sion, C. (2009). Playing in the classroom with a virtual globe for geography learning. In T. Jekel, A. Koller, & K. Donert (Eds.), *Learning with Geoinformation IV* (pp. 78–86). Heidelberg: Wichmann.

Schäfer, D. (2003). *Scalable GIS applications for schools – examples from Germany.* Paper presented at the Esri 2003, 18th European User Conference, 10. Deutschsprachige Anwenderkonferenz. Conference CD. Accessed January 2, 2010, http://www.staff.uni-mainz.de/dschaefe/pdf/14_30_Fr_Gr_schaefer.pdf

Schäfer, D., & Ortmann, G. (2004). Biotopkartierung in Rheda-Wiedenbrück. Gemeinsames Projekt zwischen Stadtverwaltung, Schule und Universität Mainz. ArcAktuell(1/2004), 44.

School Realities. (2009). *South Africa.info, the official gateway.* Department of Education. Accessed October 2009, www.southafrica.info

Schultz, R. B., Kerski, J. J., & Patterson, T. (2008). The use of virtual globes as a spatial teaching tool with suggestions for metadata standards. *Journal of Geography, 107,* 27–34.

Share, P. (1997). *Telecenters, IT and rural development: Possibilities in the information age.* Accessed March 23, 2010, http://www.csu.edu.au/research/crsr/sai/saipaper.htm#top

Sharpe, B., & Best, A. C. (2001). Teaching with GIS in Ontario's secondary schools. In D. R. Green (Ed.), *GIS: A sourcebook for schools* (pp. 73–86). London: Taylor & Francis.

Sharpe, B., & Huynh, N. T. (2005). *Geospatial knowledge areas and concepts across the Ontario Curriculum.* Technical report for the GeoSkills Program, GeoConnections. Natural Resources Canada, 22pp.

Shin, C., Jeong, Y., & Joo, S. (2002). Computational education: A development of a geographic learning courseware based on GIS. *Journal of the Korea Information Processing Society, 9-A*(1), 105–112.

Shin, E. (2006). Using geographic information system (GIS) to improve fourth graders' geographic content knowledge and map skills. *Journal of Geography, 105*(3), 109–120.

Shulman, L. S. (1986). Those who understand: Knowledge growth in teaching. *Educational Researcher, 15,* 4–14.

Siegmund, A., & Naumann, S. (2009). GIS in der Schule. Potenziale für den Geographieunterricht von heute. *Praxis Geographie, 39*(2), 4–8.

Silva, J. E. (2008). *Tendencias y futuro de los GIS en gobierno.* PowerPoint presentation, XV Conferencia de Usuarios Latinoamericanos de Esri, Santiago, Chile. http://www.esri-chile.com/lauc2008/13.proceeding_4h-2.htm. Accessed 30 October 2009.

Sinton, D. S. (2009). Roles for GIS within higher education. *Journal of Geography in Higher Education, 33*(Supplement 1), S7–S16.

SITE. (2009). *Sistema de información territorial de educación.* Ministerio de Educación, Chile. Accessed December 10, 2009, http://atlas.mineduc.cl/pmgt/

Skills Canada. (2010). *Skills Canada.* Accessed September 12, 2010, http://www.skillscanada.com/

South African Department of Education. (2003). *National curriculum statement grades 10–12 (general) geography.* Pretoria: Government Printer.

South African Department of Education. (2006). *The national policy framework for teacher education and development in South Africa.* Pretoria: Government Printer.

Spatially Speaking. (2007). *Geographical association project.* Accessed December 8, 2010, http://www.geography.org.uk/projects/spatiallyspeaking/

Specht, W. (Ed.). (2009). *Nationaler Bildungsbericht Österreich 2009, Band 1: Das Schulsystem im Spiegel von Daten und Indikatoren.* Graz: Leykam

Statistics Canada. (2010). *Statistics Canada learning resources.* http://www.statcan.gc.ca/edu/index-eng.htm. Accessed 12 September 2010.

Steiniger, S., & Bocher, E. (2009). An overview on current free and open source desktop GIS developments. *International Journal of Geographical Information Science, 23*(10), 1345–1370.

Stoltman, J. P. (1990). Geography education for citizenship. Bloomington, IN: ERIC Clearinghouse for Social Studies/Social Science Education.

Storie, C. (2000). *Assessing the role of Geographical Information System (GIS) in the geography classroom.* Unpublished Master's thesis. Wilfrid Laurier University, Waterloo, Ontario, Canada.

Strobl, J. (2008). Digital earth brainware. A framework for education and qualification requirements. In J. Schiewe & U. Michel (Eds.), *Geoinformatics paves the highway to digital earth* (pp. 134–138). Osnabrück: Universität Osnabrück (Hrsg.).

Strobl, J. (2010). Towards a Geoinformation Society. *GIS-Development, 14*(1), 102–104.

Strobl, J., & Koller, A. (1995). Das internet und Materialien für GW. *GW-Unterricht, 59.* http://www.ph-linz.at/ZIP/didaktik/gw/strobl/strobl.htm

Sui, D. Z. (1995). A pedagogic framework to link GIS to the intellectual core of geography. *Journal of Geography, 94*, 578–591.

Tan, J. (2002). Education in the early 21st century: Challenges and dilemmas. In D. da Cunha (Ed.), *Singapore in the New Millennium* (pp. 154–186). Singapore: Institute of Southeast Asian Studies.

Tapscott, D. (1998). *Growing up digital: The rise of the net generation.* Columbus, OH: McGraw-Hill.

The World Factbook: Rwanda. (2009). *General information about Rwanda.* Online resource. CIA. Accessed January 28, 2010, https://www.cia.gov/library/publications/the-world-factbook/geos/rw.html

Thurlow, C. (2010). Personal communication. GIS Training Manager, Eagle Technology.

Tierney, W. G., Bailey, T., Constantine, J., Finkelstein, N., & Hurd, N. F. (2009). Helping students navigate the path to college: What high schools can do: A practice guide (NCEE #2009-4066). Washington, DC: National Center for Education Evaluation and Regional Assistance, Institute of Education Sciences, U.S. Department of Education.

Treier R., Treuthardt Bieri C., & Wüthrich M. (2006). Geografische informationssysteme (GIS) - Grundlagen und Übungsaufgaben für die Sekundarstufe II. *Bern*, 1–150.

TÜİK. (2010). Adrese dayalı nüfus kayıt sistemi (ADNKS), 2010 Nüfus Sayımı Sonuçları. Accessed August 2, 2011, http://tuikapp.tuik.gov.tr/adnksdagitapp/adnks.zul

Tuna, F. (2008). *Taking the advantages of Geographic Information Systems (GIS) to support the project based learning in high school geography lessons,* Unpublished doctoral dissertation, Marmara University, Institute of Education Sciences, Istanbul.

UNEB. (2008). Uganda National Examination Board (UNEB). *Uganda advanced certificate of education: Regulations and syllabuses.* Kampala, Uganda.

UNESCO. (2008). *ICT competencies standards for teachers.* Accessed September 1, 2010, http://cst.unesco-ci.org/sites/projects/cst/The%20Standards/ICT-CST-Implementation%20Guidelines.pdf

UNICEF Rwanda Statistics. (2007). *Resource document.* Accessed January 28, 2010, http://www.unicef.org/infobycountry/rwanda_statistics.html

Universidad Distrital Francisco José de Caldas. (2010). Licenciatura en educación básica con Énfasis en Ciencias Sociales. http://www.udistrital.edu.co/academia/pregrado/lsocial/. Accessed 13 August 2010.

University of Malta. (2009). *SEC syllabus. (2010): Geography.* Matriculation and Secondary Education Certificate Examinations Board, Malta. Accessed January 20, 2010, http://www.um.edu.mt/matsec/docs/syllabireports/syllabi2010/sec/SEC15.pdf.

US Department of Labor. (2007). *Geospatial technology*. Accessed March 23, 2010, http://www.doleta.gov/Brg/JobTrainInitiative/.

US Department of Labor. (2010). *The President's high growth job training initiative*. Accessed March 23, 2010, http://www.doleta.gov/Brg/JobTrainInitiative/

Vadivelu, V. M. (2007). Education system and teacher training in India. *Ethiopian Journal of Education and Sciences, 3*(1), 97–102.

Vajoczki, S. (2009). Geography education in Canada. In O. Muniz-Solari & R. G. Boehm (Eds.), *Geography education: Pan American perspectives* (pp. 139–155). Austin, TX: The Grosvenor Center for Geographic Education.

van der Schee, J. (2003). New media will accelerate the renewal of geographical education. In R. Gerber (Ed.), *International handbook on geographical education* (pp. 205–214). Dordrecht: Kluwer.

van der Schee, J. A., & Scholten, H. J. (2009). Geographic Information Systems and geography teaching. In H. J. Scholten, R. Van der Velde, & N. Van Manen (Eds.), *Geospatial technology and the role of location in science* (pp. 287–301). Dordrecht: Springer.

Vera, A. L. (Geohistoria, 2003). (It works interesting links for the National Curriculum) 275 http://www.geohistoria.net/index2.asp. Geography of Spain: http://www.geohistoria.net/paginas/2bgeo.htm, 276 General geography: http://www.geohistoria.net/paginas/3eso.htm. Accessed September 1, 2010.

Vine, P., Al Abed, I., Hellyer, P., & Vine, P. (2009). *United Arab Emirates 2009*. London: Trident Press Ltd.

Vogler, R., Ahamer, G., & Jekel, T. (2010). GEOKOM-PEP – Pupil led research into the effects of geovizualisation. In T. Jekel, A. Koller, & K. Donert (Eds.), *Learning with GI V* (pp. 51–60). Heidelberg: Wichmann.

Volz, D., Viehrig, K., & Siegmund, A. (2008). GIS as a means for competence development. In: T. Jekel, A. Koller, & K. Donert (Eds.), *Learning with Geoinformation III – Lernen mit Geoinformation III* (pp. 42–48). Wichmann: Heidelberg.

Volz, D., Viehrig, K., & Siegmund, A. (2010). Informationsgewinnung mit Hilfe Geographischer Informationssysteme – Schlüsselkompetenz einer modernen Geokommunikation. *Geographie und ihre Didaktik, 38*(2), 102–108.

Wallace, R. M. (2004). A framework for understanding teaching with the internet. *American Educational Research Journal, 41*, 447–488.

Wallentin, G., Jekel, T., Rattensberger, M., & Binder, D. (2008). 'Schools on Ice' – Einbindung von Lernendenperspektiven in GI-basiertes Lernen. In T. Jekel, A. Koller, & K. Donert (Eds.), *Lernen mit Geoinformation III* (pp. 87–95). Heidelberg: Wichmann.

Wang, S. F. (2001). *The research in the geographic information ability of high school teacher*, Unpublished master's thesis, National Taiwan University, Taipei.

Wang, Z. F. (2009). *Teaching with GIS in 2006 high school geography curriculum standard*, Unpublished master's thesis, National Taiwan University, Taipei.

Warkentin, J., & Simpson-Housley, P. (2001). The development of geographical study in Canada, 1870–2000. In G. S. Dunbar (Ed.), *Geography: Discipline, profession and subject since 1870* (pp. 281–316). Dordrecht, The Netherlands: Kluwer.

Welzel-Breuer, M., Graf, S., Sanchez, E., Fontanieu, V., Stadler, H., Raykova, Z., et al., (2010). Application of computer aided learning environments in schools of six European countries. In G. Cakmakci & M. F. Taşar (Eds.), *Contemporary science education research: scientific literacy and social aspects of science* (pp. 317–326). Ankara, Turkey: Pegem Akademi.

West, B. A. (2003). Student attitudes and the impact of GIS on thinking skills and motivation. *Journal of Geography, 102*, 267–274.

West, B. A. (2006). Towards an understanding of conceptions of GIS. In K. Purnell, J. Lidstone, & S. Hodgson (Eds.), *The proceedings of the international geographical union commission on geographical education symposium, changes in geographical education: Past, present and future* (pp. 467–471). Brisbane: International Geographical Union Commission on Geographical Education and Royal Geographical Society of Queensland.

White, K. L., & Simms, M. (1993). Geographic information systems as an educational tool. *Journal of Geography, 92*(2), 80–85.

Wiegand, P. (2001). Geographical Information Systems (GIS) in education. *International Research in Geographical and Environmental Education, 10*(1), 68–71.

Wiegand, P. (2003). School students' understanding of choropleth maps: Evidence from collaborative mapmaking using GIS. *Journal of Geography, 102*, 234–242.

Wigglesworth, J. C. (2003). What is the best route? Route-finding strategies of middle school students using GIS. *Journal of Geography, 102*, 282–291.

Wu, C. S. (2006). Examination and improvement on education reform in Taiwan: 1994–2006. *Bulletin of National Institute of Education Resources and Research, 32*, 1–21.

Wu, C. S., & Kao, C. P. (2007). The analysis of the reform of secondary education in Taiwan: 1994–2007. *Bulletin of National Institute of Education Resources and Research, 34*, 1–24.

Yagoub, M. M. (2002). *Geographical information systems (GIS) education and application in the United Arab Emirates.* UAE University. Accessed December 27, 2009, http://faculty.uaeu.ac.ae/myagoub/main_GIS.htm

Yagoub, M. M. (2009). *Geographic information systems (GIS) for Schools in the UAE.* Unpublished paper.

Yang, S. (2004). *A study on developing geographic learning tool using Internet GIS.* Unpublished master's thesis, Seoul National University, Seoul.

Yap, L. Y., Tan, G. C. I., Zhu, X., & Wettasinghe, M. C. (2008). An assessment of the use of geographical information systems (GIS) in teaching geography in Singapore schools. *Journal of Geography, 107*(2), 52–60.

Yomralıoğlu, T. (2002). *GIS activities in Turkey.* Proceedings of international symposium on GIS, pp. 834–840, Istanbul, Turkey.

Yuda, M., Satori, I., & Johansson, T. (2009). Geographical Information Systems in upper-secondary school education in Japan and Finland: A Comparative Study. The Shin-Chiri (The New Geography), 57, 156–165.

Zerger, A., Bishop, I., Escobar, F., & Hunter, G. (2002). A self-learning approach for enriching GIS education. *Journal of Geography in Higher Education, 26*(1), 67–80.

Index

Printed by Books on Demand, Germany